ON CONTESTED SHORES

ON CC

U.S. troops disembarking on Utah Beach, 6 June 1944.
Official U.S. Navy photo

CONTESTED SHORES

The Evolving Role of Amphibious Operations in the History of Warfare

VOLUME 2

Edited by Timothy Heck,
B. A. Friedman, and Walker D. Mills

Marine Corps University Press
Quantico, Virginia
2024

LIBRARY OF CONGRESS CATALOGING-IN-PUBLICATION DATA

Names: Heck, Timothy, 1982- editor. | Friedman, B. A. (Brett A.), editor. | Mills, Walker D., editor. Marine Corps University (U.S.). Press, issuing body.
Title: On contested shores : the evolving role of amphibious operations in the history of warfare, vol. 2 / edited by Timothy Heck, B. A. Friedman, and Walker D. Mills.
Other titles: Evolving role of amphibious operations in the history of warfare, vol. 2
Description: First edition published 2024. | Quantico, Virginia : Marine Corps University Press, 2024. | Includes bibliographical references and index. | Summary: "The second volume of On Contested Shores serves as both manual and inspiration for its readers. While not doctrine, the organization, breadth of experience and analysis, and topics presented here should find a place on the shelves at operational units, PME schoolhouses, and on the desks of scholars looking to better understand what puts men and women in small boats to land on contested shores"—Provided by publisher.
Identifiers: LCCN 2020037806 | ISBN 9798986259581 (paperback)
Subjects: LCSH: United States. Marine Corps. | Amphibious warfare—History—Case studies. | Landing operations—History—Case studies.
Classification: LCC U261 .O5 2020 | DDC 359.9/6460973—dc23 | SUDOC D 214.513:AM 7
LC record available at https://lccn.loc.gov/2020037806

DISCLAIMER

The views expressed in this publication are solely those of the authors. They do not necessarily reflect the opinion Marine Corps University, the U.S. Marine Corps, the U.S. Navy, the U.S. Army, the U.S. Air Force, or the U.S. government. The information contained in this book was accurate at the time of printing. Every effort has been made to secure copyright permission on excerpts and artworks reproduced in this volume. Please contact the editors to rectify inadvertent errors or omissions.

Copyright for all works—journals and monographs—is retained by the author(s). Some works of authorship from Marine Corps University Press (MCUP) are created by U.S. government employees as part of their official duties and are now eligible for copyright protection in the United States; further, some of MCUP's works available on a publicly accessible website may be subject to copyright or other intellectual property rights owned by non-Department of Defense (DOD) parties. Regardless of whether the works are marked with a copyright notice or other indication of non-DOD ownership or interests, any use of MCUP works may subject the user to legal liability, including liability to such non-DOD owners of intellectual property or other protectable legal interests.

MCUP products are published under a Creative Commons NonCommercial-NoDerivatives 4.0 International (CC BY - NC-ND 4.0) license.

Published by
Marine Corps University Press
2044 Broadway Street
Quantico, VA 22134

1st Printing, 2024

ISBN: 979-8-9862595-8-1
DOI: 10.56686/9798986259581

THIS VOLUME IS FREELY AVAILABLE AT WWW.USMCU.EDU/MCUPRESS

CONTENTS

Foreword — ix
Preface and Acknowledgments — xiii
Glossary of Select Terms and Acronyms — xv

Introduction — 3
    Timothy Heck, B. A. Friedman, and Walker D. Mills

DOCTRINE AND LOGISTICS

CHAPTER ONE — 7
The Landing at Collado Beach: The Logistical Importance of the Amphibious Landing near Veracruz during the Mexican-American War
    Christopher Menking

CHAPTER TWO — 21
The Landing Craft Controversy, 1934–1942
    Jerry E. Strahan

CHAPTER THREE — 42
Red Tide over the Beach: Soviet Amphibious Warfare in Theory and Practice
    Benjamin Claremont

CHAPTER FOUR — 64
Innovative Amphibious Logistics for the Twenty-first Century
Walker D. Mills

TECHNOLOGY AND INNOVATION

CHAPTER FIVE — 89
Amphibious Juggernaut: How the Landing Ship, Tank, and Landing Vehicle, Tracked, Created the Most Powerful Amphibious Assault System of World War II
Douglas E. Nash Sr.

ORGANIZATION AND TRAINING

CHAPTER SIX — 121
The Union Defence Forces' Amphibious Invasion of German South West Africa, 1914
David Katz

CHAPTER SEVEN — 142
Operation Albion: The German Amphibious Landing on the Baltic Islands, 12–17 October 1917
Eric Sibul

CHAPTER EIGHT — 163
Beyond Cold Shores: Inland Maneuver in Historical Polar Amphibious Operations
Lance R. Blyth

CHAPTER NINE — 182
Soviet Preparations for a Naval Landing against Israel in June 1967 and Their Partial Implementation
Isabella Ginor and Gideon Remez

POLICY AND INTEROPERABILITY

CHAPTER TEN — 197
Operation Husky: The Challenges of Joint Amphibious Operations
Darren Johnson

CHAPTER ELEVEN — 213
A New Zealand-led "Commando Raid" in the South Pacific: The Green Islands, 30–31 January 1944
Shaun Mawdsley

CHAPTER TWELVE 228
PLA Amphibious Campaigns and the Origins
of the Joint Island Landing Campaign
Xiaobing Li

MILITARY MATERIEL AND PERSONNEL

CHAPTER THIRTEEN 247
U.S. Geostrategic Deterrence and A2/AD at Work
in the American Civil War, 1861–1865
Howard J. Fuller

CHAPTER FOURTEEN 276
A Groundswell of Support in the Pacific:
Deploying Small Wars Doctrine amid the Rise of Amphibious Warfare
Evan Zachary Ota

CHAPTER FIFTEEN 301
Prelude to Stalin's Third Crushing Blow: The Kerch-Eltigen Landing, 1943
Timothy Heck

CHAPTER SIXTEEN 321
Not a Carbon Copy of the U.S. Marine Corps: The Development
of the People's Liberation Army Navy Marine Corps since 1979 and
What that Means for the Chinese Power Project in the Pacific and Beyond
Edward Salo, PhD

Conclusion 333
Timothy Heck, B. A. Friedman, and Walker D. Mills

Select Bibliography and Suggested Further Reading 339
Index 353
About the Authors 361

FOREWORD

I am honored to be asked to prepare the foreword to this volume of *On Contested Shores*. The second in the series, *On Contested Shores* shows the care and effort being taken to relate the information contained in them to current day situations and circumstances. The chapters are individual contributions collected from many countries, showing the wide reach the authors have presented for completeness.

The foreword to first volume by Brigadier General Jason Q. Bohm is very well written and, for the most part, covers this volume also. That volume started with an article about an Italian special operation (June 1555) and ended with a discussion of U.S. Marine Corps expeditionary advanced base operations (EABO) concepts now and in the future. That book was chronologically organized, allowing readers to grasp some of the sweep of amphibious operations during the past 500 years.

In contrast, this volume's chapters have been organized thematically so that readers can find commonalities, intersections, and differences about a subject, concept, or event more easily. This active dialogue helps the volume meet the goal of creating an applicable history. The volume starts with the Veracruz landings during the Mexican-American War (1846–48) and concludes with a discussion of the Chinese People's Liberation Army Navy (PLAN) Marine Corps today. Additionally, the authors have addressed the technology, organizational structure, and policies needed to field modern amphibious forces.

Let me quote a short section from General Bohm: "Timothy Heck, B. A. Friedman, and Marine Corps University Press have compiled a comprehensive and well-balanced work to advance this effort. They endeavored to 'elucidate the foundations of amphibious warfare while also illuminating its future potential'."[1] To this, I add Walker D. Mills as coeditor for the 2024 volume, and comment that they have made volume 2 just as interesting and readable as a bonus for the reader.

Amphibious operations remain the essential tool for the Joint Force to conduct forcible entry operations. *Amphibious Operations*, Joint Publication 3-02, describes the role and purpose of these landings as follows:

> *Amphibious operations use maneuver principles to employ ready-to-fight combat forces from the sea to the shore to achieve a position of advantage over the enemy. During combat operations, maneuver, in conjunction with organic and supporting fires, is essential to gaining access where the enemy least expects it. It provides a position of advantage to destroy or seriously disrupt the enemy's cohesion through a variety of rapid, focused, and unexpected actions that create a turbulent and rapidly deteriorating situation with which the enemy cannot cope.*[2]

The landings, technologies, forces, and policies are examined here with an eye toward making historical analysis relevant to modern practitioners across a wide spectrum of fields and disciplines. This is, in short, not a book by Marines, for Marines, about Marines. Rather, it is a richer analysis of what is needed to enable the Joint Force to staff, train, equip, and employ amphibious forces.

In my 1987 *Marine Corps Gazette* article, "Thinking about Warfare," I argued that offensive actions are often undertaken for the specific purpose of shielding other organizations from damage and casualties.[3] This combat shield "can deny an enemy the opportunity to shoot at a force or otherwise disrupt its operations."[4] Ultimately, "the measure of success is the survival of the shielded force."[5] Amphibious operations allow commanders to place forces ashore that can then shield ships at sea and construct an airfield to extend the range of bombers and fighters, creating a new shield by which the landing force can advance like climbing the rungs of a ladder.

Many works on amphibious operations, not just this volume, focus on major force-on-force landings. Iwo Jima, Normandy, and Gallipoli all come to mind as oft-examined and documented amphibious operations. Less examined but equally important is the concept of an amphibious campaign and landing that supports operational maneuver through shielding. While the mention of U.S. Army general Douglas MacArthur might raise hackles on some, his amphibious campaign model in

---

[1] Jason Q. Bohm, "foreword," in Timothy Heck and B.A. Friedman, eds. *On Contested Shores: The Evolving Role of Amphibious Operations in the History of Warfare* (Quantico, VA: Marine Corps University Press, 2020), xi, https://doi.org/10.56686/9781732003149.
[2] *Amphibious Operations*, Joint Publication 3-02 (Washington, DC: Joint Chiefs of Staff, 2021), I-3.
[3] Philip D. Shutler, "Thinking about Warfare," *Marine Corps Gazette* 71, no. 11 (November 1987): 18–26.
[4] Shutler, "Thinking about Warfare," 20.
[5] Shutler, "Thinking about Warfare," 21.

the South Pacific during World War II and its application in Korea all bear further examination as a model for why amphibious landings are conducted and how to plan for them.

I encourage readers to look at the landing on the Green Islands, covered admirably here by Shaun Mawdsley, to get a sense of this model of amphibious operations. First, the landing was a small, combined operation conducted by U.S. Navy Pacific Fleet with the 3d New Zealand Division to accomplish a reconnaissance in force. Second, it fit in a sequence of operations that used amphibious operations to position airpower to deny the Japanese not only airpower but land and seapower as well, forcing them to abandon Kavieng, the last major base in the South Pacific theater.

The Green Islands were not the first use of a land-based force to deny enemy access to aviation. Indeed, using Marines of the 1st Marine Division to defend Henderson Field on Guadalcanal so the fighters and bombers could protect the division and the fleet offshore was the first major use by the United States of this concept. The events of 20 August 1942 are emblematic of this symbiotic shielding relationship between the infantry and aviation as part of an amphibious force:

> *On 20 August, from a point 322 kilometers south of the island, 19 planes of Marine Fighter Squadron 223 (VMF-223) flying Grumman F4F-4 Wildcats led by Major John L. Smith and 12 dive bombers of Marine Scout Bombing Squadron 232 (VMSB-232) flying Douglas SBD-3 Dauntlesses led by Lieutenant Colonel Richard C. Mangrum took off from the flight deck of the USS* Long Island *(CVE 1). Beginning at 1330, the flight ended with the safe arrival of all planes at Henderson Field by 1700. Within 8 hours of their arrival, the first great counterattack of the Japanese was thrown back; and within 12 hours, the newly arrived planes were performing their first mission in support of the ground troops—patrolling the beaches east of the Tenaru to cut off any attempt at escape by the remnants of the enemy force that Lieutenant Colonel Edwin A. Pollock's 2d Battalion, 1st Marines, had cut to pieces at the mouth of Alligator Creek.[6]*

As aircraft transitioned from ship to shore, the infantry shielded the maintainers, pilots, and aircraft who then conducted close air support strikes against the Japanese. Modern concepts of EABO and stand-in forces could learn much from these early uses of amphibious power as movable shields. The model of Guadalcanal was carried forward in the southwest Pacific theater by General MacArthur.

In his role as commanding general of the 1st Marine Division and senior officer on Guadalcanal, General A. A. Vandegrift " 'invented a new system of war—the system of seizing a beachhead on which an airfield could be constructed, setting up a cordon defense around it, then proceeding to the next step. The process was repeated in endless variations throughout the South Pacific—at Bougainville, Cape Gloucester,

---

[6] Maj John L. Zimmerman, USMCR, *The Guadalcanal Campaign* (Washington, DC: Historical Division, Headquarters Marine Corps, 1949), 64.

Hollandia, Aitape, Geelvink Bay, Mindoro.' This campaign moved the force forward 4,828 kilometers in a little more than a year."[7]

The primary maneuver element was the fighter base that was moved to the beachhead as soon as safely possible. The fighters then shielded the bombers to gain air superiority and sea control, while the surface ships gained undersea control. This allowed the next landing in the sequence to be made away from defended positions with little opposition and a lot of fire support. The new fighter base was built quickly, and the process repeated. The landings on the Green Islands were part of that sequence.

As we examine amphibious operations and the opportunities they provide the Joint and combined force, both volumes of *On Contested Shores* provide valuable insights into a form of warfare that has had comparatively little coverage in other histories. The common theme of the need for multidomain planning and cooperative execution, apparent even in the early entries, becomes more obvious in the later chapters. Future volumes could include sequences of amphibious operations that show how all U.S., allied, and coalition forces can work together across all domains—space, air, land, sea, undersea, electromagnetic, communications, intelligence, and cyber—to accomplish assigned missions with minimum casualties. Again, historical examples exist that are worth looking at, both well-known and those less studied. I look forward to reading them when they come out.

Philip D. Shutler
Lieutenant General, U.S. Marine Corps (Ret)[8]

---

[7] LtGen Matthew Glavy, USMC, and LtGen Philip D. Shutler, USMC (Ret), "Designing a Force with a Fighting Foot Ashore," U.S. Naval Institute *Proceedings* 149, no. 11 (November 2023).

[8] Philip D. Shutler was commissioned in 1948 following graduation from the U.S. Naval Academy. He served as a platoon leader and as a reconnaissance company commander in Korea, including at the Chosin Reservoir, before becoming a naval aviator in 1952. He commanded Marine Aircraft Group 31 in Vietnam and was later named the deputy commandant for aviation. He was director for Operations (J-3), Joint Staff, when he retired on 1 July 1980. He remains active in analysis and discussions of military operations, campaign planning, and strategy.

PREFACE AND ACKNOWLEDGMENTS

We started this project shortly after the first volume of *On Contested Shores* was published in 2020. We did so, in part, because the topic seems to have struck a nerve with readers, practitioners, and scholars, many of whom became potential contributors. They, and we, recognized there was much more ground (or shoreline) to be covered that had direct operational relevance to the Marine Corps, our Joint Services, and our allies and partners. Thus, a second volume was born, and Walker D. Mills was brought on board to help craft what you read today.

This volume, unlike the first, is not organized chronologically. Rather, it is grouped around common themes, namely the DOTMLPF-PI construct—doctrine, organization, training, materiel, leadership and education, personnel, facilities, policy, and interoperability—many will remember from professional military education (PME). While not all the DOTMLPF-PI categories are covered, we sought out contributions that addressed specific elements we found vitally important to the conceptualization and execution of amphibious operations. Thanks is due to the authors who allowed us to tinker with their work in a way that helped make these themes more explicit without this becoming a pedantic work aimed solely at the suffering PME student.

Ultimately, we hope this book serves as both manual and inspiration for its readers. While not doctrine, the organization, breadth of experience and analysis, and topics presented here should find a place on the shelves at operational units, PME schoolhouses, and on the desks of scholars looking to better understand what puts men and women in small boats to land on contested shores.

We would like to acknowledge the incredible assistance provided by Angela Anderson and her team at Marine Corps University Press for supporting both this volume and the previous one. Without their input, guidance, and dedication, the work you see today would exist in a dozen hard drives and half-forgotten notebooks instead of in your hands. Thank you also to Major General Jason Bohm who wrote, at Angela's request, the introduction to our first volume. We would also like to thank the United States Naval Academy for hosting the McMullen Naval History Symposium, which remains the premier conference for studying naval and amphibious operations and forces. The symposium's role as a focal point for scholarship made our process easier as we sought contributors, peer reviewers, and supporters.

To our families, friends, peers, and networks, thank you for your support as we took this volume from concept to completion. Whether it was one of our children asking us to find "more anpibian eyes" to a gracious and understanding partner allowing us to slip away to write, edit, and coordinate, we could not have done this without their support. Thank you.

GLOSSARY OF SELECT TERMS AND ACRONYMS

A2/AD	antiaccess/area-denial
ADC	Alaskan Defense Command
APD	auxiliary personnel destroyer
ARG	Amphibious Ready Group
BuC&R	Bureau of Construction and Repair
CCP	Chinese Communist Party
CNO	Chief of Naval Operations
DMO	distributed maritime operations
DOTMLPF-PI	doctrine, organization, training, materiel, leadership and education, personnel, facilities, policy, and interoperability
DUKW	six-wheeled amphibious truck
EABO	expeditionary advanced base operations
FDB	Fleet Development Board
FLEX	Fleet Landing Exercise
FMF	Fleet Marine Force
GSWA	German South West Africa
KMT	Kuomintang troops
LAW	Light Amphibious Warship
LBB	Landing Boat Board
LCAC	Landing Craft, Air Cushion
LCI	Landing Craft, Infantry

LCM	Landing Craft, Medium (Mike boat)
LCPL	Landing Craft Personnel, Large
LCP(R)	Landing Craft Personnel (Ramp)
LCT	Landing Craft, Tank
LCVP	Landing Craft, Vehicle, Personnel (Higgins boat)
LPD	Landing Platform, Dock
LSD	Landing Ship, Dock
LSM	Landing Ship, Medium
LST	Landing Ship, Tank
LVT	Landing Vehicle, Tracked
MAC	Marine Amphibious Corps
MAF	Marine Amphibious Force
MDO	multidomain operations
MEF	Marine Expeditionary Force
NATO	North Atlantic Treaty Organization
PLA	People's Liberation Army (China)
PLAAF	PLA's Air Force
PLAN	People's Liberation Army Navy
PLANMC	People's Liberation Army Navy Marine Corps
PND	Policía Nacional Dominicana
PRC	People's Republic of China
PT boat	patrol torpedo boat
recce	reconnaissance
ROC	Republic of China
SOCOM	U.S. Special Operations Command
SSF	South Sea Fleet (China)
Triple Entente	the formal association between Russia, France, and Great Britain during World War I
UAV	unmanned aerial vehicle
UDF	Union Defence Forces' (South Africa)

ON CONTESTED SHORES

INTRODUCTION

*Timothy Heck, B. A. Friedman,
and Walker D. Mills*

A Military, Naval, Littoral War, when wisely prepared and discreetly conducted, is a terrible Sort of War. Happy for that People who are Sovereigns enough of the Sea to put it into execution! For it comes like Thunder and lightning to some unprepared Part of the World.
~ Thomas More Molyneux, 1759[1]

"The Marines have landed and have the situation well in hand," a concept popularized by correspondent Richard Harding Davis at the end of the nineteenth century, has served as a buzz phrase, recruiting slogan, and catchphrase signifying that the United States is taking decisive action in response to a crisis somewhere in the world.[2] While stirring and captivating, the phrase implies a simplicity, an almost mathematical certainty to amphibious operations: problem + Marine Corps = problem solved.

The reality of landing Marines, or any amphibious force, however, is a decidedly more complex process than just crossing the beach. A successful amphibious force is far from something that happens overnight or in an ad hoc manner. At the least, forces need to be raised, equipped, trained, provided with doctrine, transported, landed,

---

[1] Quoted in *Expeditionary Operations*, Marine Corps Doctrinal Publication 3, with change 1 (Washington, DC: Headquarters Marine Corps, 2018), 4-1.
[2] "Famous Quotes," Marine Corps History Division, accessed 20 December 2023.

supplied, evacuated, and supported. They need to communicate, conduct reconnaissance, interact with local populations, administer their own population, and prepare to repeat the operation again. Similarly, the forces needed to repel amphibious landings must also be carefully planned for and prepared. For any situation to be well in hand, an amphibious force needs a planned starting point and a decided end state.

The littoral battlespace, focus of Molyneux's opening epigraph, remains just as vital a battlespace today as it did in 1759; and amphibious forces, vessels, and concepts remain crucial to understanding how war can and is being prosecuted. In the Indo-Pacific, the Philippine Navy is engaged in an ongoing struggle with the Chinese People's Liberation Army Navy over sovereignty and control of the islands making up the South China Seas. Most notably, in 1999, the Filipino government ran the BRP *Sierra Madre* (LT 57) aground on the Second Thomas Shoal in the Spratly Islands to help bolster its claim against Chinese expansionism.[3]

While seemingly part of a nonamphibious operation, the *Sierra Madre* was built in World War II for just that purpose. Commissioned into the U.S. Navy as a Landing Ship, Tank (LST) in 1944, it served in the Western Pacific until mothballed at war's end. After nearly two decades, the ship was recommissioned and served as part of the logistics basing for the Mobile Riverine Force in the Mekong Delta before being turned over to the Republic of Vietnam Navy in 1970 as the *My Tho*. With the collapse of Saigon imminent, the *My Tho* set sail loaded with refugees, eventually docking at the American naval station at Subic Bay, where diplomatic agreements transferred it to the Philippine Navy in 1976.[4] Now aground, the ship serves as an outpost and visible reminder of Filipino sovereignty in these contested seas. It is resupplied by at least one other World War II-vintage LST, the BRP *Benguet* (LS 507).[5] The Philippine military has almost taken a page directly from then-Commandant of the Marine Corps General David H. Berger's June 2023 *Force Design 2030 Annual Update*: "Amphibious warfare ships are the cornerstone of maritime crisis response, deterring adversaries, and building partnerships. They persist forward, are globally deployable, and offer fleet and joint force commanders flexible and tailorable force options in competition and conflict."[6]

As Douglas Nash writes in his chapter, the development of the LSTs, like all purpose-built amphibious technology, was one fraught with progress and setbacks. That these two ships, laid down in southern Indiana in 1944, continue to play a significant role in global politics is a tribute to their designers, builders, crews, and the

---

[3] Jon Hoppe, "The Measure of the Sierra Madre: The Extensive History of the Sierra Madre, Originally the USS LST-821," *Naval History Magazine*, vol. 36, no. 1, February 2022.
[4] Hoppe, "The Measure of the Sierra Madre."
[5] Camille Elemia, "How a Decaying Warship Beached on a Tiny Shoal Provoked China's Ire," *New York Times*, 11 November 2023.
[6] Gen David H. Berger, *Force Design 2030: Annual Update, June 2023* (Washington, DC: Headquarters Marine Corps, 2023), 4.

enduring value of amphibious forces and technology. The *Sierra Madre* and *Benguet* have witnessed and served in the evolution of a variety of amphibious operations since being laid down, including full-scale landings, low-intensity conflict, humanitarian operations, and now, for the *Sierra Madre*, as a focal point for regional and global strategic competition.

Work on this volume started before the Russian invasion of Ukraine in February 2022. With Russia's seizure of Crimea in 2014, Russian naval dominance in the littoral waters around Ukraine should have provided ample opportunity to employ Russia's Naval Infantry (*Morskaya pekhota Rossii*) in an attempt to strike well behind Ukrainian front lines. Instead, the Russian Naval Infantry seems to have conducted only a few small-scale landings and elements of it have been soundly defeated by the defending Ukrainians.[7] While this lack of D-Day-style landings might be shocking to Western observers, Soviet amphibious doctrine, which the Russians are heir to, categorized landing operations in a variety of ways, only two of which were operational or strategic in nature.[8] The Russian Naval Infantry's presence alone, combined with Russian naval reach, provides a valuable service and capability to Russian commanders, giving them assets to conduct tactical and operational maneuver from the sea while requiring Ukrainian planners to calculate a potential Russian landing into defensive considerations.[9]

But amphibious operations, even in the Molyneux or Davis version, are more than just boats crashing over the surf to discharge troops. The chapters in this volume reflect that expansion and are divided into the following sections:

Doctrine and Logistics
Technology and Innovation
Organization and Training
Policy and Interoperability, and
Military Materiel and Personnel

Each chapter largely nests in its selected theme but, as with the blended and combined nature of amphibious operations, elements bleed from one to the other. New Zealand's landing on Green Island in 1944, for example, could not have been facilitated without technology provided by or organizational lessons learned by others previously.

Underlying all of them, though, is belief that amphibious operations remain relevant. The Marine Corps, which has started a massive organizational redesign to ad-

---

[7] See Michael Schwirtz et al., "Putin's War," *New York Times*, 16 December 2022.
[8] V. I. Achkasov and N. B. Pavlovich, *Sovetskoe voenno-morskoe iskusstvo v Velikoĭ Otechestvennoĭ voĭne* [Soviet Naval Operations in the Great Patriotic War, 1941-1945], trans. U.S. Naval Intelligence Command Translation Project (Annapolis, MD: Naval Institute Press, 1981), 97.
[9] For more, see Walker Mills and Timothy Heck, "What Can We Learn about Amphibious Operations from a Conflict that Has Had Very Little of It? A Lot," Modern War Institute, 22 April 2022.

Introduction

dress the expected future battlefield, still sees them as central to purpose and identity. We hope this volume provides ideas, inspiration, and debate about the application of amphibious power, reinforcing the idea that the Marines will be landing and soon have the situation well in hand.

CHAPTER ONE

The Landing at Collado Beach

The Logistical Importance of the Amphibious Landing near Veracruz during the Mexican-American War[1]

Christopher Menking

The Mexican-American War is the United States' first war of expansion against a large foreign nation. It represents the nation's first large-scale invasion of another county, mobilizing armies and fighting in three separate foreign theaters of war, and maintaining logistical networks to support these armies in the field across the North American continent. During the war, the U.S. military grew to meet the new demands of a foreign conflict. The Regular Army, the U.S. Army Quartermaster Department, and U.S. Marine Corps saw permanent expansion during the war, laying the foundation for future growth.

The campaign for central Mexico presented the most significant logistical challenge of the war. The Quartermaster Department would have to transport supplies and soldiers from New England manufacturing depots to New Orleans then to the various ports on the Rio Grande and in Mexico. The campaign brought not only the risk of the gulf but also the added obstacle of springtime diseases that plagued the Mexican coast each season. Further complicating this endeavor was the fact that the campaign began with the first major joint amphibious operation for the U.S. Army and Navy, which caused additional logistical hurdles that needed to

---

[1] Much of this chapter is based on Christopher Menking, "Remembering the Forgotten D-Day: The Amphibious Landing at Collado Beach during the Mexican War" (thesis, University of North Texas, October 2013).

be surmounted to achieve success. With utmost preparation, the Quartermaster Department helped coordinate the largest amphibious invasion of the war and up until World War II.

The true heart of the invasion's success lay with the interdepartmental cooperation between the Army under General Winfield Scott, the Navy led by Commodore David Conner, and the Army Quartermaster Department commanded by General Thomas Sidney Jesup. The ability and willingness of these three men to cooperate at a time when the United States military was often rife with internal conflict and political intrigue is truly unique. Their cooperation was not perfect. However, when it mattered, each set aside their pride to assure success of the operation. Scott and Jesup buried issues from previous wars, Jesup and the quartermasters executed the onerous demands of the invasion, and Scott subordinated Army troops to Conner and the Navy to ensure the landing was successful. During the months of planning and movement, these three men and their subordinates worked surprisingly well together and achieved one of the most important victories of the war that led to Mexico's ultimate surrender.

On 27 October 1846, General Scott submitted a memorandum proposing an invasion of Mexico from the coast titled "Vera Cruz and Its Castle." He discussed what would be necessary to capture the port city of Veracruz and its protecting castle, San Juan de Ulúa. President James K. Polk, Secretary of War William L. Marcy, and the rest of the cabinet had been debating the best course to bring the war to a close. They knew that the Army must take possession of Mexico City to force the Mexican government to admit defeat and come to the negotiating table. In the early months of the war, it became clear that it would not be logistically feasible for General Zachary Taylor to march his army to Mexico City from the north. There simply were not enough roads; the terrain was extremely hostile, being mostly desert; and the distance to maintain the supply lines to the Army would have been too great.[2]

Scott's memorandum argued cogently that the capture of Veracruz without an advance inland would be meaningless. With the expectation that the capture would be "a step towards compelling Mexico to sue for peace," Scott outlined what forces he believed would be needed to capture Veracruz, including "an army of at least ten thousand men, consisting of cavalry (say) 2,000, artillery (say) 600, and the remainder infantry." The full memorandum outlined the preliminary expectations Scott had regarding what forces were needed to land in the face of what he expected would be staunch opposition. Not only did Scott believe that the landing would meet Mexican resistance on the beach, but he "did not doubt meeting at [the] landing the most formidable struggle of the war. No precaution was therefore neglected." Ten thousand troops, custom built landing craft, and support from the Navy were all essential com-

---

[2] K. Jack Bauer, *The Mexican War, 1846-1848* (New York: Macmillan, 1974), 233.

ponents to success in Scott's mind. Time would deprive him of much that he wanted, but he would receive enough of each of these three components to execute a successful landing, which fortunately proved to be unopposed.[3]

On 16 November 1846, after four more days of prodigious activity, Scott produced yet another memorandum summarizing the needed troops, supplies, and ships for the operation, which he gave to Marcy:

> *For transporting 14,000 men to Veracruz, with horses, artillery, stores, and boats, 50 ships, of from 500 to 750 tons each.*
>
> *The Boats of the blockading squadron are not, I learn, capable of putting ashore, at once, more than (say) 500 men—only one have the number to be drawn from that fleet.*
>
> *We should therefore require (say) 140 flat boats, to put ashore at once, say 5,000 men, with 8 pieces of artillery. Horses might follow in the second or their trip of boats.*
>
> *The form of the boats, &c., shall be determined by to-morrow, when orders may be given for their purchase, (probably) construction. Colonel Stanton, chief quartermaster, is expected back to-night.*
>
> *The ships need not (to avoid demurrage) be chartered until the troops are known to be nearly in position to embark.*
>
> *P.S.—Orders should be given at once, to have in readiness to be shipped, ordnance and ordnance stores for the water expedition.*[4]

The Quartermaster Department was already working at full capacity to supply both the Army and Navy with necessary supplies and ships. Scott's memorandum placed a whole new burden on the department. While maintaining its already high level of production and procurement, the department now had to supply, move, and support an additional army in the field. Beyond the daunting new task of Scott's expedition was the short time frame the general placed on the production of materiel and the movement of troops. Springtime in the Gulf of Mexico brought malaria and yellow fever to the Mexican shore. In Spanish, yellow fever was called the vomito negro because of the black, tar-like vomit that its victims expelled. The disease is transmitted by mosquito and can debilitate a person within a day of infection and roughly 25–50 percent of all victims die. Scott hoped to land, capture Veracruz, and

---

[3] K. Jack Bauer, *Surfboats and Horse Marines: U.S. Naval Operations in the Mexican War, 1846–48* (Annapolis, MD: U.S. Naval Institute, 1969), 63–64, 66; and Winfield Scott, "Vera Cruz and Its Castle," in *Messages of the Presidents of the United States, with the Correspondence, therewith Communicated, between the Secretary of War and Other Officers of the Government, on the Subject of the Mexican War*, House Executive Documents no. 60, 30th Cong., 1st Sess., Serial Set 520 (Washington, DC: Wendell and Van Benthuysen, 1848), 1268–74, hereafter *Messages of the Presidents of the United States*.

[4] Winfield Scott, "Memoranda for the Secretary of War," in *Messages of the Presidents of the United States*.

move inland before his army succumbed to the ravages of disease. It was at this point in the war that the Quartermaster Department truly came into its own and stepped up to meet the challenges placed before it.[5]

On 23 November 1846, Scott received his orders from Marcy:

Sir: The President of the United States desires you to repair to the lower Río Grande, in order to take upon yourself the general direction of the war against Mexico from this side of the Continent, and more particularly to organize and conduct an expedition (with the co-operation of the navy) against the harbor of Vera Cruz.[6]

With this order, Scott began his journey from Washington, DC, to Veracruz, where he and Commodore Conner would become the first soldiers to successfully invade the coast of Mexico at Veracruz since Hernán Cortés, conquering again the "Halls of the Montezuma."[7]

Scott requested enough custom-built boats to put ashore 5,000 troops, including light artillery batteries, in the first wave of landings. After receiving Marcy's orders to construct the surfboats, Assistant Quartermaster General Henry Stanton wrote, "The Department has been recently required to provide, at an embarrassingly short notice, one hundred and fifty boats or barges, of the description indicated in the drawings and specifications handed you yesterday, by the 1st of January!" The success of delivering these boats proved to be one of Stanton's greatest achievements during the war. Lieutenant George M. Totten, a Navy officer, designed the surfboats, which were built near Philadelphia. The boats were double-ended, broad-beamed, and flat-bottomed, with frames built of well-seasoned white oak. They were built in three sizes so as to nest together for transport: 40 feet to could hold at least 45 troops, 37 feet to hold approximately 40 troops, and 35 feet to hold a maximum of 40 troops. Each surfboat carried a crew of six oarsmen, one coxswain, and a skipper, and ranged in cost between $795 and $950 per boat. These vessels, given their nesting feature, could be stacked to fit into ships with oversized hatches and be stored in their holds. The boats were completed in the 30 days as Scott had requested, though according to Stanton it was "one of the most difficult orders which has ever been imposed on me." Timely delivery of the surfboats proved to be almost as difficult as their rapid production. The 141 boats in 47 stacks were shipped partly in Army vessels, whose decks had been cut to admit them into the hold, and partly on the decks of vessels chartered by the Quartermaster Department. Only 65 of the 140 finished boats made it to Scott by the time of the landing. Though this was only one-half of the requested amount, it proved

---

[5] Winfield Scott to William Marcy, 16 November 1846, in *Messages of the Presidents of the United States*, 1274.

[6] William L. Marcy to Scott, Projét, 23 November 1846, in *Messages of the Presidents of the United States*, 1275–76.

[7] William L. Marcy to Scott, Projét, 23 November 1846, in *Messages of the Presidents of the United States*, 1275–76.

to be enough to accommodate the original 2,500-troop first wave Scott called for in his first memorandum.[8]

General Jesup, while operating out of New Orleans, directed the new volunteer regiments to be outfitted and put on transports. Steadily new waves of soldiers made their way to Mexico for the continuation of hostilities in the new theater of war. By 10 December, Stanton sent out a circular stating that "instructions have been given to muster the Volunteers into service, by companies, as they report themselves ready without waiting for the enrollment of the entire Regiments." These units would be supplied and ready to embark for Mexico as soon as able rather than delay waiting for the full compliment. Scott's landing needed as many troops as possible before executing the landing, but he had a deadline set by the seasons and the threat of disease.[9]

The Mexico City campaign last touched American soil at Brazos Santiago, Texas, en route to Veracruz. Scott's army of invasion gathered as they awaited troops from New Orleans and pulled regulars from Taylor's forces to make the army that would land on Collado Beach, south of Veracruz. This depot became one of the main coaling stations for the Army transports and Navy vessels on their journeys south to Tampico or Veracruz. In late 1846, Jesup traveled to Brazos Santiago to help coordinate Scott's landing at Veracruz. The bulk of supplies from New Orleans traveled through the harbor at Brazos Santiago. The growing port became the key forward logistical center for the entire war.[10]

During early January, the department continued working to move the new volunteers from across the United States equipped and transported to Mexico. Most of the new recruits mustered into service either traveled to join Scott in the invasion at Veracruz or to reinforce Taylor's forces in northern Mexico. Transfers of experienced Regular Army troops to Scott's expedition left Taylor with depleted forces. The new recruits easily filled the gaps in the ranks for the armies remaining in northern Mexi-

---

[8] "Boats: Surf Boats of Mexican War," 31 December 1846, John Lenthall Papers (1794-1865), Independence Seaport Museum, Philadelphia; Bauer, *Surfboats and Horse Marines*, 63-64, 66; Chester L. Kiefer, *Maligned General: The Biography of Thomas Sidney Jesup* (San Rafael, CA: Presidio Press, 1979), 285; K. Jack Bauer, "The Veracruz Expedition of 1847," *Military Affairs* 20, no. 3 (Autumn 1956): 164; Winfield Scott, "Vera Cruz and Its Castle," in *Messages of the Presidents of the United States*, 1268-74; Ivor D. Spencer, *The Victor and the Spoils: A Life of William L. Marcy* (Providence, RI: Brown University Press, 1959), 147, 164; and William G. Temple, "Memoir of the Landing of the United States Troops at Veracruz in 1847," in Philip Syng Physick Conner, *The Home Squadron under Commodore Conner in the War with Mexico, Being a Synopsis of Its Services, 1846-1847* (n.p., 1896), 60-62.

[9] Henry Stanton to B. Alvoro, 3 December 1846, Henry Stanton to John Goolrick, 3 December 1846, Letters Sent by the Office of the Quartermaster General, microfilm no M745, Roll 21, 309-10; and Henry Stanton to Thomas Jesup, New Orleans, 3 December 1846, United States Department of War, Letters Sent, Roll 21, 311.

[10] Edward J. Nichols, *Zach Taylor's Little Army* (Garden City, NY: Doubleday, 1963), 194; William H. Samson, ed., *Letters of Zachary Taylor from the Battle-Fields of the Mexican War* (Rochester, NY: Genesee Press, 1908), 104; J. Jacob Oswandel, *Notes of the Mexican War, 1846-1848* (Knoxville: University of Tennessee Press, 2010), 26; W. L. Marcy to MajGen Z. Taylor, in *Messages of the Presidents of the United States*, 365-66; and Thomas T. Smith, *The U.S. Army and the Texas Frontier Economy, 1845-1900* (College Station: Texas A&M University Press, 1999), 16, 20, 24, 71, 112, 138.

The Landing at Collado Beach

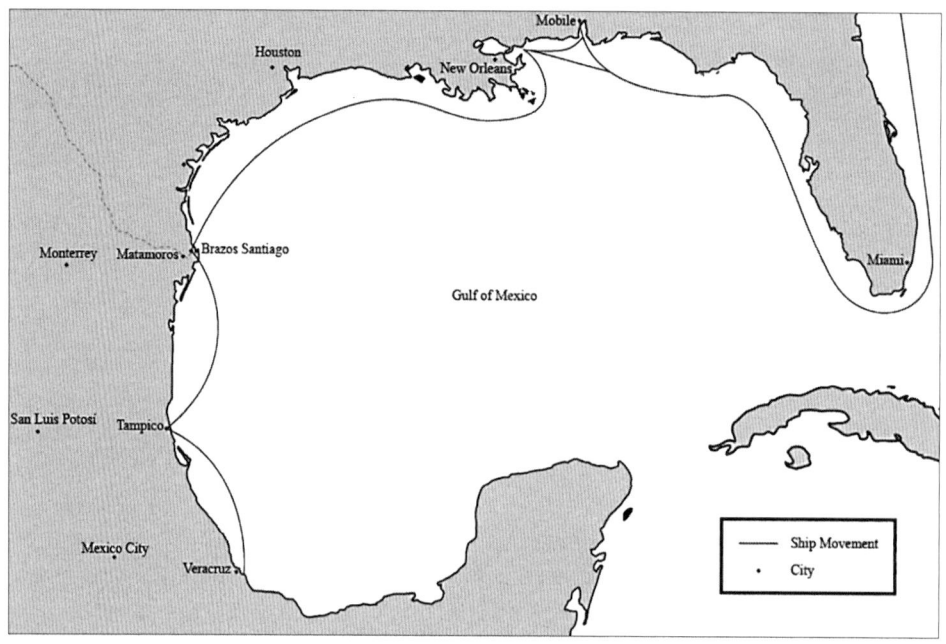

MAP 1
Troop and materiel movement to Mexico, 1846.
Source: sourtesy of the author, adapted by MCUP

co. Since it seemed likely most of the fighting was moving to central Mexico, inexperienced volunteer recruits would be far less of a liability in the northern theater than in active combat during the Mexico City campaign.[11]

Scott reached Brazos Santiago from New Orleans on 27 December 1846. Within a week, on 3 January, he called for the detachment of a portion of General Zachary Taylor's troops—1,000 cavalry, 4,000 regulars, and 4,000 volunteers, less those already headed to Tampico—for departure from that port. On 4 January, Secretary Marcy noted that their intelligence had reported no large covering army being assembled at Veracruz, and he expressed hope that Scott would be able to take the city and castle before such a force could be assembled.[12]

Despite lingering differences from the Second Seminole War, Scott supported Jesup's proposal to Secretary Marcy for an increase of the Quartermaster Department's assigned officers. During the previous war Jesup wrote a series of misunderstood and poorly conceived letters that created unnecessary tension between himself and Scott,

---

[11] Henry Stanton to Samuel Dusenbery, 2 January 1847, Henry Stanton to John Goolrick, 4 January 1847, Henry Stanton to R. F. Loper, 4 January 1847, Henry Stanton to D. H. Vinton, 9 January 1847, Letters Sent by the Office of the Quartermaster General, microfilm no M745, Roll 21, 368–70, 382.

[12] Bauer, *The Mexican War, 1846–1848*, 238; W. L. Marc to MajGen Winfield Scott, in *Messages of the Presidents of the United States*, 391; and Henry Stanton to R. F. Loper, Philadelphia, 4 January 1847, United States Department of War, Letters Sent, Roll 21, 370.

significantly straining their long friendship. Jesup wrote a letter to President Andrew Jackson complaining of Scott's "foot dragging," which resulted in Jackson ordering Scott back to the capital to explain his failures before a military court of inquiry. Scott's wrath fell on Jackson, Jesup, and Brevet Major General Edmund Pendleton Gaines, the latter also wrote a letter placing the blame for the military failures in Florida on Scott's shoulders. Ultimately, Scott received vindication from the court with Jesup and Gaines receiving slaps on the wrists for comments made to the press. With Scott's help, Jesup received the first department expansion in almost 25 years. Jesup recommended and received the appointment of 4 additional majors and 10 additional assistants from the Army. Scott worked with Jesup and the Quartermaster Department to quickly achieve ambitious goals. Given the confines of time, the Department performed admirably, due in large part to Jesup's hard work and effort. Such cooperation and commitment to the invasion contributed to its ultimate success.[13]

Stanton informed Jesup on 14 January 1847 that the 140 boats requested by General Scott were completed. More importantly, all of the ships carrying the nesting landing crafts were en route to Brazos Santiago. The transport ships carrying the surfboats had been purchased for this sole purpose, which added to the expeditious nature of their shipping. In addition, many of the ships that were carrying volunteers and ordnance destined for Mexico were already at sea or would be leaving shortly. Transports containing ordnance that had yet to set sail received direction to Isla de Lobos, roughly 340 kilometers from Collado Beach, rather than Brazos Santiago due to their late departure. The department achieved the task of maintaining Taylor's forces while building a separate army for invasion of the Mexican coast. Though undoubtedly stressful for the quartermasters involved, the achievement of both material production and manpower transportation is one that surpassed military efforts up to this point in United States history.[14]

At Brazos Santiago, every brigade commander was exceedingly anxious to avail themselves to Scott to ensure that their brigade would participate in the amphibious assault; "but General Scott, with his usual military diplomacy, met all such application with the stereotyped assurance that there would be more work to do than he had troops to accomplish, and that before they reached the City of Mexico they would have all the fighting they wanted." Brevet Brigadier General William J. Worth arrived

---

[13] Kiefer, *Maligned General*, 119-22; John S. D. Eisenhower, *Agent of Destiny: The Life and Times of General Winfield Scott* (Norman: University of Oklahoma Press, 1999), 163-66; Th. S. Jesup to Wm. L. Marcy, in *Messages of the Presidents of the United States*, 568-69; and Erna Risch, *Quartermaster Support of the Army: A History of the Corps, 1775-1939* (Washington, DC: Quartermaster Historian's Office, Office of the Quartermaster General, 1962), 286.

[14] Henry Stanton to Thomas Jesup, New Orleans, 14 January 1847, United States Department of War, Letters Sent, Roll 21, 384-85; Henry Stanton to D. H. Vinton, New York, 19 January 1847, United States Department of War, Letters Sent, Roll 21, 393; and Henry Stanton to Thomas Jesup, New Orleans, 29 January 1847, United States Department of War, Letters Sent, Roll 21, 415-16.

with his division of regulars at the mouth of the Río Grande on 22 January, providing his troops with the honor of serving as the vanguard of the invasion.[15]

General Scott had hoped to sail from Brazos Santiago by the beginning of February to avoid *vómito negro* season, but delays pushed the departure of the expedition force back until mid-February. Despite delayed ships, Scott ordered Captain A. R. Hetzel—a Brazos Santiago quartermaster—to charter enough vessels locally to get the troops at Brazos Santiago afloat by 10 February and those at Tampico afloat five days later. On 15 February 1847, Scott set sail from Brazos Santiago aboard the steamer USS *Massachusetts* (1845), destined for Tampico. With his departure, he left orders that after replenishing their water tanks, all ships with troops or supplies destined for the landing were to rendezvous on the ocean side of the barrier island of Isla de Lobos.[16]

General Scott arrived at Tampico on 16 February, leaving two days later for Isla de Lobos.[17] Colonel George T. M. Davis—aide-de-camp to General Scott—noted that "the arrival on the 16th of General Scott and his staff was strong as proof of Holy Writ that the hour of action was at hand, and the enthusiasm and military demonstration with which his advent was hailed at Tampico must have convinced him that he enjoyed the unlimited confidence of the citizen-soldier composing the brigades of Generals Quitman, Shields, and Pillow." Scott stopped in Tampico to monitor embarkations at Tampico of some of the remaining troops there, namely regulars. Once his orders were issued, he continued to Isla de Lobos.[18]

Isla de Lobos is located about 97 kilometers south of Tampico, roughly 13 kilometers east of Tamiahua Lagoon.[19] General Scott arrived on 21 February, bringing troops with him to join the troops already on the island. The day after his arrival, Scott informed Commodore Conner that he was sending two vessels with ordnance supplies, two with surfboats, and some transports ahead to Antón Lizardo, about 26

---

[15] George T. M. Davis, *Autobiography of the Late Col. Geo. T.M. Davis, Captain and Aide-de-camp Scott's Army of Invasion (Mexico), from Posthumous Papers* (New York: Jenkins and McCowan, 1891), 121; and Winfield Scott to William Marcy, Brazos San Iago, 24 January 1847, in *Messages of the Presidents of the United States*, 856.

[16] Winfield Scott to William Marcy, Brazos San Iago, 12 January 1847, in *Messages of the Presidents of the United States*, 844-46; Winfield Scott to Commodore Connor, at Sea, 26 December 1846, in *Messages of the Presidents of the United States*, 846-47; Winfield Scott to W. O. Butler, Camargo, 3 January 1847, in *Messages of the Presidents of the United States*, 851-852; and Bauer, *Surfboats and Horse Marines*, 71-72.

[17] There is some confusion about when Gen Scott actually arrived in Tampico. Some sources say 16 February 1847, others record it as 18 February. Given the distance of approximately 483 kilometers, Scott aboard the *Massachusetts* should have been able to make the trip within a day. Given this information, 16 February will be the date used here.

[18] Davis, *Autobiography of the Late Col. Geo. T.M. Davis*, 121; Winfield Scott to General Brooke, Brazos San Iago, 12 January 1847, in *Messages of the Presidents of the United States*, 855-56; and Bauer, *Surfboats and Horse Marines*, 72.

[19] During the war, Isla de Lobos was chosen for its good harbor. In more recent years, the island has become a favored spot for tourism, especially for divers and fishermen due to the wildlife in the surrounding reefs.

kilometers southeast of Veracruz just off the coast. Scott requested to have the troops land and encamp ashore. On 26 February, Scott informed Conner that once the regulars, one-third of his siege train, and more surfboats arrived, he would leave for Antón Lizardo and attempt a landing.[20]

The vanguard of transports reached Antón Lizardo on 4 March 1847. On 5 March, Scott, arrived aboard the *Massachusetts*. Lieutenant Raphael Semmes records that "our hitherto quiet headquarters, in which we had stagnated all winter, became daily more animated, until Antón Lizardo was crowded with a magnificent fleet of steamers and sail-vessels; all bearing at their gaff-ends the proud flag of the republic." While at Antón Lizardo, Scott issued General Order No. 45, which assigned the three landing waves to their respective transports and specified which units would be in each line. The first line was under the command of General William Worth. The second under the command of General Robert Patterson, and the third included the reserves placed under General David E. Twiggs that were made up of the 2d Brigade of Regulars. An amphibious landing in the line of battle presented a daunting task.[21]

After arriving at Antón Lizardo, Scott joined Conner on the steamer USS *Petrita* (1846) to reconnoiter the beaches between Anton Lizardo and Veracruz for a suitable location to land the surfboats. Accompanying them were Worth, Twiggs, Patterson, and Major General Gideon Pillow, as well as Scott's staff, including Captains Robert E. Lee and Joseph E. Johnston and Lieutenants Pierre G. T. Beauregard and George G. Meade. After discussion, they decided on a sandy stretch of shore almost five kilometers south of Veracruz, beyond the range of its guns. They chose "a gently curving strip of sand paralleled by a line of sand hills about 150 yards inland, Collado Beach lies behind Sacrificios Island, two and one-half miles southeast of Veracruz." This would soon prove a fortuitous choice for the troops making the landing.[22]

At daylight on 9 March 1847, the troops assembled. General Scott could not have chosen a better day. Historian K. Jack Bauer later poetically described it as "a brilliant sun sparkled in the cloudless blue sky and illuminated the snowcapped grandeur of distant Mount Orizaba once again looking upon a conqueror landing at Veracruz." Lieutenant Semmes noted that "if we had had the choice of weather, we could not have selected a more propitious day." Many of the soldiers and officers in their journals mentioned a feeling or connection to the time of Hernan Cortez, as if this invasion force were walking in the conquistadors' footsteps. Scott felt that "the sun

---

[20] Bauer, *Surfboats and Horse Marines*, 75.
[21] Bauer, *The Mexican War, 1846–1848*, 240; Bauer, *Surfboats and Horse Marines*, 76–77; Adm Raphael Semmes, *Memoirs of Service Afloat during the War between the States* (Baltimore, MD: Kelly, Piet, 1869), 125; Conner, *The Home Squadron under Commodore Conner in the War with Mexico*, 19; Roger G. Miller, "Winfield Scott and the Sinews of War: The Logistics of the Mexico City Campaign October 1846–September 1847" (master's thesis, North Texas State University, 1976); and General Orders No. 45, 7 March 1847, Adjutant General's Office General Orders, Record Group 94.
[22] Bauer, *The Mexican War, 1846–1848*, 241; Bauer, *Surfboats and Horse Marines*, 77; Conner, *The Home Squadron under Commodore Conner in the War with Mexico*, 19; and Temple, "Memoir of the Landing of the United States Troops at Veracruz in 1847," 64.

dawned propitiously on the expedition." As if predestined, the landing took place on the 33d anniversary of Scott's promotion to the rank of general. The auspicious day was enhanced by calm seas with little surf on the beach, a condition Scott felt was necessary for the landing.[23]

At 0945 that morning, the covering force hoisted anchor and sailed for the landing area. The USS *Reefer* (1846), *Bonita* (1846), *Petrel* (1846), *Tampico* (formerly USS *Pueblano*), and *Falcon* (formerly USS *Isabel*), which formed the inshore covering force, hoisted anchor and stood out for the landing area. Fifteen minutes later, the USS *Raritan* (1843) signaled the main body of ships to prepare to get under way. According to William G. Temple, serving under Conner,

> *all preliminary arrangements having been made, between 11:00 a.m. and 12 o'clock noon, the fleet—Commodore Conner leading, in the flag-ship* Raritan *under Captain Forrest, whose decks, like those of the other ships, were crowded with troops, and General Scott following at a short distance, in the steamer Massachusetts—got underway, in gallant style, and filed, one by one, out of the narrow pass leading from the anchorage.*[24]

General Scott wrote that "the whole fleet of transports—some eighty vessels, in the presence of many foreign ships of war, stood up the coast, flanked by two naval steamers and five gunboats to cover the movement. Passing through them in the large propeller, the *Massachusetts*, the shouts and cheers from every deck gave me assurance of victory, whatever might be the force prepared to receive us." Even though the beach did not have defenses built on it, Scott believed Worth and his troops would face Mexican forces that would try to throw the Americans back into the sea.[25]

At 1215 that afternoon, the inshore covering force moved offshore of Collado Beach. The next three hours were filled with the movement of the larger vessels as they appeared and moved to their assigned posts. At 1245, the *Reefer* and accompanying gunboats arrived off Isla Sacrificios, directly across from the city of Veracruz and less than 10 kilometers southeast of San Juan de Ulúa. The rest of the ships soon arrived and took their assigned places with little disorder or confusion. Once they were safely anchored, the steamers cast the surfboats loose, whose oarsmen propelled them to the troop ships to embark their passengers. At 1530, the steamers USS *Spitfire* (1846) and *Vixen* (1846), with five schooner gunboats of the inshore force, closed to within 90 yards of shore. During this preparation, the flotilla of gunboats attached to the squadron under Commander Josiah Tattnall as senior officer took position within

---

[23] Bauer, *The Mexican War, 1846–1848*, 242; Semmes, *Memoirs of Service Afloat during the War between the States*, 126; Bauer, *Surfboats and Horse Marines*, 79; and Winfield Scott, *Memoirs of Lieut.-General Winfield Scott*, ed., Michael Gray and Timothy D. Johnson (Knoxville: University of Tennessee Press, 2015), 418–19.
[24] Temple, "Memoir of the Landing of the United States Troops at Veracruz in 1847," 67.
[25] Bauer, *The Mexican War, 1846–1848*, 242; Bauer, *Surfboats and Horse Marines*, 80; Semmes, *Memoirs of Service Afloat during the War between the States*, 126; Conner, *The Home Squadron under Commodore Conner in the War with Mexico*, 19; and Scott, *Memoirs of Lieut.-General Winfield Scott*, 419.

grapeshot range of the beach, so as to cover the landing with its guns, as previously ordered by Commodore Conner.[26]

Three flags were hoisted on the main truck of the *Massachusetts*, signaling Worth's division to prepare for the landing. Soldiers clambered down into the surfboats. Lieutenant Semmes remembered that "the surfboats, 67 in number, and each one manned by experienced seamen of the navy, were hauled alongside of the ships; the soldiers, with their arms and accoutrements, were passed into them; and as each boat received her complement, she shoved off, and laid on her oars at a little distance, until the others should be ready." When each detachment was ready, it formed up in the line of battle parallel to Collado Beach and abreast to the acting naval transports some 450 yards offshore. The strong currents that swirled around Isla Sacrificios and its reef threw the surfboats into confusion. The units became mixed up, but rather than sort them out boat by boat, General Worth ordered that each regiment pull for the boat with its regimental colors hoisted. The perfect line of battle was lost, but each surfboat landed next to others in their regiment.[27]

While the surfboats formed up parallel to the shore, Mexican cavalry could be seen in the dunes behind the beach. In response, the mosquito fleet, a U.S. Navy squadron detachment, under Commander Tattnall ran close into the beach and kept up constant shelling. At 1700, the *Tampico* hurled a 24-pound shell at cavalry who could be seen on the dunes behind the beach. The shot had no visible effect on the cavalry. For the anxious Americans, this cemented their fear that the landing force would have to fight strong Mexican opposition to claim the beach.[28]

At 1730, the *Massachusetts* fired a shot, signaling the beginning of the landing. The cannon silenced the murmur among the fleet; all eyes were fixed on the surfboats as the sailors pulled hard to cover the 450 yards to the beach. The setting sun behind the dunes silhouetted the walls and castle of Veracruz. While the small surfboats closed in on the beach, not a single crack of musket fire was heard from the shore. Then, just before the surfboats touched the sand, a figure leaped out of one of the craft into water up to his armpits. He waded ashore. It was General Worth. His staff followed him onto the beach, and surfboats began hitting the sand all around them.[29]

In a matter of moments, the first wave followed Worth, 2,595 troops in all, onto the beach without a single casualty. Oswandel watched from his ship and remembered that "as soon as the surf boats struck the beach the soldiers instantly jumped on shore,

---

[26] Bauer, *Surfboats and Horse Marines*, 8, 80; Bauer, *The Mexican War, 1846–1848*, 242; and Conner, *The Home Squadron under Commodore Conner in the War with Mexico*, 19. Grapeshot refers to a type of cannon charge using round pellets that when fired spread in an effect much like a shotgun blast. This type of shot was particularly devastating against infantry.

[27] Bauer, *Surfboats and Horse Marines*, 80–81; and Semmes, *Memoirs of Service Afloat during the War between the States*, 126–27.

[28] Capt William Harwar Parker, *Recollections of a Naval Officer, 1841–1865* (New York: Charles Scribner's Sons, 1883), 84; and Bauer, *Surfboats and Horse Marines*, 81.

[29] Bauer, *The Mexican War, 1846–1848*, 242, 244; and Bauer, *Surfboats and Horse Marines*, 81–82.

The Landing at Collado Beach

some in the water. We are now looking for the Mexicans to attack our men, but on they rushed in double quick time until they came to a sand hill. Here they planted the flag of our country with three hearty cheers, responded to with great enthusiasm by every soldier on board the ships." At 1740, Worth's troops planted the American flag on the dunes, as "the troops debarked in good order; and in a few minutes afterward a detachment, which had wound its way up one of the sand-hills, unfurled the American flag, and waving it proudly over their head, planted it in the land of Cortez."[30]

When the American soldiers reached the top of the sand hills, they realized that the Mexicans had fled back behind the safety of the city walls. After the first assault, the remaining United States forces landing at the beach no longer tried to land in the order of battle. In less than five hours, more than 10,000 troops landed at Collado Beach without a single loss of life.[31] Extra care had to be taken in landing the siege train. At first, they tried to land the heavy batteries from two surfboats lashed together, but this did not work. The remaining guns were lowered into the surfboats carefully due to their fragile pine bottoms.[32]

This landing positioned the American forces to besiege and take the city of Veracruz, beginning the march to Mexico City. If the Mexican soldiers had met the Americans on Collado Beach, the Army would have been in far worse shape. This decision not to resist the landing by the Mexican commander changed the landing from a hazardous amphibious assault to a perfect example of how to execute such an operation flawlessly for future American military leaders. During the next week, Scott directed his forces to take up positions around Veracruz to begin the siege. General Scott chose to besiege the city rather than assault it, as was the popular idea among his men. He did so to save American lives and those of citizens in the city. As the investment around the city continued, the Mexicans sent cavalry to find soft points in the American lines. Brigadier General Juan Morales—the Mexican commander at Veracruz—chose to hold his small garrison within the walls.[33]

The U.S. Navy and Marines participated in the landing first as escorts commanding the landing craft and then fighting alongside the Army once ashore. Sailors and Marines served in naval batteries under the command of the Navy. Some Marines also served with the Army as more traditional soldiers. Eventually, a Marine battalion arrived in Mexico under the command of Lieutenant Colonel Samuel E. Watson. The

---

[30] Oswandel, *Notes of the Mexican War, 1846-1848*, 35-36; Bauer, *The Mexican War, 1846-1848*, 244; Davis, *Autobiography of the Late Col. Geo. T.M. Davis*, 125; Bauer, *Surfboats and Horse Marines*, 82; and Semmes, *Memoirs of Service Afloat during the War between the States*, 128.

[31] The number of troops landed at Collado Beach during these five hours varies between 8,600 and around 13,000, depending on the source.

[32] Bauer, *Surfboats and Horse Marines*, 82, 419-20; Conner, *The Home Squadron under Commodore Conner in the War with Mexico*, 20; and Temple, "Memoir of the Landing of the United States Troops at Veracruz in 1847," 68-69.

[33] Bauer, *The Mexican War, 1846-1848*, 245-48.

battalion was attached later to Major General John A. Quitman's 4th Division. They went on to serve alongside the Army with distinction during the Battle of Chapultepec. Beginning with the landing at Veracruz and the culminating with the victory at Chapultepec, the Marines demonstrated their value to the United States Armed Services.[34]

On 22 March 1847, Scott called for the formal surrender of Veracruz, which Morales rejected. Gun batteries, both ashore and afloat, continued to bombard the walls of the city, trying to force its capitulation. Finally, on 29 March, the formal surrender of Veracruz took place. Scott achieved the surrender by tempering his demands and allowing the Mexican forces to save face. He agreed to parole the whole garrison and allowed civilians free movement around the city. General Worth assumed charge of Veracruz as military governor.[35]

The landing was a success and the city was taken. General Scott and Commodore Conner deserved the accolades given to them for this operation. It was a positive example of what could be accomplished with joint operations. The Army and Navy had worked in unison to achieve a herculean feat at Collado Beach. Midshipman William H. Parker stated that "whatever may be said of Commodore Conner's management of affairs up to this time, the arrangements for this service were perfect."[36]

Commodore Conner must be credited with successfully conducting an incredibly complicated operation. He suggested the landing place, proposed the method of transporting troops to the debarkation point, and handled the details of the landing. General Scott deserved credit for conceiving and planning such an audacious operation. Moreover, Scott managed to land on a hostile shore without much logistical support and not quite the number troops that he deemed minimal to execute the operation. General Jesup and his quartermasters achieved a monumental success by supplying three armies in the field, while also transporting one of those armies to execute the largest amphibious assault to date. With the landing complete, General Scott took Veracruz and began his march to capture Mexico City, the first foreign capital ever occupied by the United States Army.

The U.S. Army Quartermaster Department, under the guidance of Jesup, provided Scott with the manufacturing, transportation, and manpower he needed to undertake one of the most important battles of the war. The supply networks established the department crossed the United States East Coast, the Gulf of Mexico, and inland to Mexico City after Veracruz's capture. The quartermasters overcame the unpredictable northers of the gulf, the risk of disease along the Mexican coast, and partisan bandits raiding American supply lines in central Mexico. The Quartermaster

---

[34] Gabrille M. Neufeld Santelli, *Marines in the Mexican War*, ed., Charles R. Smith (Washington, DC: History and Museums Division, Headquarters Marine Corps, 1991), 31–33, 36–39.
[35] Bauer, *The Mexican War, 1846–1848*, 249–53.
[36] Parker, *Recollections of a Naval Officer, 1841–1865*, 84.

Department more than met the expectations demanded of them during the war and the experience gained during the Mexican-American War shaped how quartermasters waged the coming American Civil War on a much grander scale.

The war finally came to an official end on 2 February 1848, with the signing of the Treaty of Guadalupe Hidalgo, a small suburb of Mexico City where the Mexican government had fled during the occupation of the city. Diplomat Nicholas P. Trist served as the U.S. representative in the negotiations. The final provisions of the treaty dealt with many of the territorial issues between the neighboring countries, including finalizing the Texas-Mexico border and the ceding of a vast portion of Mexico's far northern land. This treaty led to bitterness on both sides of the war, but the conflict came to an end in a relatively short time as a result of the central Mexico campaign that started with the amphibious operation at Veracruz.[37]

---

[37] The Treaty of Guadalupe Hidalgo, 2 February 1848, National Archives and Records Administration, Washington, DC.

CHAPTER TWO

The Landing Craft Controversy, 1934–1942

Jerry E. Strahan

In his book, *First to Fight*, Marine Corps lieutenant general Victor H. Krulak declared that the American landing craft of 1937 "had not advanced far beyond" what they were "during the Revolutionary War."[1] This lack of advancement can be attributed to three factors. First, extremely limited Navy budgets—funds simply were not available for the development of such craft. Second, there was a belief that advancements in air power had made successful amphibious assaults impossible. Third, during the interwar period, many of America's prewar planners believed that if war broke out in Europe, the French would hold back the invading forces and their ports would remain open as they had during World War I.[2] As a result of this type of thinking, in January 1939, just eight months prior to the start of the war in Europe, the United States had a total of 19 personnel landing craft.[3] This situation would rapidly change, but not without controversy and competition involving the Navy's Bureau of Ships and boatbuilder Andrew Jackson Higgins.

In the late 1920s, Higgins owned a small boatyard in downtown New Orleans where he built rugged workboats for oil exploration and timber companies. These

---

[1] LtGen Victor H. Krulak, USMC (Ret), *First to Fight: An Inside View of the U.S. Marine Corps* (Annapolis, MD: Naval Institute Press, 1984), 90.
[2] VAdm Daniel E. Barbey, USN (Ret), *MacArthur's Amphibious Navy: Seventh Amphibious Force Operations 1943–1945* (Annapolis, MD: Naval Institute Press, 1969), 12.
[3] "Report on Landing Boat Program of the Navy Department," n.d., Senate Documents, Record Group 46, Senate 79A-F30, 33, National Archives, hereafter Senate Report.

FIGURE 1
Andrew J. Higgins.
Source: courtesy the Higgins family

companies operated deep in the Louisiana swamps and needed a boat capable of passing over floating logs, crossing submerged sandbars, pulling up on a riverbank, and then retracting with ease. In addition, these maneuvers had to be accomplished without damaging the boat's hull or propeller.

In 1931, Higgins announced the development of just such a boat—the "Eureka."[4] By necessity and coincidence, the shallow-draft Eureka possessed many of the same characteristics required of future landing craft. Higgins had attempted to interest the Navy in his boats as early as 1927. Of his first meeting, he recalled, "They were very nice, but definitely not interested." He continued calling on the Navy, but claimed that "they did not lend an attentive ear."[5]

---

[4] "Drawing Number Book, no. 1," n.d., Higgins Industries, New Orleans, LA, author's collection, 8; and "23-foot 'Eureka' Model," n.d., advertising letter, Higgins Industries, New Orleans, LA, author's collection. The letter's heading, "Higgins Industries, Inc.," establishes that the letter was written post 26 September 1930–the day Higgins started the company.

[5] "A Revisal by A. J. Higgins Sr., President, Higgins Industries, Inc., of Transcript of Hearing before the Navy Department Price Adjustment Board," 7 October 1943, Sen 79A-F30, OP5, Box 185, U.S. Senate, Record Group 46, National Archives, 19, hereafter Higgins Revisal.

FIGURE 2
An early Eureka workboat designed for oil-field companies, timber companies, and trappers for use in Louisiana's shallow water swamps. The Eureka later evolved into the Navy's LCP and then its 36-foot LCPL and LCVP.
Source: courtesy the Higgins family

It was not until 1934 that he found a group who showed a strong interest in his boat—the U.S. Marine Corps. The Corps' leadership at Quantico, Virginia, were impressed with his Eureka, but the Corps, like the Navy, lacked the funds to purchase such craft.[6] Fortunately, the following year the Navy received a budget increase, thus allowing its Bureau of Construction and Repair (BuC&R) to solicit bids. The BuC&R was looking for a boat of a specific length and weight. What it was hoping to find was an existing commercial boat capable of serving as both a standard launch and a personnel landing craft.[7] Nine New England companies responded, five entries were selected for testing. Four were wooden deep-vee hull Eastern seaboard fishing skiffs and the fifth was a steel commercial craft that was quickly eliminated. From August through October 1936, the fishing boats underwent sea trials at Cape May, New Jer-

---

[6] Higgins Revisal, 19.
[7] LtCol Kenneth J. Clifford, USMCR, *Progress and Purpose: A Developmental History of the United States Marine Corps, 1900–1970* (Washington, DC: History and Museums Division, Headquarters Marine Corps, 1973), 48–49. On p. 3 of Senate Report, the date given for the bids being advertised is early 1936.

FIGURE 3
A Eureka climbs Lake Pontchartrain's concrete seawall to exhibit the strength of its hull.
Source: courtesy the Higgins family

sey, where it was discovered that all four boats had disadvantages that alterations could not correct.[8] Even so, three were chosen for additional testing.

Higgins never submitted a bid. It appears he was not officially notified of the process. Later, he wrote to the BuC&R, "We have been aware that the Navy Department has a need and have been investigating suitable types of boats for parties to land through the surf.... We know that we have designed, perfected, and are building the very type of boat best fitted for this purpose."[9] He then reminded the BuC&R that his company had written them on several occasions and even included specifications and drawings. In closing, he requested that a representative be sent to New Orleans to test a Eureka.[10] No BuC&R representative was dispatched.

Three and a half months after the Cape May trials, the Navy took a major step

---

[8] Senate Report, 6.
[9] "Andrew J. Higgins to the Bureau of Construction and Repair," 1 October 1936, National Defense Committee Files, OP-5, Navy Department Matters Ships, Shipbuilding and related Matters, Box 182, Record Group 46, National Archives. The Navy Department existed until the passage of the National Security Act of 1947, which created the Department of the Navy, officially replacing the Navy Department.
[10] "Andrew J. Higgins to the Bureau of Construction and Repair," 1 October 1936.

in the development of landing craft. The secretary of the Navy created the "Navy Department's Continuing Board for the Development of Landing Boats for Training in Landing Operations," generally referred to as the Landing Boat Board (LBB). Simultaneously, the commander in chief of the United States Fleet established the Fleet Development Board (FDB). The LBB and the FDB were intended to work jointly—the LBB developing the boats and the FDB overseeing their testing and making recommendations to the BuC&R.[11]

Approximately two weeks after the formation of these boards, the BuC&R and the Navy's Bureau of Engineering, authorized the Philadelphia Navy Yard to build a prototype landing boat. The design was intended to incorporate the best features of three of the sea skiffs tested at Cape May. The newly formed FDB objected, reasoning that "little would be gained by constructing a boat so similar in design" to the unacceptable skiffs. In spite of the objection, the BuC&R had the Philadelphia Navy Yard move forward with building a 30-foot landing boat. Additionally, the three skiffs continued to be tested.[12]

At a later meeting of the LBB, various potential landing boats were discussed. Included in the discussion was the design of a 33-foot Eureka submitted by Higgins Industries. The board rejected the Eureka, claiming the design failed to show "sufficient promise" as a landing boat.[13]

In complete contrast to the LBB's actions, when Lieutenant Commander Ralph S. McDowell, the officer responsible for landing craft development in the BuC&R, learned of Higgins's Eureka he wrote to the boatbuilder and invited him to Washington. Higgins accepted and the two men spent a week discussing the Eureka's design and capabilities.[14] Little else could be accomplished because funds were still limited, and landing craft were low on the Navy's list of priorities.

Tests continued to be run on the fishing skiffs and the BuC&R's Philadelphia boat. In early 1938, all four craft participated in Fleet Landing Exercise 4 (FLEX 4). Despite the fact that the Philadelphia boat was considered "the least suitable" of all of the boats tested, the BuC&R ordered five additional boats of the same design. The senior member of the FDB "urged" that the order be canceled. The Chief of Naval Operations (CNO) responded, "Until a more suitable boat can be developed, their completion is considered justified."[15]

During this same period, McDowell once again contacted Higgins. This time to inform him that the Navy had $5,200 available to purchase a 30-foot experimental

---

[11] VAdm George Carroll Dyer, USN (Ret), *The Amphibians Came to Conquer: The Story of Admiral Richmond Kelly Turner*, vol. 1 (Washington, DC: Department of the Navy, 1969), 205.
[12] Senate Report, 8–11.
[13] Senate Report, 14.
[14] LtCol Frank O. Hough, USMCR, Maj Verle E. Ludwig, USMC, and Henry I. Shaw Jr., *History of the U.S. Marine Corps Operations in World War II*, vol. 1, *Pearl Harbor Guadalcanal* (Washington, DC: Historical Branch, G-3 Division, Headquarters Marine Corps, 1958), 26–27.
[15] Senate Report, 18–20.

landing craft. If he would agree to furnish a boat of his own design, at the specified price, and not exceeding 30-feet in length, he would be awarded a contract.[16]

Higgins vehemently opposed the 30-foot requirement. In his opinion, the boat's beam was too wide for its length. He believed the boat should be at least 39-feet long. However, despite the 30-foot limit and the fact that the boat would cost considerably more than $5,200 to build, he accepted McDowell's offer.[17]

After approximately 11 years of calling on the Navy, on 5 May 1938, Higgins received his first contract. Within weeks, he had a 30-foot Eureka ready for shipment. The boat cost more than $12,500 to build. There was also the additional expense of transporting it and the cost of sending a retired captain, Bert Oakley, to properly demonstrate its capabilities.[18]

On 27 May 1938, Oakley sent Higgins a telegram describing the preliminary trials as "very spectacular and a sensation." According to the captain, the chief boat builder of the Norfolk Navy Yard commented, "The boat was doing the impossible and [he] could hardly believe what he had actually seen." Oakley then declared that the members of the board, the Coast Guard representatives in attendance, and the crew, were all, "astonished and pleased with the trials."[19]

McDowell was so impressed by the Eureka's performance that he suggested the boatbuilder contact U.S. Navy commander M. W. Powers, an officer assigned to the Construction Corps of the U.S. naval mission in Lima, Peru. McDowell was aware that the Peruvian government was interested in purchasing several shallow draft workboats and he believed the Eureka would be ideal for their purpose.[20]

Lieutenant Commander George H. Bahm, head of the special board responsible for conducting the Eureka's Norfolk trials, was also impressed by the boat's performance. He reported, "The Higgins boat is considered generally the best of the Experimental Landing Boats thus far tested for the purpose intended."[21] Following Bahm's report, the LBB recommended to the CNO that Higgins be awarded a contract to build four experimental 30-foot landing boats. Two were to be constructed of wood and two were to be fabricated of metal.[22]

---

[16] Higgins Revisal, 20.
[17] Higgins Revisal, 20; and Andrew Higgins to Gen Holland M. Smith, 3 February 1948, Coll/2949 Holland M. Smith Collection, 1905-67, Box 1, Series 1.2, Personal Correspondence 1917-65, Folder 9, Personal Correspondence 1947-48, Archives Branch, Marine Corps History Division, Quantico, VA, hereafter Higgins to Smith, February 1948; and Higgins to Whitt and Chambers, Ltd., July 8, 1940, author's collection.
[18] Higgins Revisal, 20. Bert Oakley may refer to Robert B. Oakley, though all sources simply refer to Bert.
[19] Oakley to Higgins, 27 May 1938, author's collection. The Norfolk Navy Yard's name was changed in 1945 to Norfolk Naval Shipyard.
[20] LtCdr R. S. McDowell to Andrew J. Higgins, 1 June 1938, author's collection.
[21] "Higgins Experimental Landing Boat: Report of Tests," USS *Arkansas*, 7 June 1938, National Defense Committee Files, OP-5, Navy Department Matters Ships, Shipbuilding and Related Matters, Box 182, Record Group 46, National Archives.
[22] Senate Report, 31-32.

Higgins was appreciative of the order, but still frustrated by the 30-foot requirement. In a postwar letter to Marine Corps lieutenant general Holland M. Smith, Higgins wrote, "I got some experimental orders, again for the goddamned 30' length boat. I built these more or less under protest." When he questioned the BuC&R as to why the boats had to be 30-feet in length, he was told that the existing davits on military and commercial ships could not handle anything longer. During a meeting with the Navy Department, Higgins exclaimed, "To hell with designing a boat to fit the davits. . . . They should design their davits to fit a proper size boat."[23]

There was also another reason why Higgins was upset. Shortly after the BuC&R began testing his Eureka, its Design Division's Small Boat Desk came out with a new set of plans. According to Higgins, their plans incorporated several of his boat's features. In his opinion, the Small Boat Desk was attempting to steal his design, but "they missed the point and the features they tried to copy were defeated by malformed under-water sections."[24]

In the fall, the BuC&R followed up on a previous recommendation of the FDB and the LBB. It awarded a contract to build three experimental landing boats that would be similar to an earlier BuC&R-designed metal landing boat built by Welin Davit Corporation; however, they were to incorporate changes recommended by the FDB. The records do not show whether any of the changes included features copied from the Eureka.[25] What is known is that, in June 1938, the LBB recommended that two wooden and two metal Eurekas be purchased. Following two design modifications and approval by the CNO, on 1 December 1938, the Navy purchased the four Higgins boats.[26]

Two months later, FLEX 5 began its naval exercises in the West Indies. As a result of their poor showing during the exercises, the three sea skiffs and the Philadelphia boats were eliminated as potential landing boats. Also eliminated from consideration were the original BuC&R boat built by Welin Davit, once hailed as the biggest advancement thus far in a landing boat; two other metal BuC&R boats; and the original 30-foot Eureka. This left the three modified BuC&R boats built by Welin Davit and the four newly purchased Eurekas. The FDB suggested several modifications to both the BuC&R and Higgins's designs. The LBB then recommended that one BuC&R and one Higgins boat be constructed incorporating the changes.[27]

The modified bureau and Higgins boats were retested during FLEX 6. The official report determined that the Higgins boat was "considered to be the best all-round boat for the purpose intended."[28] In reference to the bureau's metal boat, the report con-

---

[23] Higgins to Smith, February 1948.
[24] Higgins to Smith, February 1948.
[25] Senate Report, 24–25.
[26] Senate Report, 32.
[27] Senate Report, 34–39.
[28] Senate Report, 41.

FIGURE 4
A 36-foot Eureka LCPL during a training exercise on Lake Pontchartrain.
Source: courtesy the Higgins family

cluded, "About the only advantage offered by this type of boat is the cheapness and speed with which they can be manufactured."[29]

Completely disregarding the report's findings, the LBB and FDB recommended continued development of the bureau boats. Later, during the spring and summer of 1940, the bureau awarded contracts to build 41 30-foot metal bureau boats and 62 of Higgins's 30-foot wooden Eureka landing boats.[30]

Higgins was excited about the order, but again frustrated by the 30-foot limitation. He later recalled, "I got so exasperated that on my own, and without an order, and at my own expense, I built a boat 36-foot of length, and bore all the expense of shipping it to Norfolk, demanding that it be tested."[31]

On 11 September, the CNO ordered competitive trials between the 36-foot Higgins boat, a metal bureau boat, and a landing boat built by Chris-Craft Corporation. Tests were conducted on 17 September, and a full report was forwarded to the secretary of the Navy William Franklin Knox. The report indicated that Chris-Craft's twin-engine entry performed excellently but had difficulty retracting. Because of the

---

[29] Senate Report, 42.
[30] Senate Report, 45.
[31] Higgins to Smith, February 1948.

retracting issue, the FDB recommended against further development of the boat. Of the modified 36-foot Higgins entry, the FDB reported that it was "by far the most superior" and "exceeded in performance any other landing boat that the members of the Board had ever seen."[32] As for the metal bureau boat, it was considered, "the least satisfactory of the three tested."[33]

Despite the findings, six days after the test, the Bureau of Ships, which had been established on 1 July 1940 to assume the combined functions of the BuC&R and the Bureau of Engineering, awarded a contract to Gibbs Gas Engine Company of Jacksonville, Florida, to build 16 bureau-type metal landing boats. The bureau simply refused to give up on its Small Boat Desk's design in spite of its deficiencies.[34]

Based on the test results, on 19 September, the LBB recommended to the CNO that Higgins be awarded a contract to build 335 36-foot Eureka landing boats, now designated by the Navy as a Landing Craft Personnel, Large (LCPL).[35] Prior to going into production, Higgins built two new 36-foot Eureka landing boats and absorbed all costs. He wanted to ensure that his company would be giving the military the best possible boat. Each boat was constructed with slightly different hull modifications. Informal tests were held on 22 October 1940 at Virginia Beach, and the Eureka with the slightly flatter hull design was determined to be the superior of the two craft. It had better retracting capabilities and surpassed the Navy's speed requirements. Because of its length, the 36-foot Eureka could carry more troops and materiel. Also, because of its improved hull design, it was nine miles per hour faster than a 30-foot boat with the same engine.[36] On 30 November, the bureau officially awarded Higgins Industries the contract. Approximately five months later, on 30 April 1941, Higgins received a second contract for an additional 188 36-foot Eureka landing boats.[37]

It had been 13 years since Higgins first approached the Navy. Approximately two years had passed since he shipped his first boat to Norfolk. The competition between the boatbuilder and the Bureau's Small Boat Desk over the design of the personnel landing craft had finally come to an end.

According to General Smith, "through the unfathomable process whereby the official mind finally emerges from the darkness into the light, the Navy eventually decided to standardize on the 36-foot Higgins boat."[38] In Smith's opinion, Higgins "won the opening phase of the boat battle singlehanded, with loud Marine applause."[39]

---

[32] Senate Report, 46.
[33] Senate Report, 46.
[34] Senate Report, 45.
[35] Senate Report, 50.
[36] Higgins to Whitt and Chambers, 8 July 1940, author's collection.
[37] "The Chief of the Bureau of Ships to the Under Secretary of the Navy (Clearing Office)," 15 September 1942, National Defense Committee Files, OP-5, Navy Department Matters Ships, Shipbuilding, and Related Matters, Box 182, Record Group 46, National Archives.
[38] Gen Holland M. Smith, USMC (Ret), and Percy Finch, *Coral and Brass* (New York: Charles Scribner's Sons, 1948), 91.
[39] Smith and Finch, *Coral and Brass*, 90.

FIGURE 5
City Park Plant.
Source: courtesy the Higgins family

FIGURE 6
City Park LCVP production line.
Source: courtesy the Higgins family

Higgins might have won the opening phase, but his battle with the bureau was far from over.

In July 1940, as Higgins was trying to interest the British in his Eurekas, he purchased the Albert Weiblen Marble and Granite Works and converted it into a multistory, $1.5 million boat building facility, known as the City Park Plant. According to Higgins, Rear Admiral Claude A. Jones, the assistant chief of the bureau, and Captain Norborne L. Rawlings, procurement officer for the bureau, warned him against constructing additional plants for landing craft production.[40]

Higgins ignored their advice and moved forward.[41] During the early stages of his new plant's construction, he determined that as presently designed, it would never be capable of delivering the thousands of landing craft he believed it would be called on to produce once the United States entered the war. His solution—immediately redesign and enlarge the facility. The problem was he lacked available land. Delgado Trade School was on one side of the Louisiana plant. On the other side was the Southern Railway tracks, which would be vital for bringing material in and transporting finished boats out. In front of the plant ran City Park Avenue. Bordering the back of the plant was Holt Cemetery. Higgins "knowingly and willingly" enlarged his plant onto

---

[40] Andrew J. Higgins Jr., interview with author, 25 March 1975, author's collection; Higgins Revisal, 36; and Statement by Andrew Higgins, 10 January 1947, Statler Hotel, Washington, DC, typescript in Papers of Harry S. Truman, File 633, Harry S. Truman Library, Independence, MO, 7, hereafter Statler Hotel statement.

[41] Higgins Revisal, 33; and House Documents, 77th Cong., 2d. Sess., Serial no. 10600, vol. 23, no. 281, 49.

an unused portion of the cemetery. By the time the plant was complete, 40 percent sat on property to which he held no title. He reasoned, once the war was won, the legal issues could be resolved.[42]

The City Park Avenue plant, after completion, held the honor of being the world's largest boat-building facility housed under one roof dedicated to the production of landing craft.[43] It also had the distinction of being the first boat-building plant to implement assembly line production techniques. Additionally, it was the only boat-building plant to produce landing craft on the second floor and then lower them by a huge elevator to waiting railroad cars below.

According to Marine Corps General Robert E. Hogaboom, this was typical of Higgins. The general once wrote that, "in discussing problems of the Corps and details of boat requirements he [Higgins] was quick to grasp an idea and seemed to be able to mentally translate it into practical design. . . . Quick decision he gave and immediate action he demanded. He was the sort who tended to knock down anything that stood in his way."[44] A series of events starting in March 1941 would prove Hogaboom's characterization of Higgins to be true.

As the country drew closer to entering the war, it appeared the United States might have to seize the island of Martinique from the French to halt its use as a German submarine base.[45] Such action would require an amphibious assault and General Smith had no faith in the bureau's ability to design and build the landing boats required for such an operation. Therefore, he turned to the one person he believed could accomplish it—Andrew Higgins.

Smith sent Captain Victor H. Krulak and Major Ernest E. Linsert, secretary of the Marine Equipment Board, to New Orleans to meet with Higgins. They showed him photographs, taken by Krulak in 1937, of ramped Japanese landing boats participating in the invasion of China. Afterward, they informally asked him if he could install a ramp in the bow of a Eureka. Higgins immediately accepted the challenge and, at his own expense, had his men begin working on the design.[46]

Linsert later returned to New Orleans, accompanied by Navy commander Ross B. Daggett, representing the bureau, and on 26 May they observed three ramped Eurekas undergo tests on Lake Pontchartrain.[47] Linsert reported to Brigadier General Charles D. Barrett, director of plans and policies at Marine Corps headquarters, that

---

[42] Statler Hotel statement, 7.
[43] City Park Plant Dedication Booklet, author's collection, 19.
[44] Gen Robert E. Hogaboom to Benis M. Frank, 3 November 1975, copy in author's collection courtesy of Frank.
[45] LtGen Victor H. Krulak (Ret) to Jerry Strahan, 5 November 1975, author's collection; and Krulak, *First to Fight*, 94.
[46] Krulak, *First to Fight*, 94.
[47] Chief of the Bureau of Ships to Under Secretary of the Navy, 15 September 1942, National Defense Committee Files, OP-5, Navy Department Matters Ships, Shipbuilding and Related Matters, Box 182, Record Group 46, National Archives.

he considered both boats to be quite satisfactory.[48] On 27 May, the LBB informed Higgins via telephone that a special board of Marine Corps and bureau representatives would arrive in New Orleans in three days to officially test the new ramped Eurekas. The LBB also requested that he have preliminary drawings of a 45-foot tank landing craft ready for review.[49]

For more than a decade the bureau's Small Boat Desk had been trying to design a craft capable of landing a tank over an open beach, but had not been successful. During the Caribbean exercises of 1941, the Marines had been supplied with three 45-foot BuC&R-designed lighters, which according to General Smith, "were unmanageable and unseaworthy in heavy surf."[50] During the exercise, one of the lighters capsized. The bureau was now desperate.

Higgins agreed to accept the LBB's request, but under one condition. He would only work with the Marine Corps.[51] Once that was established, he informed the caller that when the board arrived, instead of drawings, he would have a completed craft in the water ready for testing. The caller failed to take him seriously.[52]

Higgins had on hand a partially completed towboat and dredge tender. When a proposed contract with the Mobile District, U.S. Army Corps of Engineers failed to materialize, he was left with the unfinished boat.[53] The towboat's length and width were the approximate size required for the tank lighter. Immediately, he had his workers begin converting the towboat into a tank landing craft.

When the special board members arrived in New Orleans, they found not only the new 36-foot ramped Eurekas but also a completed 45-foot tank landing craft. The lighter had been designed, fabricated, and put in the water in 61 hours.[54] Higgins had accomplished in less than three days what the BuC&R's Design Division had been unable to accomplish in more than two decades. The board considered the trials of the Eurekas and the tank lighter to be highly successful.[55]

Previously, the senior member of the LBB had instructed the CNO that if the ramped Eurekas proved acceptable, the contract for the 188 spoonbill-bow Eureka landing boats was to be modified. All remaining boats were to be the new ramped Eureka design, now designated by the Navy as Landing Craft Vehicle, Personnel (LCVP). Also, Higgins was to immediately begin construction of 49 tank lighters,

---

[48] Clifford, *Progress and Purpose*, 51.
[49] Andrew J. Higgins Jr., interview with author, 26 June 1975, author's collection.
[50] Smith and Finch, *Coral and Brass*, 92.
[51] Higgins to Smith, 3 February 1948, Archives Branch, Marine Corps History Division, Quantico, VA.
[52] Andrew J. Higgins Jr., interview with author, 20 April 1973, author's collection. Higgins served as vice president of Higgins Industries during the war and took over as president when his father died in 1952.
[53] Higgins Industries Inc., draftsman's drawing no. 2583, 28 May 1941, and draftsman's drawing no. 2511-A, 31 May 1940, author's collection; Graham Haddock, interview with author, 26 June 1975, author's collection; and J. A. Dovie, "Drawing Number Book," author's collection, 66.
[54] Higgins to Smith, February 1948.
[55] Higgins to Smith, February 1948; and Hough, Ludwig, and Shaw, *Pearl Harbor to Guadalcanal*, 28.

FIGURE 7
Troops disembark from a Higgins LCVP.
Source: courtesy the Higgins family

now classified as a Landing Craft, Mechanized (LCM). There was one stipulation: as many of both boats were to be delivered to Norfolk, Virginia, within 14 days.[56]

Higgins now had a contract, but he lacked engines, steel, bronze rods, and a place to produce the LCMs and LCVPs. His City Park Plant was already dedicated to the mass production of 36-foot Eureka landing boats, but it could be converted to manufacture LCVPs. That left him still needing a place to produce the LCMs. His St. Charles Avenue plant lacked the space to fabricate the 45-foot steel craft. To solve the problem, he purchased an old carriage barn on Polymnia Street. The barn's left wall ran directly behind the rear wall of the St. Charles Avenue plant, making its location ideal. Additionally, its size was adequate for assembling the lighters. However, Higgins still needed a place to fabricate parts, so he took matters into his own hands. He barricaded off the block of Polymnia Street bordering one side of his plant, covered it with canvas, and turned the street into a temporary warehouse and fabrication yard. Residents living on the closed off block could not drive their cars home, gar-

---

[56] Graham Haddock, interview with author, 26 June 1975, author's collection; Senate Report, 35; and "Statement by A. J. Higgins, of Higgins Industries Inc., New Orleans," 24 September 1942, 4, author's collection, hereafter Higgins statement.

The Landing Craft Controversy

FIGURE 8
The first LCMs being fabricated in the converted Polymnia Street carriage barn.
Source: courtesy the Higgins family

bage trucks could not pick up trash, and a brothel owner complained the noise was destroying romance and killing business.[57]

Now that Higgins had a place to produce the craft, he quickly gathered the necessary components. Higgins Industries served as an outlet for Gray Marine motors. Therefore, it had some of the required engines on hand, but not nearly enough. With the factory unable to supply additional motors in time, Higgins contacted other Gray Marine dealers nationwide and purchased their stock. To expedite delivery, he sent his company trucks across the South to pick up engines and rush them back to New Orleans.[58]

As for the lack of steel, the industrialist discovered a barge load of the required type moored near Baton Rouge. He sent a fleet of chartered trucks and armed plant guards to persuade the consignee to release the material to Higgins Industries. To

---

[57] Haddock June interview; and "The Boss," *Fortune*, July 1943, 214.
[58] Graham Haddock, interview with author, 10 November 1975, author's collection.

get additional material, he had a Birmingham, Alabama, steelmaker called off a golf course on a Sunday morning. He then persuaded him to furnish the necessary metal plating. Next, he contacted President Ernest E. Norris of Southern Railway and requested that the flatcars loaded with steel be attached to the first possible passenger train headed to New Orleans. Norris informed Higgins that regulations prohibited such an action. Higgins contacted the Navy; the regulations were temporarily suspended and the steel was soon headed south.[59]

At this point, Higgins lacked one last critical item: bronze rods to be used as propeller shafts. When he discovered the mills could not provide the rods in time, he searched for other sources. Rods were located at an oilfield depot in Texas, but the owner refused to sell.[60]

The CNO had instructed "that every practicable means be taken to expedite completion and delivery" of the boats.[61] That was all Higgins needed to know. Since there was no time for the Navy to expropriate the material, Higgins sent his son, Andrew Higgins Jr., with some plant workers to Texas. Accompanying them was a pair of wire cutters. After dark, the crew "borrowed" an ample supply of rods from the oilfield depot and loaded them in the back of their company truck. With Texas police in pursuit, their truck crossed the Louisiana state line, where Louisiana State Police cars were waiting to escort the shipment to New Orleans. Shortly thereafter, the benevolent depot owner received full payment for the material.[62]

At the end of the 14-day time limit, Higgins delivered 26 LCVPs and 9 LCMs. The LCMs had to be partially painted as the train rolled down the tracks toward Norfolk. In spite of incredible difficulties, the Navy received their boats on time.[63]

As the LCVPs and LCMs rolled east, it appeared the competition between the bureau and Higgins Industries had finally come to an end. Such was not the case. The bureau's Small Boat Desk was determined to design the LCVP and LCM that would serve as the Navy's standardized landing craft, and they had an ace in the hole: the bureau was in charge of awarding landing craft contracts.

The Amphibious Force, U.S. Atlantic Fleet, had requested that all future personnel landing boats be the new Higgins ramped LCVP design. The bureau ignored their request and awarded a contract to produce 200 Higgins-designed LCPLs. However, the contract was not awarded to Higgins but to Chris-Craft Corporation. Later, it was revised, and the final 162 boats produced were LCVPs. But the bureau chose not

---

[59] "The Boss," 214; *Ideas for United Nations*, film, n.d., Gayle Higgins Jones Collection, New Orleans, LA; and Andrew J. Higgins Jr., interview with author, 20 April 1973, author's collection.
[60] Gen Robert E. Hogaboom to Benis M. Frank, 3 November 1975, copy in author's collection courtesy of Frank.
[61] Secretary of Naval Operations to Secretary of the Navy, 29 May 1941, Office of the Secretary General Correspondence, 1940–42, S82-3 (1) W4-3 (410527), Record Group 80, National Archives.
[62] Gen Robert E. Hogaboom to Benis M. Frank, 3 November 1975; and Andrew J. Higgins Jr., interview with author, 25 March 1975, author's collection.
[63] Haddock interview, June 1975; and Higgins statement, 4.

The Landing Craft Controversy

to produce the Higgins-designed LCVP. Instead, its Small Boat Desk had recently come out with its own version and the bureau contracted with Chris-Craft to mass produce it.[64]

The bureau's boat featured a 3.5-foot-wide ramp and was given the designation Landing Craft Personnel (Ramp) (LCP[R]). Higgins's LCVPs had a 6-foot 2-inch-wide ramp that allowed small vehicles to be carried by the boat. The bureau's narrower ramp on the LCP(R) was incapable of transporting a vehicle. The Higgins LCVP could debark 36 troops in 19 seconds. The LCP(R) took 32 seconds to debark the same number of soldiers. The LCP(R)'s narrower ramp also limited its use as a materiel carrier. Plus, the LCP(R) proved to be bow heavy. The Amphibious Force of the Atlantic Fleet had not been consulted about the LCVP's design changes. An LCP(R) pilot model was never produced or tested to uncover potential flaws. Instead, the Small Boat Desk's design went straight into production with 1,587 LCP(R)s being manufactured.[65]

As negative reports came in from the commanding general of the Amphibious Force and from additional testing, production of the narrow-ramped boat was halted.[66] All future boats were to be of the Higgins's design. Thus, for the second time, a Higgins-designed boat had beaten out a bureau designed boat to become a standardized Navy landing craft. The Small Boat Desk had lost its second battle, but it was not yet ready to concede the war. In General Smith's opinion, "in the Navy, tradition never dies while there is a shot left in the locker."[67] In the summer of 1941, the bureau was about to fire its third and final shot.

Officers assigned to the Small Boat Desk were insistent that, given enough time, the defects in their tank lighter could be corrected. From May to July 1941, they focused their attention on redesigning it. The result was a 47-foot lighter that was extremely similar to their previously unsatisfactory 45-foot craft. Again, no model-basin test had been run and no prototype had undergone sea trials. The forces afloat had previously recommended that no additional bureau-type lighters be produced. Yet, in August, when the bureau received a directive to build 131 additional tank landing craft, it only requested bids on its newly modified untested lighter.[68]

As the bureau pushed forward with its bid process, on Sunday morning 7 December 1941, Higgins and members of the New Orleans Dock Board held a roadside meeting near the Industrial Canal (a.k.a. Inner Harbor) in eastern New Orleans. The industrialist was interested in leasing land from the board so he could build a large boat building plant. The plant would help turn out the massive orders the

---

[64] Senate Report, 61–65.
[65] Senate Report, 74.
[66] Senate Report, 69–71.
[67] Smith and Finch, *Coral and Brass*, 93.
[68] Senate Documents, 78th Cong., 2d Sess., no. 71, Report 10, Part 15, *Investigation of the National Defense Program*, 154, hereafter Senate Report 10.

FIGURE 9
Industrial Canal Plant.
Source: courtesy the Higgins family

boatbuilder believed would soon be forthcoming. In fact, the British Admiralty had already notified the bureau that it wanted to purchase 150 Higgins-designed LCMs.[69] Higgins saw that as simply the beginning. As he and board members negotiated terms, a broadcast from a nearby car radio announced that Pearl Harbor had been attacked. Terms were immediately agreed on and, by that afternoon, Higgins had crews clearing the site.[70]

Because of the now urgent requirement for lighters, the bureau suddenly needed Higgins's production capabilities.[71] Therefore, on 26 December, the 131 tank lighter bid was modified. The bid now called for 10 47-foot bureau lighters and 20 45-foot Higgins lighters. The remaining 101 lighters were to be 50-foot in length to accommo-

---

[69] Senate Report 10, 157.
[70] Statler Hotel statement, 12.
[71] After the attack on Pearl Harbor, there was urgent need for an increase in the production of landing craft. According to George E. Mowry's report, "Landing Craft and the War Production Board," Special Study no. 11 (first issued on 15 July 1944 specifically for the War Production Board), there were two major landing craft production programs. The first began in April 1942 in preparation for the invasion of North Africa and ended in the spring of 1943. The second major production program in preparation for the invasion of Western Europe and the Pacific operations began in August 1943 and peaked in May 1944. The problem, even during the peak production periods, was that often after a directive from the CNO to produce landing craft it might be months before the Bureau of Ships awarded a contract.

The Landing Craft Controversy

date the Army's new 30-ton M4 Sherman medium tank. Seventy-six of the 101 were to be Higgins's lighters and 25 were to be the bureau-designed boats.[72]

During a meeting at the White House on 4 April 1942, the bureau was instructed to provide 600 50-foot lighters by 1 September for "imminent military operations."[73] The bureau increased the order to 1,100 lighters and decided that all 1,100 would be its design. This decision was made in spite of the results of two separate tests. On 20 April 1942, during trials held at Ipswich, Massachusetts, the bureau lighter demonstrated it had "no directional control" when in reverse. Later, during trials held at Philadelphia, it was discovered the boat could not be run at full speed and retain "seaworthiness."[74]

When the Army learned of the bureau's decision to produce its lighters instead of Higgins's, it strongly objected. Conferences concerning the tank lighters were held between the Army and the bureau in early May 1942. The Army continued to insist on the Higgins lighter and the Navy was adamant that its tank landing craft was capable of handling Army needs. The Army reached out for help. They requested that Higgins lend them his chief naval architect George Huet, and also a member of his engineering department, Graham Haddock, to serve as consultants in the forthcoming inter-Service meetings.[75]

As the meetings took place in the Navy building on Constitution Avenue, the Army positioned Huet and Haddock in its headquarters nearby. If a question arose that their representatives could not answer, they quickly sent a messenger to obtain the needed information from their hidden experts.[76] During the discussions, Higgins was in New Orleans being honored by the city as part of its Maritime Day celebration. As soon as the festivities were over, he immediately headed to Washington to join Huet and Haddock.

After arriving, he discovered the bureau still planned to produce its tank lighter. In response, Higgins visited Senator Harry S. Truman (D-MO), head of the Senate's Special Committee to Investigate the National Defense Program.[77] After two additional meetings, Truman ordered the Navy to have its tank lighter compete one-on-one against Higgins's LCM.[78] The competition took place on 25 May 1942, near

---

[72] Senate Report 10, 157.

[73] Senate Report 10, 157. During the Senate hearings, the questioner instructed those testifying that specific geographic locations were not to be used when referencing future military actions. Instead, those answering were to use the term *imminent military operations*. This also seems to be the case at the White House meeting on 4 April 1942.

[74] Senate Report 10, 159-61.

[75] Haddock interview with author, 8 January 1993, author's collection.

[76] Haddock interview, June 1975.

[77] The Senate's Special Committee to Investigate the National Defense Program was commonly referred to, even in government documents, as the "Truman Committee." See Special Committee to Investigate the National Defense Program (1 March 1941) in "Chapter 18. Records of Senate Select Committees, 1789-1988," National Archives.

[78] Krulak, *First to Fight*, 97-98.

FIGURE 10
Bureau of Ships' tank lighter during 25 May 1942 competition against Higgins Industries' LCM. The bureau's crew was prepared to abandon ship if necessary. Higgins's LCM easily handled the mildly choppy seas and successfully landed its tank.
Source: courtesy Graham Haddock from original Higgins files

Norfolk, Virginia. Senior officers from the Navy Department, the Bureau of Ships, the Army, the Marine Corps, and an administrative assistant from the Truman Committee were all on hand. By the end, the results were clear[79].

The following day, Major Howard W. Quinn, from the Operations Division, U.S. Army Transportation Services, wrote to the commanding general of the Services of Supply that "as we neared the net it became apparent that the Navy Bureau-type tank lighter was in trouble.... It appeared that the lighter was going to overturn."[80]

Quinn described the crew as "straddling" the sides of the lighter and the coxswain as "steering the vessel from the rail." He concluded, "As far as comparison of characteristics of the types of tank lighters are concerned, it may be stated that on May 25 tests there was no comparison."[81] The official report concluded that "the Bureau-type lighter was unseaworthy and that the Higgins lighter performed excellently."[82]

---

[79] Senate Report 10, 162.
[80] Senate Report 10, 163.
[81] Senate Report 10, 163.
[82] Senate Report 10, 163.

The Landing Craft Controversy

As a result of the 25 May competition, the Bureau of Ships notified all Navy yards building bureau lighters under the 1,100-boat contract, that they were to convert their production to the Higgins-designed tank lighters.[83] None of the 126 produced 50-foot bureau lighters were ever assigned to combat. The CNO reported to committee investigators that the bureau's lighters were "restricted to service for training purposes or for miscellaneous utility lightering [sic] work."[84]

The competition between Higgins and the bureau was finally over. However, the investigation into the Bureau of Ships was just beginning. On 8 June 1942, the Truman Committee officially opened hearings concerning the landing craft program. Approximately two months later, the committee forwarded its findings to secretary of the Navy Knox. The report concluded that the bureau, "for reasons known only to itself, stubbornly persisted for over five years in clinging to an unseaworthy tank lighter of its own." It then claimed that in the bureau there was "an inherent reluctance on the part of its personnel to accept any design but, its own, even though this involves a flagrant disregard for the facts, if not also for the safety and success of American troops."[85] In the committee's opinion, "If a better design had not been available, persons in the Design Division of the Bureau, responsible for the lighter program, might be deemed merely incompetent."[86] As a result of their findings, the committee recommended that Knox "reorganize the sections of the Bureau's design division that had been responsible for the tank lighter program."[87]

Also, on 5 August 1942, Truman sent a letter to Secretary Knox stating, "I cannot condemn too strongly the negligence or willful misconduct on the part of the officers of the Bureau of Ships entrusted with this vital matter, involving as it did both the success of our military forces and the lives of American marines, sailors, and soldiers." As for the bureau's treatment of Higgins Industries, the senator found it to be "biased and prejudiced." Truman claimed "that the war effort has not suffered an irreparable injury is due largely to the ability and energy of Higgins Industries, Inc. and to its repeated criticisms of the shortcomings of the designs prepared by the Bureau of Ships." In his opinion the boatbuilding company "should be commended for doing this without fear of the results which such criticisms might incur with the agency on which it was dependent for contracts."[88] The following day, Secretary Knox informed the committee that he was authorizing an examination of the tank lighter program be made on behalf of the Navy Department.[89]

On 18 August 1942, Knox authorized Yale University professor Herbert. L. Seward to conduct the Navy's investigation. As Seward began his inquiry, Knox was

---

[83] Senate Report 10, 164.
[84] Senate Report 10, 136.
[85] Senate Report 10, 167–68.
[86] Senate Report 10, 167.
[87] Senate Report 10, 168.
[88] Harry Truman to Secretary Frank Knox, 5 August 1942, from Higgins's 1942 copy, author's collection.
[89] Senate Report 10, 133.

already initiating a reorganization of the bureau. In early November, Seward submitted his findings to Knox. He had discovered the "slow-going" peacetime practices of the bureau had continued, that "rifts and chasms between factions" existed, and that no existing procedures allowed for "proper consideration of suggestions submitted from outside sources." Just as troubling, his investigation found no process for allowing recommendations from the forces afloat to be quickly brought to the attention of those in charge of design and procurement.[90]

Seward's report also verified that the bureau's lighter had been given undue preference, while Higgins Industries had received piecemeal orders, thereby making a steady production flow difficult to maintain. As for the treatment accorded Higgins, Seward described it as "unfortunate." His report concluded that "the Higgins lighter is superior to the Bureau type lighter."[91]

It had taken the Marine Corps, the Army, the Truman Committee, the Navy Department's own investigation, personnel changes in the Small Boat Desk, and the 1 November appointment of Captain Edward L. Cochrane as the new chief of the Bureau of Ships to correct the pre-1942 problems and prejudices of the bureau. It had also taken an industrialist willing to call out the unfair practices of some officers in the Bureau of Ships, even though such an action could have been detrimental to his company.

In a postwar letter to General Smith, Higgins wrote, "I would not care to appear as if I was disgruntled with the Navy, for after 1942, we got along excellently."[92] By late 1942, he had no reason to be resentful. His LCP, LCPL, LCVP, and LCM had all become the Navy's standardized landing craft.

When the Allies invaded Normandy in 1944, Higgins Industries consisted of eight plants and employed more than 20,000 workers. The company produced 20,094 boats and ships for the Allied war effort. A remarkable achievement for a company that in 1937 employed approximately 50 workers and operated from a single small boatyard not located on the waterfront. An achievement that almost never occurred because of a biased few in the Small Boat Desk of the Bureau of Ships.

---

[90] Senate Report 10, 133–34.
[91] Senate Report 10, 133–34.
[92] Higgins to Smith, February 1948.

CHAPTER THREE

Red Tide over the Beach

Soviet Amphibious Warfare in Theory and Practice

Benjamin Claremont

The term *amphibious operations* generally does not bring to mind the Soviet military. If it does, the image is likely influenced by the work of Tom Clancy and Larry Bond, whose dramatic Soviet invasion of Iceland featured heavily in their bestselling *Red Storm Rising*.[1] Even Cold War-era American intelligence of the Soviet *Morskaya Pekhota* (Naval Infantry) was limited.[2] Allen E. Curtis, the liaison between the U.S. Army's Soviet Army Studies Office and the National Training Center's opposing force in 1989–2000, called Defense Intelligence Agency efforts "pathetic," noting there was one unclassified report from 1979 that was never updated.[3] Indeed, even for the Soviet Navy, the Naval Infantry, along with *Morskaya Aviatsiya* (Naval Aviation), was seen as something of an unwanted and often neglected distraction from the Navy's priorities. However, the Soviet Union had a long history of amphibious operations, especially during and after the Second World War. The study

---

[1] Tom Clancy and Larry Bond, "Operation Polar Glory," in *Red Storm Rising* (New York: G. P. Putnam's Sons, 1986).
[2] The abbreviation *MorPekh*, short for Морская Пехота (*Morskaya Pekhota*) or Naval Infantry, will be used throughout. In addition, the acronyms MPBn and MPBr, meaning Naval Infantry Battalion and Brigade, will be used.
[3] Allen E. Curtis, "Soviet Marines in the 70s–80s," Miniatures Page, 20 May 2007; the document is in LtCol Louis N. Buffardi, *The Soviet Naval Infantry*, DDB-1200-148-80 Defense Intelligence Report (Warsaw: DIA Soviet Warsaw Pact Division, Directorate for Research, 1980).

of Soviet amphibious warfare offers a unique perspective that contrasts with Western experience.

Given Russian performance in the Russo-Ukrainian War, one might question the utility of understanding Russian and Soviet amphibious warfare, especially for the United States, an insular power heavily invested in expeditionary amphibious warfare and Joint forcible entry. Russia, like the Soviet Union before it, is a continental land power. Their navy, and by extension their Naval Infantry, exist to support the activities of the Ground Forces (Soviet Army). This is a fundamentally different perspective of Joint warfare than in the United States. However, it is one which both challenges core assumptions of American thinking on amphibious warfare and has been consistently understudied despite the criticality of coastal and littoral regions during and after the Cold War.

Unlike the Western allies with histories of colonial campaigns and marine expeditions, the Soviet Naval Infantry really only began conducting amphibious assaults during the Second World War. The material conditions of the Nazi-Soviet war meant that victory or defeat would be decided by the large-scale land campaigns in the Soviet Union and bordering states. Geography determined that the main water obstacles were riverine, on the great rivers such as Don, Dniepr, Vistula, Oder, and Volga.

The amphibious assaults of the Western allies moved toward applying an unstoppable force of operational-strategic air interdiction, close air support, and a volume of naval gunfire only possible when the world's two largest naval powers—the United States and United Kingdom—focused their might on a few kilometers of beach.[4] They were often strategic assaults, crossing oceans and breaking into a continental theater with forces numbering in the tens of divisions. In contrast, Soviet amphibious assaults were small (battalion-regimental scale), at shallow depths (often less than 150 km from friendly forces), rarely had anything larger than destroyers for fire support, and were made to insert forces to outflank defenses or to insert a forward detachment.

Despite its alien context, the Soviet/Russian perspective is useful for three major reasons. First, understanding the theory and practice of the probable enemy empowers leadership at all levels to shatter mirror imaging and work forward probable enemy courses of action on the basis of battlefield conditions, and the Russians build off the foundation of Soviet theory and practice. Second, it provides perspective on how amphibious warfare can support and enable successful large-scale, high-intensity ground forces operations in continental theaters. Third, and perhaps most importantly, studying an external approach to amphibious warfare forces reassessment of what preconceptions and assumptions are taken for granted.

Before diving into historical vignettes and Soviet amphibious theory, some context is necessary. The Soviet study of war was a fully articulated academic field with its

---

[4] The term *United Kingdom* here includes Dominion, Commonwealth, and Empire forces.

own language, subfields, methodological structures, and lively debates.[5] The Soviets used very precise terminology when discussing military science, and there are many false friends with English language terms. To deconflict, this chapter will put the terms in English when discussing the Western understandings and in Russian or italic translation or transliteration when using Soviet definitions.[6] The following introduction will hopefully orient the rest of this chapter within Soviet terminology and their intellectual framework. This, of course, does not imply that the author or publisher condones or supports the ideology of the USSR in any way. It is, however, important to understand an organization through their own eyes and in their own words.

In the Soviet understanding, *war* (война, *voyna*) was distinct from *armed conflict* (вооружённая борьба, *vooruzhonnaya bor'ba*).[7] *War* was a broad sociopolitical phenomenon that is defined based on Vladimir Lenin's Clausewitzian articulation as an expression of the politics of the warring powers and the classes within them.[8] The Soviets considered *war* to be total, a struggle by the whole of a country (coalition) in which *armed conflict* was only currently the main form of struggle, alongside economic, diplomatic, and ideological conflict.[9] As the Soviets viewed *war* as encompassing the totality of the state, *war* fell under the purview of civilian leadership.[10] However, and most importantly, the Soviets did not see *war* as a failure of diplomacy or policy but as one tool among many to achieve policy aims, one which carried great risk and so was dangerous and undesirable, but one which may be forced on the USSR.[11] Only a fool would desire war, but to the Soviets it was something that must be prepared for, endured should it come, and its opportunity not squandered.[12]

In contrast, *peace* (мир, *mir*) was primarily defined as the conduct of foreign

---

[5] Peter H. Vigor, "The Function of Military History in the Soviet Union," and Christopher N. Donnelly, "The Soviet Use of Military History for Operational Analysis: Establishing the Parameters of the Concept of Force Sustainability," in Col Carl W. Reddel, USAF, ed., *Transformation in Soviet and Russian Military History: Proceedings of the Twelfth Military History Symposium*, AFD-101028-004 (Washington, DC: U.S. Air Force Academy, Office of Air Force History, U.S. Air Force, 1986), 117–40, 243–72; and Christopher Donnelly, *Red Banner: The Soviet Military System in Peace and War* (London: Jane's Information Group, 1988), 182–83.

[6] For example, in the case of strategy, operations, and tactics, Western definitions will work from the paradigm established by B. A. Friedman in *On Operations: Operational Art and Military Disciplines* (Annapolis, MD: Naval Institute Press, 2021) or U.S. Joint doctrine like *Strategy*, Joint Doctrine Note 2-19 (Washington, DC: Joint Chiefs of Staff, 2019) and *Joint Warfighting*, Joint Publication 1 (Washington, DC: Joint Chiefs of Staff, 2023), while Soviet definitions will universally be taken from two authoritative Soviet reference works: Советская военная энциклопедия [Soviet Military Encyclopedia] [SVE], 8 vols. (Moscow: Voenizdat, Soviet Ministry of Defense, 1979–89), hereafter SVE volume number, and Военный Энциклопедический Словарь [*Military Encyclopedic Dictionary*] [VES] (Moscow: Sovetskaya Entsiklopediya, 1986), hereafter VES 1986. Definitions will be given in footnotes.

[7] VES 1986, 151, for война, 157 for Вооружённая борьба.

[8] VES 1986, 151.

[9] VES 1986, 151.

[10] VES 1986, 151.

[11] VES 1986, 151.

[12] Donnelly, *Red Banner*, 104; and VES 1986, 151.

policy without the use of *armed conflict*.¹³ Indeed, the Soviet *Military Encyclopedic Dictionary* defines *mir* as explicitly including competition below the threshold of *armed conflict*, and notes that "*peace* without weapons and violence, *peace* in which every people chooses the path of their development, their way of life, is the ideal of Socialism."¹⁴ To the Soviets, *peace*, in the Western sense of "peace with goodwill," could only come when the fundamental antagonisms inherent to capitalism were resolved by its elimination.

Finally, *armed conflict* being the primary mode of struggle in *war*, but able to exist outside of it, was the sum of military actions taken to achieve political and military goals.¹⁵ As it related primarily to the activity of the armed forces, it was managed by military leadership.¹⁶

These subjects and definitions fell under the category of *military affairs* (Военное дело, *Voyennoye delo*), the term for all issues relating to the theory, practice, and construction of armed forces, and more particularly in the USSR the system of knowledge required for service personnel to successfully fulfill their military duty.¹⁷

The highest level of *military affairs* in the USSR was *military doctrine* (доктрина военная, *doktrina voennaya*), the official policy statement (system of views) of the civilian government of the USSR espousing the scientifically based, officially ordained system of understanding *war* and the use of the armed forces within it, in present and future.¹⁸ It was both military-technical and sociopolitical. The sociopolitical aspect of *military doctrine* set the policy objectives, methods, and force posture, and was the product primarily of the civilian leadership of the USSR. This broad approach was then refined by the military-technical aspect of *military doctrine*.¹⁹

Military doctrine's military-technical aspect laid out the scientifically supported state-approved theory and practice of warfare. It was derived from theoretical research, practical assessments of military and economic capabilities, and political policy and goals to create a logically sound and coherent *military doctrine* that reflected rigorous and objective research, not simply the preferences of any Soviet general or marshal.²⁰

The Soviets studied *war*, *peace*, and *armed conflict* as part of a rigorous academic field: *military science* (военная наука, *voyennaya nauka*).²¹ *Military science* was "the system of knowledge about the laws of war, military strategy, the nature of war,

---

[13] VES 1986, 448.
[14] VES 1986, 448.
[15] VES 1986, 157.
[16] *The Soviet Army: Operations and Tactics*, FM 100-2-1 (Washington, DC: Department of the Army, 1990), 1-5.
[17] VES 1986, 139.
[18] SVE, vol. 3, 225-29; and James M. McConnell, *Analyzing Soviet Intentions: A Short Guide to Soviet Military Literature* (Alexandria, VA: CNA, 1989), 2.
[19] Donnelly, *Red Banner*, 106.
[20] Col David M. Glantz, *Soviet Military Operational Art: In Pursuit of Deep Battle* (Abingdon, UK: Frank Cass, 1991), 2-5; and *The Soviet Army: Operations and Tactics*, 1-8.
[21] VES 1986, 135-36.

the construction and preparation of the armed forces and the country for war, and methods of conducting armed conflict."[22] This broad field was subdivided into a number of other subfields, though this chapter will focus on *military art* (Военное искусство, Voyennoye iskustvo).[23]

Military art was the theory and practice of the preparation and conduct of military action (*armed conflict*) on land, sea, and in the air.[24] It further broke down into the fields of:

- *Military strategy*, the theory and practice of planning for, preparing for, and fighting *armed conflict* at the national or TVD level.[25] Due to this scale, it was definitionally joint and combined arms.
- *Operational art*, the theory and practice of planning and conducting combined arms (common fleet), joint and independent operations (combat actions) by various formations of the armed forces.[26]
- *Tactics* is the theory and practice of preparing and conducting combat by *subunits*, *units* and *formations* of various branches of the armed forces, combat arms (forces) and special forces. It is subdivided into general tactics and branch tactics of the armed forces, combat arms and *special troops*.[27]

Having oriented this chapter within the Soviet understanding of military theory and the terminology they used to describe it, the time comes to examine the Soviet Naval Infantry and their concepts for use. Soviet *MorPekh* (Naval Infantry) existed to support and enable the Ground Forces, and so it is important to understand them within the context of Soviet *military art* and that of the Ground Forces in particular.

By the late 1980s, the Soviet military had adopted an iterated and modernized derivative of the military concept it had pioneered before World War II and refined

---

[22] VES 1986, 135–36; and McConnell, *Analyzing Soviet Intentions*, 2–4, notes the distinction between systems of knowledge (sciences) characterized by roughly free theoretical exploration and systems of views (policy/doctrine) characterized by official authoritative statements.

[23] VES 1986, 136. The other fields include general theory, theory of the construction of the armed forces, theory of military training and indoctrination, theory of the military economy and rear of the armed forces, theory of command and control, branch-specific theory, and military history.

[24] VES 1986, 139–40. By the late 1980s, this definition appears to have expanded to include space in the classified literature, but this cannot be confirmed from primary sources yet.

[25] SVE, vol. 7, Стратегия военная, 555–65. TVD is often translated as "theater of military activity."

[26] SVE, vol 6, Оперативное искусство, 53–57.

[27] V. G. Reznichenko, *Taktika* [Tactics: A Soviet View] (Moscow: ВОЕННОЕ ИЗДАТЕЛЬСТВО, 1987), introduction. Branch is used to translate "вид Вооруженных Сил," while "arm" is used to translate "Род войск." *Special troops* is a translation of "специальных войск," which is a term encompassing most logistics, combat support, and combat service support functions. Do not confuse it with "особого назначения" or "специального назначения" (OsNaz/OsN or SpetsNaz/SpN, meaning Special Purpose), which refer to Special Operations Forces (SOF). This distinction is generally unclear in the English-language literature, likely due to translation issues.

during that conflict: *deep operations*.[28] It was the ideal to which the Soviet military strove, much as the U.S. military sought to execute its concept of AirLand Battle.[29] Deep operations was an integrated military concept that discussed warfare from under the sea to above the atmosphere, from the level of national military decision-making to small-unit tactics.[30] Note that "deep" does not refer to overall depth of advance or to the distance from jump-off points or how far from the initial forward edge of the battle area (FEBA) forces reach, but the separation between advanced forces and the main body.[31]

The 1980s theory of deep operations was typified by a robust and integrated joint and combined arms approach using modern technology to improve on the concept. Its defining feature is that rather than the stereotypical "Soviet steamroller," an enemy defense is split by several "finger-like penetrations controlled by a single powerful hand."[32] These fingers are the advanced forces, tasked with critical assets such as enemy airfields, preempting the enemy's ability to form a coherent defense by seizing key terrain or interdicting the flow of reinforcements, or collapsing planned defenses by seizing them before the enemy can establish a position.[33]

While the Operational-scale Advanced Force, a.k.a. the OMG (Operational Mobile Group, *Operativnaya Podvizhnaya Pruppa*), has received far greater attention, in the realm of amphibious operations the relevant concept is the much more common and less discussed *PO* (Forward Detachment, *Peredvoi Otriad*).[34] A typical *PO* would be a battalion reinforced with attachments to act as a task-organized, self-sufficient combined arms group capable of independent action.[35] Acting as a forward detachment in support of ground forces would be a very likely role for *MorPekh* (Naval Infantry) in a coastal direction.

---

[28] David M. Glantz, *The Military Strategy of the Soviet Union: A History* (London: Routledge, 1992), 200-8, https://doi.org/10.4324/9781315035666. There are a number of key differences, especially in echelonment, but these are largely outside the scope of this chapter.

[29] Vincent H. Demma, *Department of the Army Historical Summary, Fiscal Year 1989*, ed., Susan Carroll (Washington, DC: U.S. Army Center of Military History, 1998), 45-50.

[30] It was expressed in a classified 1980s General Staff Directive, which carried with it the weight of law. Though this document does not currently exist in the open literature, there are several references to it by those involved in its production. MajGen Yuri Kirshin quoted in John G. Hines and Ellis Mishulivich, *Soviet Intentions, 1965-85*, vol. 2, *Soviet Post-Cold War Testimonial Evidence* (Washington, DC: Office of Net Assessment, Department of Defense, 1993), 104; and LtGen Gelii Viktorovich Batenin, quoted in Hines and Mishulovich *Soviet Intentions, 1965-85*, vol. 2, 7-8. This work also features extensive interviews with Col Gen A. A. Danilevich, who was the leader of the author-collective on this work.

[31] *The Soviet Army: Operations and Tactics*, 1-48. The implications of such a concept in sea, air, and space are outside the scope of this chapter, but they do carry over.

[32] *The Soviet Army: Operations and Tactics*, 1-48.

[33] *The Army Field Manual*, vol. 2, pt. 2, *A Treatise on Soviet Operational Art* (London: British Army, 1991), 6-16-6-17.

[34] OMG is often mistranslated as Operational Maneuver Group. Also note that PO are not the only advanced force relevant here, raiding detachments and other units are also relevant.

[35] David M. Glantz, *The Soviet Conduct of Tactical Maneuver: Spearhead of the Offensive* (London: Frank Cass, 1991), 10-13.

The Soviet Union, like Russia before and after it, was a continental land power. Its security concerns were primarily focused on its land borders—NATO, the Middle East, and after 1960 the People's Republic of China.[36] Their significant continental holdings required defense, and this commitment absorbed the bulk of the Soviet military's attention.[37] The Soviet Navy was officially less important than the ground forces, and the priority for the Soviet Navy through the bulk of the Cold War was Admiral Sergey G. Gorshkov's *Withholding Strategy*, a modernized nuclear "fleet in being," which saw the Soviet fleet committed to protecting its nuclear ballistic missile submarines (SSBN).[38] These SSBNs would serve a crucial role in intrawar deterrence and conflict termination, so their survival was an incredibly high priority for the Soviet military as a whole.[39]

The Soviet Navy thus did not place a high priority on expeditionary amphibious warfare. Amphibious operations were to take place at relatively shallow depths in support of ground forces actions. Indeed, the Soviets reported 114 amphibious landings during the Nazi-Soviet war, of which only 4 were large-scale operations.[40] In contrast, the Western Allies conducted 22 major and hundreds of minor landings during the war.[41]

The Soviet Naval Infantry, like the Soviet airborne forces, spent most of the war fighting as ground troops. They came to prominence during the sieges of Sevastopol and Odessa, earning the moniker of "Black Death."[42] At Leningrad, the Soviet Navy committed more than 87,000 sailors as Naval Infantry, and large numbers were employed in the defense of Moscow and Stalingrad, as well as assisting in crossing the Don, Dnepr, Danube, and Amur.[43] By the end of the war, the Soviet Naval Infantry numbered approximately 500,000 personnel, of which approximately 300,000 had been cumulatively landed.[44] During the Second World War, Soviet naval development was,

---

[36] In the post-World War II period. For the interwar period, this would shift to British and Japanese Imperial holdings and the USSR's capitalist neighbors.

[37] Alongside the strategic nuclear forces (SNF). For more on SNF, see John Hines, *Soviet Intentions, 1965-1985*, vol. 1, *An Analytical Comparison of U.S.-Soviet Assessments during the Cold War* (McLean, VA: BDM Federal, 1995).

[38] James McConnell, *Admiral Gorshkov on "Navies in War and Peace*," ADA003071 (Arlington, VA: Center for Naval Analyses, 1974), 76-81.

[39] Brad Dismukes, "The Return of Great Power Competition: Cold War Lessons about Strategic ASW," *Naval War College Review* 73, no. 3 (2020): 3, 5-7.

[40] John J. Carroll, *Soviet Naval Infantry*, ADA047604 (Leavenworth, KS: Army Command and General Staff College, 1977), 42.

[41] Carter A. Malkasian, *Charting the Pathway to OMFTS: A Historical Assessment of Amphibious Operations from 1941 to the Present* (Alexandria, VA: CNA, 2002), 10, 19.

[42] LtCol Donald K. Cliff, USMC, "Soviet Naval Infantry: A New Capability" (master's thesis, School of Naval Warfare, Naval War College, 1971), 14.

[43] Cliff, "Soviet Naval Infantry," 15. The use of dedicated amphibious forces in support of river crossings merits further study.

[44] Norman Polmar, Thomas A. Brooks, and George E. Federoff, *Admiral Gorshkov: The Man Who Challenged the U.S. Navy* (Annapolis, MD: Naval Institute Press, 2019), 49.

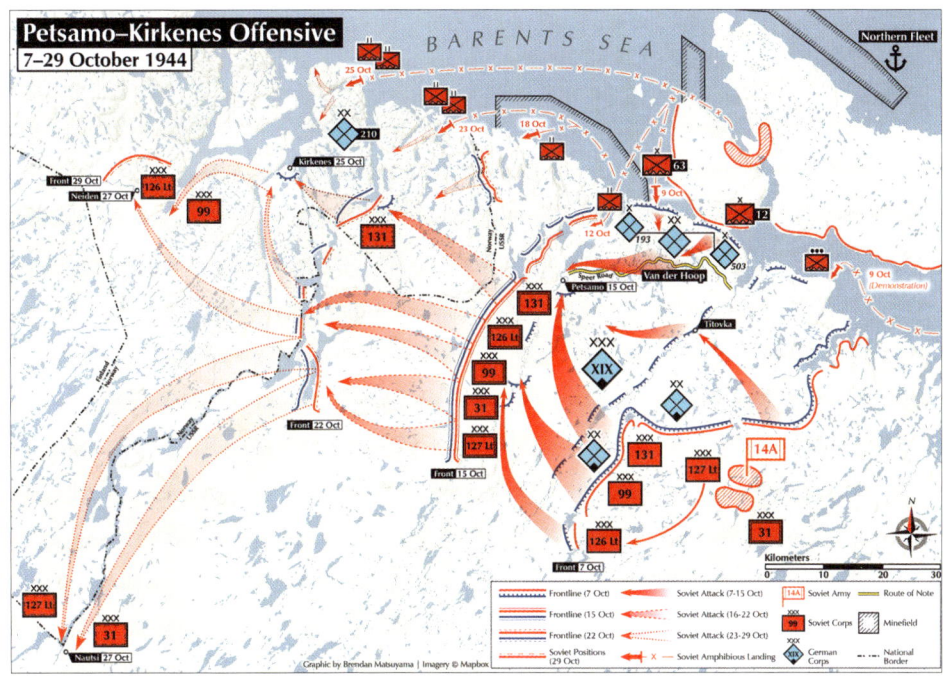

MAP 1
Petsamo–Kirkenes offensive operation, 7–29 October 1944.
Source: courtesy of Brendan Matsuyama, adapted by MCUP

by their own admission, focused "on developing ways for the Navy to assist the ground forces in the defense and attack on seaside directions."[45] Despite this supporting role, the Black Death had earned a reputation as relentless and formidable combatants. To further explore Soviet Naval Infantry and amphibious operations during World War II, the 1944 Petsamo-Kirkenes operation will be offered as a vignette.[46]

Despite austere beginnings, the Soviet Northern Fleet was a sizable force by October 1944. At the same time that U.S. and Allied forces were landing and fighting on Leyte, the Soviets were conducting five amphibious landings across Finnmark during northern Norway's arctic autumn. The first and largest of these landings was at the bay of Bukhta Maativuono (now Guba Malaya Volokopaya), where approximately 3,000 Soviet Naval Infantry landed around midnight between 9 and 10 October 1944.[47]

---

[45] VES 1986, 140.
[46] Petsamo is the Finnish name. In Norwegian, it is Petsjenga; in Russian, it is Pechenga. The town is currently located in Murmansk Oblast, Russia. There is quite a lot of room for scholarship on the subject of Soviet amphibious warfare. Even the four large landings—Kerch-Feodosiya in 1941, Novorossiysk in 1943, Kerch-Eltigen in 1943, and Moon Sound in 1944—have relatively little written on them in English.
[47] James Gebhardt, "Petsamo-Kirkenes Operation (7–30 October 1944): A Soviet Joint and Combined Arms Operation in Arctic Terrain," *Journal of Slavic Military Studies* 2, no. 1 (1989): 58.

Red Tide over the Beach

With no dedicated amphibious vessels, the Soviets successfully put this force ashore in three waves supported by light craft and shore batteries.[48] The only opposition to the landing came from German coastal artillery, but these had been located during Soviet preparations for the landing and were suppressed by Soviet fires. This force, the bulk of the 63d Naval Infantry Brigade (MPBr), was tasked with hindering the withdrawal of Nazi forces, especially *Division Group van der Hoop*, alongside forces of the 12th MPBr, who attacked overland from the Sredni peninsula.[49] These two naval infantry brigades formed up by midday on 10 October and began pursuing Nazi forces down the "Speer Road" running between Pechenga and Titovka, less than 30 km directly.[50]

The Soviet landing was successful for several reasons. First, the landing site was lightly defended and German defenses were targeted robustly. The Soviet Naval Infantry had approximately 275 aircraft of the Northern Fleet's Naval Aviation on standby. The Northern Fleet, under Admiral Arseni Golovko, conducted extensive hydrographic surveys and navigational support, photoreconnaissance of German positions and potential landing sites, as well as presurveying Soviet shore battery positions and locating German shore defenses for suppression or destruction. Furthermore, German shore batteries were suppressed by coordinated attack: small craft would locate the batteries by drawing their fire and then vector Naval Aviation aircraft to strike the German guns.[51]

It is worth noting the command relationship throughout these amphibious operations. Naval Infantry forces were subordinate to Admiral Golovko and the Northern Fleet, and the forces in general were subordinate to the Karelian front under Kirill A. Meretskov.[52] The two organizations were formally coordinated through Stavka (General Headquarters), though Golovko and Meretskov reportedly had an effective and congenial working relationship.[53] However, this lack of formal command relationships provided ample opportunity for friction: there were separate Northern Fleet and Karelian front forward command posts that had no direct communications, only through the Northern Fleet headquarters in Polyarny, more than 100 km to the east.[54]

The Northern Fleet landings during the Petsamo-Kirkenes offensive operation were surprisingly successful for how ad hoc they were. As James Gebhardt notes, "The [Northern] Fleet had no amphibious landing craft," and one landing (that of

---

[48] Maj James F. Gebhardt, *The Petsamo-Kirkenes Operation: Soviet Breakthrough and Pursuit in the Arctic, October 1944*, Leavenworth Paper no. 17 (Leavenworth, KS: Combat Studies Institute, U.S. Army Command and General Staff College, 1989), 90.
[49] Gebhardt, *The Petsamo-Kirkenes Operation*, 90.
[50] Gebhardt, *The Petsamo-Kirkenes Operation*, 38–39. The "speer road" appears to be roughly the same route as the modern A138/E-105.
[51] Gebhardt, *The Petsamo-Kirkenes Operation*, 89.
[52] Gebhardt, *The Petsamo-Kirkenes Operation*, 87.
[53] Gebhardt, *The Petsamo-Kirkenes Operation*, 86–87.
[54] Gebhardt, *The Petsamo-Kirkenes Operation*, 88. The graphic is located on 27.

12 October at Liinakhamari) was conducted by an improvised force of 500 sailors, fleshing out a cadre of 150 naval infantrymen.[55] Whereas the Western Allies used purpose-designed landing craft—Landing Ship, Tank (LST); Landing Craft, Tank (LCT); Landing Ship, Medium (LSM); Landing Craft, Support (LCS); Landing Craft, Assault (LCA); and iconic Landing Craft Vehicle, Personnel (LCVP, better known as the Higgins Boat)—Soviet landings generally relied on motor torpedo boats, motor gun boats, minesweepers, submarine chasers, and other nonspecialized craft.[56] The Soviet Naval Infantry of WWII had no analog for the American Landing Vehicle, Tracked (LVT). Their ship-to-shore connectors were wooden gangplanks, running the vessel up to the shore, or whaleboats.[57] As a result, the actual landing was generally a lengthy and vulnerable process. The Soviets were aware of this and carefully chose landing sites to avoid robust German beach defenses, while remaining in range of air and artillery support. Furthermore, the lack of robust Soviet amphibious lift capacity significantly hindered the landing of both follow-on forces and heavy equipment. Soviet authors such as Admirals Ivan Isakov and K. A. Stalbo candidly spoke to these shortcomings, but noted that, in spite of the improvised landing craft, lack of heavy equipment, sustainment issues and the knock-on effects thereof, Soviet landings were often successful.[58]

There were practical, geographical, and economic factors that caused the lack of specialized landing vessels during the Nazi-Soviet war. Practically speaking, the USSR had no large force of amphibious warfare ships, much less personnel to use them. As Admiral Stalbo described it:

> *In order to land forces in the war years, we had to resort to using warships, and poorly-suited ships and boats. . . . The lack of specialized landing ships often led to considerable losses of landing forces and made weather conditions of special significance.*[59]

To develop such a force during the war would have been extremely wasteful, given the strain on the Soviet state. Both the USSR and Nazi Germany were continental powers, the bulk of whose combat power was found in their ground and air forces. For the USSR to defeat Nazi Germany, a necessity given the war of genocide and conquest the Nazis unleashed on the Soviet people, it first had to liberate the occupied regions of the USSR and destroy Nazi Germany.[60] Given the geography of the region,

---

[55] Gebhardt, *The Petsamo-Kirkenes Operation*, 92.
[56] Gebhardt, *The Petsamo-Kirkenes Operation*, 94.
[57] Gebhardt, *The Petsamo-Kirkenes Operation*, 92.
[58] Adm I. S. Isakov, *Red Fleet in the Second World War* (London: Hutchinson, 1947); RAdm K. A. Stalbo, "The Naval Art in Landings of the Great Patriotic War," *Morskoi Sbornik*, no. 3 (1970): 3; and quoted in Carroll, *Soviet Naval Infantry*, 39.
[59] Carroll, *Soviet Naval Infantry*, 38-39.
[60] This is not to minimize the violence inflicted by the USSR on the various peoples living in the Soviet Union, of Eastern and Central Europeans, or the peoples of the various areas illegally annexed in 1939-45.

amphibious warfare would either be river crossings or on the coastal periphery, universally in support of the ground forces. In addition, limited Soviet economic resources during the interwar period had been primarily focused on ground and air forces modernization. The result of these and other factors was, as S. G. Gorshkov noted,

> [before the war] neither the building of landing ships nor the training of special landing troops were given due attention. All our fleets came into the war without having a single specially constructed landing ship. . . . All this limited the potential of the [Navy] in solving the tasks of assisting land forces and made it harder for it to stage landings from the sea . . .[61]

Despite these handicaps, the landings during the Petsamo-Kirkenes offensive were tactically successful, but they lacked the overall joint coordination to turn a successful landing into a successful amphibious operation. For example on 9–10 October, the Soviet Naval Infantry landed well after the beginning of the Soviet offensive and more than 30 hours after the Nazi *Division Group van der Hoop* was authorized to retreat toward Pechenga/Petsamo.[62] While the Naval Infantry was able to engage van der Hoop's forces and prevent their redeployment, they were unable to force an encirclement or prevent their retreat.[63] This is typical of Soviet issues with coordinating multiple front (fleet)-level entities prior to the adoption of the theater command in the late summer of 1945.[64]

More information on the Petsamo-Kirkenes landings can be found in James Gebhardt's *The Petsamo-Kirkenes Operation*, whose bibliography includes much of the Soviet-era historiography and analysis of the operation, while the pair of articles by Sven Holtsmark in *Journal of Slavic Military Studies* cites a robust overview of Soviet contemporary primary sources.[65] The Petsamo-Kirkenes offensive is an excellent example of how the Soviets used *naval desant* to insert critical forces into the enemy rear to support and enable larger ground forces offensives. Through the end of the Cold War, the Petsamo-Kirkenes operation was held up by authoritative Soviet publications as a decisive and important historical model for the use of amphibious operations to support ground forces.[66] While it has seen more research in recent years,

---

[61] S. G. Gorshkov, *The Sea Power of the State* (Oxford, UK: Pergamon, 1979), 140.

[62] Gebhart, *The Petsamo-Kirkenes Operation*, 94.

[63] Gebhart, *The Petsamo-Kirkenes Operation*, 97, 116.

[64] LtCol David M. Glantz, *August Storm: The Soviet 1945 Strategic Offensive in Manchuria*, Leavenworth Papers no. 7 (Leavenworth, KS: Combat Studies Institute, U.S. Army Command and General Staff College, 1983), 37.

[65] Sven G. Holtsmark, "Improvised Liberation, October 1944: The Petsamo Kirkenes Operation and the Red Army in Norway. Part I," *Journal of Slavic Military Studies* 34, no. 2 (2021): 271–302, https://doi.org/10.1080/13518046.2021.1990554; and Sven G. Holtsmark, "Improvised Liberation, October 1944: The Petsamo Kirkenes Operation and the Red Army in Norway, Part 2," *Journal of Slavic Military Studies* 34, no. 3 (2021): 426–58, https://doi.org/10.1080/13518046.2021.1992707.

[66] Gebhart, *The Petsamo-Kirkenes Operation*, 116.

Claremont

the English language historiography would benefit from robust modern work on the subject, especially that incorporating post-Soviet archival material and historiography on the topic.[67] Furthermore, there is room to examine the causal forces behind Soviet force development during the interwar and wartime periods, especially between the doctrinal avoidance of amphibious warfare and lack of landing means.[68]

Despite their excellent combat record and relatively large size at the end of the Great Patriotic War, the Soviet Naval Infantry was quickly cut down post war due to shifting views of the character of warfare. With the rise of Nikita S. Krushchev and the ouster of Admiral Kuznetsov, the Soviet Navy focused on submarines and nuclear strikes, while the responsibility for amphibious warfare was quietly shifted to the army.[69] Krushchev was politically opposed to expeditionary amphibious warfare, which he saw as a tool of the warmongering imperialists.[70] The Naval Infantry was successively downsized, folded into the coastal troops, and retired without fanfare.[71]

However, contemporaneous with the removal of Krushchev in 1964 came the rebirth of the naval infantry.[72] This appears to be related to the rising prominence of Admiral S. G. Gorshkov, made deputy minister of defense in 1962, who had a special interest in amphibious operations due to his service during the Great Patriotic War.[73] Gorshkov was a prominent Soviet naval commander during the war and led approximately one-quarter of all Soviet amphibious landings during the war.[74] Much in the way that the British Royal Navy of the First World War was the product of Admiral John A. Fisher, the Soviet Navy was shaped by Gorshkov's concept of maritime warfare during his tenure 1956–85.[75]

In contrast to the continental Soviet Union, the United States is and historically has been an insular maritime power.[76] America is protected from attack by significant maritime borders, and its security is thus contingent on command of the sea. Possessing a large navy, its primary mode of military activity is projecting power

---

[67] The Soviet (now Russian) military history journal *VIZh*, as well as magazines like *Sovietskiy Morpekh* or *Morskoye Pekhotinets* or the journals *Morskoi Desant* and *Morskoi Sbornik* are generally available and underutilized.

[68] That is to say, were landing means not procured because they were not needed in doctrine, or were they not needed in doctrine because none were likely to be procured? Not to mention the ideological-political and bureaucratic-political influences.

[69] Carroll, *Soviet Naval Infantry*, 51–53.

[70] N. K. Krushchev, *Krushchev Remembers: The Last Testament*, ed. and trans. Strobe Talbott (Boston, MA: Little, Brown, 1974), 26; and quoted in Carrol, *Soviet Naval Infantry*, 53.

[71] Carroll, *Soviet Naval Infantry*, 53. The retirement was so subtle, the author was unable to find a specific date in any source.

[72] Carroll, *Soviet Naval Infantry*, 54.

[73] Polmar, Brooks, and Federoff, *Admiral Gorshkov*, 135–37.

[74] Polmar, Brooks, and Federoff, *Admiral Gorshkov*, chaps. 4, 5, and 6 provide a solid biographical picture; and see Cliff, "Soviet Naval Infantry," 54, for the number of landings commanded by Gorshkov.

[75] Polmar, Brooks, and Federoff, *Admiral Gorshkov*, 202–3.

[76] *Naval Warfare*, Naval Doctrine Publication 1 (Washington, DC: U.S. Navy, Marine Corps, and Coast Guard, 2020), 1.

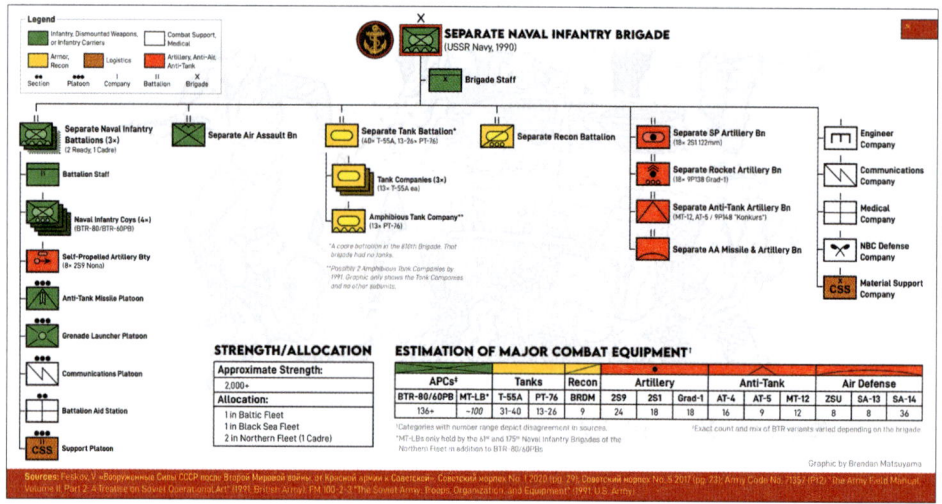

FIGURE 1
Soviet Naval Infantry Brigade table of organization and equipment, ca. 1990.
Source: courtesy of Brendan Matsuyama, adapted by MCUP

from the sea to the land.[77] Therefore, expeditionary amphibious warfare is a critical capability. As such, the amphibious forces of the U.S. Navy and Marine Corps have been tailored to these requirements, especially since the Second World War.

Thus, to the American audience, and indeed many other audiences, amphibious warfare is almost inherently expeditionary. This was not the case for the Soviets, and they tailored their force structure and military thought accordingly. The Soviet Union did not possess any "Big Deck" amphibious warfare ships, compared to the U.S. Navy's 13 in 1989.[78] Instead, they had a large fleet of smaller amphibious ships, with more than 250 LSTs, LPDs, LCACs, and LCUs.[79] Indeed, some of the capabilities they pursued greatly exceeded Western requirements. The *Zubr*-class LCAC, for example, is able to travel 300 nautical miles at 55 knots and land a mechanized infantry company.[80] In addition, the Soviets made extensive use, in both exercise and theory, of civilian

---

[77] *Naval Warfare*, 1–3.
[78] In the late 1980s, the Soviet General Staff actually had a requirement for an LHD (Landing Helicopter Dock/UDK in Soviet parlance), resulting in Project 11780 *Kherson*, but it was killed by the navy because it could not be made at the same time as Project 11437 *Ulyanovsk* carriers due to lack of shipyard space. Alexander Karpenko's *Nevskii-Bastion* blog is a convenient, albeit Russian language, source and not authoritative.
[79] Of the *Ropucha*, *Alligator*, *Ivan Rogov*, *Polnocny*, *Tsaplya*, *Lebed*, *Gus*, *Zubr*, and *Aist* classes, as well as the Project 106 small landing ship (LCU). Compare to ~175 U.S. Navy vessels of comparable role and capability in 1989 per Navsource/DANFS.
[80] Yuri Apalkov, *Ships of the Soviet Navy Handbook*, vol. 4, *Landing and Minesweeping Ships* (St. Petersburg: Morkniga, 2007), 48–56.

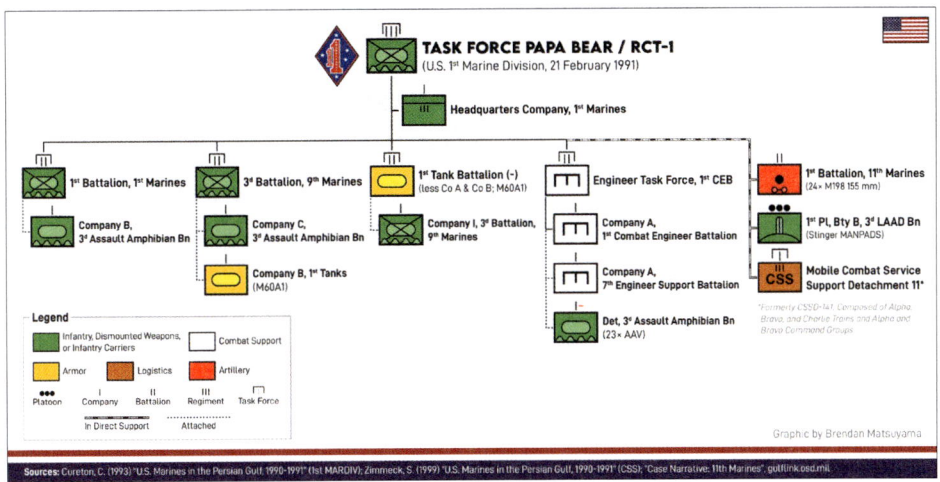

FIGURE 2
Task Force Papa Bear/Regimental Combat Team 1 (RCT-1), 1991.
Source: courtesy of Brendan Matsuyama, adapted by MCUP

roll on/roll off (RORO) ships, a practice alluded to by the use of MV *Yulius Fucik* in Clancy and Bond's *Red Storm Rising*, and taken to an extreme in "Sea Control in the Arctic: A Soviet Perspective."[81]

For example, the Soviet amphibious lift capability in the Northern Fleet ca. 1987–90 was able to move approximately one brigade, though a more likely employment scenario would be multiple reinforced battalion task forces.[82] In the late 1980s, the Soviet Northern Fleet had the capacity to simultaneously land three such naval task forces.[83] In total, it fielded two naval infantry brigades, the 61st and 175th

---

[81] Cdr Dennis M. Egan, USCG, and Maj David W. Orr, USMCR, "Sea Control in the Arctic: A Soviet Perspective," *Naval War College Review* 41, no. 1 (Winter 1988): 51–80. Egan and Orr propose that the Soviets would send multiple divisions to northern Alaska via the Arctic route to attack oil and gas infrastructure mainly using ROROs. Questions of the utility and viability of sustaining a campaign in northern Alaska for any length of time are not well explored. It is, however, thought provoking. It is worth noting that Soviet merchant shipping, like Soviet civilian aviation, was openly viewed as a mobilization asset.
[82] This would force an adversary to confront multiple dilemmas simultaneously. It also would fit into the training patterns and logistical capacity of Soviet amphibious forces, especially in the northern theater. *SSRC Soviet Amphibious Warfare* (The Hague: Soviet Studies Research Center, 1985), 55–58, touches on this. It also adds redundancy should any single landing fail.
[83] By the late 1980s, the Soviet Navy in total had four naval infantry brigades and one division: the 61st and 175th Naval Infantry Brigades in the Northern Fleet, the 810th Naval Infantry Brigade in the Black Sea, 336th Guards in the Baltic, and the 55th Naval Infantry Division in the Pacific Fleet. "Военно-Морской Флот (ВМФ)," Navy (VMF), accessed 1 September 2023.

Naval Infantry Brigades.⁸⁴ Each brigade was approximately equivalent to one of the regimental combat teams formed by the 1st Marine Division during Operation Desert Storm (1990–91).⁸⁵

The Northern Fleet also possessed two distinct types of special operations forces: Отряды Специального Назначения по Борьбы с ПДСС, Detachments of Special Purpose (SOF Detachment), for combating underwater sabotage forces and means, or PDSS, and Отдельные морские разведывательные пункты специального назначения, Separate Naval Reconnaissance Point for Special Purposes, or OMRp SpN.⁸⁶ PDSS were primarily tasked with defending Soviet naval bases from enemy divers, and they were armed with underwater firearms and a number of specialized antidiver grenades and launchers.⁸⁷ OMRp SpN filled a much more traditional over-the-beach deep reconnaissance, sabotage, and direct action role, and had a history going back through the Second World War.⁸⁸ Sources are limited and unclear, but it appears that OMRp SpN were focused more on deep reconnaissance than the Western naval SOF emphasis on beach reconnaissance and obstacle clearance coming out of the Underwater Demolition Team/Special Boat Service (UDT/SBS) tradition. Soviet and Russian naval SOF are, however, a relatively understudied topic, and one that merits further research. There is notable lack of clarity in the exact ways in which they would be used, and how those would dovetail with conventional forces.

RED WAVES WASHING ASHORE: THE MECHANICS OF LANDINGS

In Soviet terminology, landings were defined by scale and, to an extent, purpose. Soviet definitions ranged from the multiarmy operational-strategic naval landing (OSMD) through the multidivisional operational naval landing (OMD) to the tactical naval landing of reinforced company to reinforced regiment scale.⁸⁹ However, while they categorized a wide scale of landings, the Soviets only rarely conducted or exercised OMDs, with the vast majority of exercises being Тактический морского Десант (tactical naval landings, or TMD), tending toward reinforced battalion

---

⁸⁴ See 61st Independant Naval Infantry Brigade and 175th Independent Naval Infantry Brigade, "Военно-Морской Флот (ВМФ)."
⁸⁵ LtCol Charles H. Cureton, USMCR, *U.S. Marines in the Persian Gulf, 1900-1991: With the First Marine Division in Desert Shield and Desert Storm* (Washington, DC: History and Museums Division, Headquarters Marine Corps, 1993), 20. Task Forces Papa Bear, Ripper, and Taro (Regimental Combat Team 1/7/3) to be specific.
⁸⁶ A clearer but less literal translation of these units might be "Counter-Frogman Detachment" for PDSS and "Separate Naval Reconnaissance Team" for OMRp SpN.
⁸⁷ Information on Soviet naval SOF is limited, but it is possible PDSS had organizational control of the DP-62 Damba jet bombing system, a BM-21 Grad modified to fire depth charge rockets queued by sonar.
⁸⁸ James F. Gebhardt, "Soviet Naval Special Purpose Forces: Origins and Operations in the Second World War," *Journal of Slavic Military Studies* 2, no. 4 (1989): 563–64, https://doi.org/10.1080/13518048908429964.
⁸⁹ Milan Vego, *Soviet Naval Tactics* (Annapolis, MD: Naval Institute Press, 1990), 287–88. There are also diversionary and reconnaissance landings.

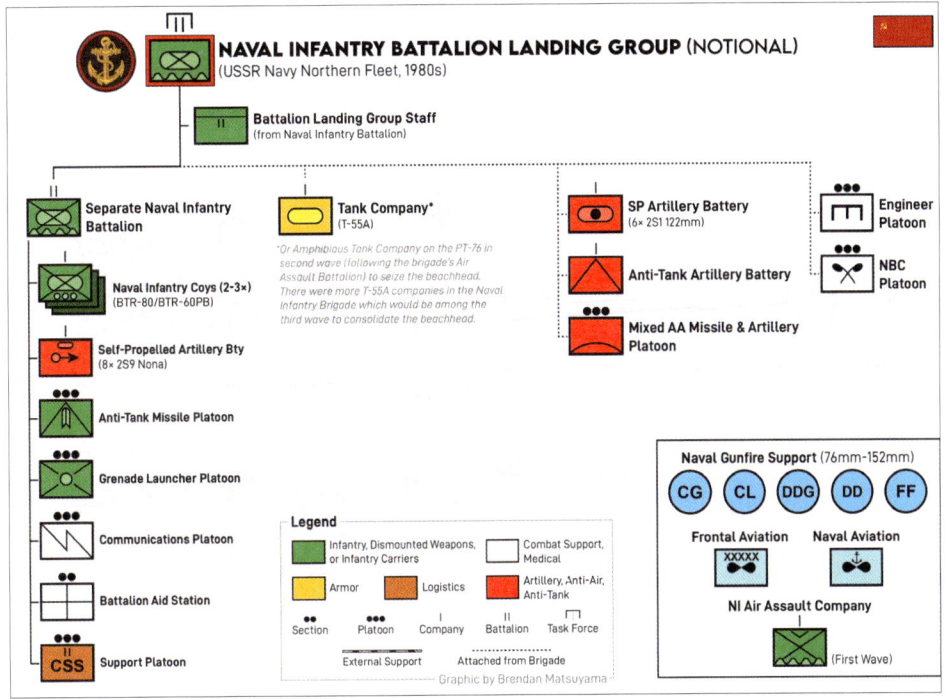

FIGURE 3
Notional Northern Fleet battalion-strength landing group, ca. 1987.
Sources: based on author's analysis, courtesy of Brendan Matsuyama, adapted by MCUP

scale.[90] Soviet writings emphasized flexible and quick-reacting TMDs coordinated with ground forces, rather than set piece and/or expeditionary OSMD.[91] Indeed, the distinction between Soviet TMD and OMD was the scale of the landing force not its depth or mission.[92]

Note that this emphasis on TMD does not preclude multiple simultaneous landings of reinforced battalion scale, which was within Soviet capability and dovetails well with their desire to overload the adversary decision-making system.[93] Imposing so many problems on the enemy as to induce paralysis was a hallmark of the Soviet concept of warfare, which can be found at least as far back as the concept of deep battle.[94] Confronting a potential adversary with multiple task organized reinforced-battalion scale combined arms groupings would be a classic mission for

---

[90] *SSRC Soviet Amphibious Warfare*, 46; and Vego, *Soviet Naval Tactics*, 299.
[91] James F. McConnell, Robert G. Weinland, and Michael K. McGwire, *Admiral Gorshkov on "Navies in War and Peace"* (Arlington, VA: CNA, 1974), 70.
[92] *SSRC Soviet Amphibious Warfare*, v. This does not preclude brigade-size landings, but they were assessed as less likely, especially in the Barents/Norwegian Seas.
[93] *The Soviet Army: Operations and Tactics*, 1-42.
[94] Note that this is a designed byproduct and not the end goal.

Soviet Naval Infantry during the conduct of armed conflict within a continental TVD.

By the 1980s, *naval landings* occupied "an important, if limited role" in the Soviet concept of warfare, especially in maritime or coastal theaters.[95] Naval infantry forces would act as advanced forces for a larger Soviet ground force in roles such as raiding detachments, forward detachments, and other forces to destabilize the enemy's scheme of deployment/maneuver.[96]

The TMD was divided into several stages for planning purposes. First came preparation and embarkation, during which forces were trained, planning and staff work conducted, and landing forces and naval assets organized and loaded. After this came the sea transit, debarkation (including the battle for debarkation), where troops moved from ship to shore, and finally the battle ashore and, if necessary, reembarkation.

Preparation and embarkation were two separate phases, but were closely linked as the point of embarkation was in part decided by the target and what preparations were necessary. Preparation consisted of conducting the required reconnaissance, staff work, and planning to select the port of embarkation, landing site, further tasks, command and control, the force necessary, and timing of a landing.[97] Available sources indicate shore reconnaissance and obstacle clearing would be conducted by "diver demolition teams" or "assault frogmen specially trained in underwater demolition, engineers, reconnaissance and communications personnel," but the precise designation, chain of command, attachment or subordination of these personnel is unclear.[98] These forces, known as the advance detachment, would typically land between H-hour and H+5 minutes, often by a mix of small boats, hovercraft, and rotary-wing aviation.[99]

Preparation occured as continuously as possible until the point of embarkation, including rehearsals and other typical measures. Typical timing allotments by the late 1970s would have been (approximately): two hours for elaboration of the commander's decision; one hour for route reconnaissance between assembly and embarkation areas, typically separated by 8-15 km; one hour for coordination; three hours for final material preparations; one hour for watertight integrity checks; and a final hour for party-political work, which consisted of efforts to improve morale, unit cohesion, and combatant motivation.[100] After this had been accomplished, the forces moved, typically by company (with reinforcements attached), to the embarkation point and loaded onto vessels.[101] Timing was ideally such that the amphibious vessels and the

---

[95] *The Army Field Manual*, vol. 2, pt. 2, *A Treatise on Soviet Operational Art* (London: British Army, 1991), 10-1.
[96] *The Army Field Manual*, vol. 2, pt. 2, 10-1.
[97] Vego, *Soviet Naval Tactics*, 302-3.
[98] Vego, *Soviet Naval Tactics*, 306-7; and *SSRC Soviet Amphibious Warfare*, 50.
[99] Vego, *Soviet Naval Tactics*, 311-12.
[100] Vego, *Soviet Naval Tactics*, 50-51, 303, 8-15 km from *The Army Field Manual*, vol. 2, pt. 2.
[101] *SSRC Soviet Amphibious Warfare*, 44; and Vego, *Soviet Naval Tactics*, 303-4.

landing force were only static at the embarkation point for as long as it took to load.[102] Generally, an embarkation area would have two alternative embarkation points in case the primary is disabled.[103]

Once the landing force had loaded onto the transports, they began the sea transit, taking constant precautions to prevent enemy means of reconnaissance, as well as to obfuscate the time, place, and scale of the landing until as close to when it occurred as possible.[104] The landing force was generally escorted by a close screen of fast attack craft and antisubmarine warfare vessels, preceded by mine warfare vessels, accompanied by a fire support ship detachment, and protected by antiair warfare combatants pushed 30–50 km down the likely threat axis.[105] The Soviets desired to conduct the embarkation and sea transit during one period of darkness, arriving at the debarkation area.

The debarkation area is chosen following mine countermeasures and assault diver sweeps of the debarkation area, the landing force anchors and the MPBn uses its amphibious assets to assault the shore. For an unprepared beach, the typical norm for a Soviet MPBn was a landing area 400–600 m wide, from which the MPBn would establish a beachhead 3,000–4,000 m by 1,500–2,000 m.[106] The Soviets called this process the battle for debarkation.[107] Western readers might know it better as the amphibious assault.[108] It was a combat action fought by joint air, naval, and ground forces to "break enemy anti-landing defenses, destroy enemy forces on the coast, and establish a beachhead."[109]

Vertical envelopment was a key tool in the Soviet amphibious landing playbook, though the Soviet Navy had relatively meager capability for the task organically.[110] Indeed, by the 1980s, the Soviets "consider[ed] . . . that an *amphibious assault alone would be most unusual.*"[111] Accompaniment by vertical envelopment, whether heliborne or parachute landed, was ubiquitous by the late 1970s, and a percentage of the Soviet Naval Infantry went through airborne training.[112] Typically, a vertical

---

[102] Vego, *Soviet Naval Tactics*, 303–4.
[103] *The Army Field Manual*, vol. 2, pt. 2, 10-5.
[104] Vego, *Soviet Naval Tactics*, 304–5.
[105] *The Army Field Manual*, vol. 2, pt. 2, 10-6.
[106] Vego, *Soviet Naval Tactics*, 307.
[107] Vego, *Soviet Naval Tactics*, 307.
[108] *The Army Field Manual*, vol. 2, pt. 2, 10-6; and *SSRC Soviet Amphibious Warfare* use this term for the combat phase.
[109] Vego, *Soviet Naval Tactics*, 308.
[110] The Soviets never built an aviation-focused amphibious warfare vessel. The *Kiev*-class ship was unsuited to the task and needed for its intended role, while the *Moskva*-class ship was a particularly poor design by any measure. For more background on aviation surface combatants, see Benjamin Claremont, "Why the *Moskva*-Class Helicopter Cruiser Is Not the Best Naval Design for the Drone Era," CIMSEC, 13 October 2021.
[111] *SSRC Soviet Amphibious Warfare*, 61, emphasis original.
[112] *The Army Field Manual*, vol. 2, pt. 2, 10-5; and Carroll, *Soviet Naval Infantry*, 84–85. Approximately one battalion per brigade is the commonly cited ratio.

envelopment would land either alongside the main body of forces or 10–20 minutes before the advanced detachment.[113] In addition to organic fixed- and rotary-wing vertical envelopment capabilities, Soviet Naval Infantry and Airborne Forces often worked in close cooperation. The Soviet Airborne Forces (*Vozdushno-desantnye voyska*, VDV) were equipped as mechanized infantry, albeit in lightly armored vehicles.[114] They possessed a full suite of parachute-capable infantry fighting vehicles, armored personnel carriers, artillery, multiple rocket launchers, and self-propelled antitank guns.[115] The mobility and combat power of the VDV allowed the Soviet joint force to inject forward detachments or other advanced forces into the enemy depth simultaneously to an amphibious assault, enhancing the ability of the Soviet military to rapidly undermine the coherence of an adversary's defensive structure. Vertical envelopment could also be used in the more traditional "bite and hold" role of light infantry airborne forces, using the mobility of the helicopter to avoid the exhausting marches that had incapacitated Soviet light infantry during the Petsamo-Kirkenes offensive.[116]

The Soviet tendency to never throw away equipment, no matter how outdated, left them with a surprisingly strong naval gunfire support capability in the 1980s. The Northern Fleet's 37th Naval Landing Division had two *Sverdlov*-class light cruisers attached through the end of the Cold War.[117] These would be supplemented by smaller Soviet surface combatants with 130mm, 100mm, and 76mm guns.[118] In addition to this, the ground forces might support a landing, if it was conducted within the range of the long-barrel 203mm, 152mm, and/or 130mm guns.[119] Whether in range of ground forces artillery support or not, Soviet TMD would, as a rule, occur within range of Soviet air support.[120] This could take the form of naval aviation aircraft, such as Sukhoi Su-17 Fitter attack aircraft, or fighters like the Sukhoi Su-27 Flanker, but it was also not uncommon for naval aviation Tupolev Tu-22M Backfire medium bombers to take part in strikes.[121]

The Soviet approach to amphibious warfare is alien in many details compared to

---

[113] Vego, *Soviet Naval Tactics*, 314–15.
[114] The BMD (*Boyevaya Mashina Desanta* or roughly airborne combat vehicle) family—BMD-1, BMD-2, BTR-D and variants—are only resistant to infantry small arms and light artillery fragmentation.
[115] "Whatismoo's Unclassified Soviet Army Field Guide," YouTube video, 4 pts., provides a handy quick reference to Soviet vehicles and equipment.
[116] Gebhardt, *The Petsamo-Kirkenes Operation*, 43–44. The Light Rifle Corps saw mixed success, consistently suffering from exhaustion due to the grueling requirements of walking long distances under severe noise, light, and engineering discipline to preserve operational security. The 70th Naval Infantry Brigade in particular exhausted itself reaching the objective and was unable to block the road it was assigned to.
[117] "37-я дивизии морских десантных сил, Military Unit: 51309," 37th Naval Landing Division, accessed 1 September 2023. *Sverdlov*-class ships are roughly equivalent to the U.S. Navy's *Cleveland*- or *Fargo*-class gun cruisers.
[118] Vego, *Soviet Naval Tactics*, 300, references the 1981 use of *Kara*- and *Krivak*-class ships in this role.
[119] *The Army Field Manual*, vol. 2, pt. 2, 10-3.
[120] Vego, *Soviet Naval Tactics*, 301, 308–9.
[121] Vego, *Soviet Naval Tactics*, 309.

the Anglo-American school of thought. While the broad strokes are similar, taking military personnel and moving them from sea to shore, the Soviets had a unique methodology from the highest conceptual levels to the precise timing and order of tasks.

To an uncharitable Western eye, the Soviet approach seems both rigid and slapdash, an overaggressive and underresourced way to put a small force not very far behind enemy lines with little provision for further supply over the beach. Such a judgment would not be incorrect, but not because the Soviets were unaware of alternative approaches. Soviet authors examined contemporary foreign amphibious operations throughout the Cold War and integrated their findings where they felt appropriate.[122] The limitations of Soviet amphibious forces were intentional choices made to optimize the force for the distinct role of amphibious warfare within their understanding of the theory and practice of the conduct of and preparation for war.

The Soviets viewed expeditionary amphibious warfare as inherently imperialist and so pursued no extensive capability for it. The continental nature of the USSR meant that naval activity would act in support of a ground campaign, with amphibious assaults acting more as a horizontal envelopment than a forcible entry. Therefore, logistics over the shore were not necessary. By leaning on a flexible and aggressive approach to landing with a short turnaround from deciding on a landing to troops ashore, the Soviets hoped to get inside the enemy's ability to react and to minimize the temporal length of the vulnerable period of transit and disembarkation.

The USSR consistently chose to have amphibious forces focused on battalion-to-brigade scale landings done at short notice over short distances in support of ground forces in a coastal axis. One of the best examples of this is Project 11780 *Kherson*, a 1980s Soviet LHD program to produce two ships: *Kherson* and *Kremenchuk*.[123] The ships were to approximate a 60-percent scale *Tarawa*-class ship to the point that designers reportedly called them "*Ivan Tarava*."[124] The ships were designed for a mix of Yakovlev Yak-38 Forger and Yakovlev Yak-141 Freestyle jumpjets, Kamov Ka-29 Helix-B assault helicopters, and *Tsaplya*-class LCACs.[125] They were designed for transporting two naval infantry battalions a range of 12,875 km at 18 knots.[126] The Soviet General Staff supported the LHD program even at the expense of aircraft carriers, but the navy refused to abandon the aircraft carrier program. The design bureau in charge

---

[122] Jacob W. Kipp, *Naval Art and the Prism of Contemporaneity: Soviet Naval Officers and the Lessons of the Falklands Conflict*, Stratech Studies Series (College Station: Center for Strategic Technology, Texas A&M University, 1983), 22–33.

[123] Alexei Sokolov, Альтернатива. Непостроенные корабли Российского Императорского и Советского флота [Alternative: Unbuilt Ships of the Russian Imperial and Soviet Fleets] (Moscow: Военная книга, 2008), 43; see also, "Фотогалерея Pilot'a Модели авианесущих крейсеров проект 11780," for images from the project.

[124] Alexander Karpenko, "Project 11780 Universal Landing Ship," Nevskii-bastion.ru, accessed 1 September 2023.

[125] Karpenko, "Project 11780 Universal Landing Ship."

[126] Karpenko, "Project 11780 Universal Landing Ship."

of Project 11780 eventually released a design update that shifted the 130mm twin-gun mount and 3K95 Kinzhal (SA-N-9 Gauntlet) surface-to-air missile system to the middle of the flight deck, which cascaded into the program's termination.[127]

The resources available to the Soviet Navy and their priorities meant that even when the technical capacity to pursue a robust expeditionary amphibious capability existed, the institution would choose to maintain the existing paradigm and further support the primary mission of aggressive bastion defense. The Soviet political repudiation of expeditionary warfare likely assisted the navy in this debate with Soviet General Staff. While the General Staff desired the capability, it would have come at the cost of handicapping the carrier fleet and expending a great deal of political capital with the Politburo to obtain a capability the Soviet government was ideologically opposed to.[128] By 1986, the program was canceled and with it the only serious effort by the USSR to pursue expeditionary amphibious capabilities.

MARINES WITHOUT LANDINGS

Despite a robust capability supporting a coherent, albeit alien, concept of amphibious operations, the most pitched battles fought by Soviet Naval Infantry, and post-Soviet Naval Infantry in Russia and Ukraine, have all been fought ashore. At Sevastopol (Crimea, now part of Ukraine), Odessa (Ukraine), and Leningrad (Russia), in Afghanistan and Chechnya, and in Mariupol (Ukraine) and, ironically, in Kherson (both in Ukraine and the namesake of the previously mentioned abortive Soviet LHD effort), Soviet, Russian, and Ukrainian naval infantry fought protracted campaigns and battles where they were singled out as notably skilled combatants but rarely conducted amphibious landings.[129]

The lack of Soviet post-WWII amphibious landings leads to the first and largest caveat: this chapter cannot judge the effectiveness of Soviet concepts. It can say the Soviets had a robustly provisioned capability that suited their understanding and intentions, but of course being well suited to a concept of use and way of war is not inherently a recipe for success.[130] In addition, the sourcing for this chapter is broadly imperfect. The most accessible sources are not recent, and due to accessibility issues, this chapter is largely interacting with Soviet professional literature as interpreted by secondary sources.

These secondary sources are high quality, but few in number and lack exploitation

---

[127] Karpenko, "Project 11780 Universal Landing Ship."
[128] That this all happened against the backdrop of Gorbachev's rise and the war in Afghanistan should not be forgotten.
[129] There were a fairly large number of amphibious landings in the Black Sea and Azov region in WWII, including at Mariupol, but Mariupol is mentioned here for the participation of the Ukrainian 36th OBrMP and Russian 810th Gv. OBrMP. The Ukrainian 35th OBrMP fought in Kherson Oblast during late October early November 2022, and at least four Ukrainian Naval Infantry Brigades have taken part in the Ukrainian summer 2023 counteroffensive.
[130] Notably, the Soviets put little effort into developing field rations.

of post-Soviet access to archival material. They also focus quite heavily on the Northern Flank, the Barents and Norwegian Seas from the Kola around the North Cape and down to southern Norway. Fundamentally, Soviet and Russian naval infantry have been understudied. There is great room for further research and writing to be done on the topic, and on non-Soviet Warsaw Pact (NSWP) and Soviet-aligned amphibious forces. For example, Poland especially had significant naval infantry forces. Nor should Soviet theory/concepts be directly extrapolated to NSWP or "Soviet Pattern" forces such as Vietnam without careful assessment of these countries using their own primary sources and within their own context.

This chapter should be read as the start of a conversation not the final word. There is much work to be done on the history of Soviet Naval Infantry, especially with the greater access afforded to materials and sources that had been trapped behind the Iron Curtain since the dissolution of the USSR. The history and evolution of Soviet Naval Infantry is a fascinating contrast to the more familiar Western school of thought. Starting from fundamentally different assumptions about the relationship between the sea and the state, and with a radically different combat record, the Soviet Navy and naval infantry articulated and procured a relatively large, coherent and well-resourced amphibious force. While the Soviet concept of amphibious warfare would not make a good fit for the needs and missions of a force like the U.S. Navy or Marine Corps, its study does demand that one interrogate their own core assumptions about the nature of combat on contested shores.

CHAPTER FOUR

Innovative Amphibious Logistics for the Twenty-first Century

Walker D. Mills

I don't know what the hell this "logistics" is that [General George C.] Marshall is always talking about, but I want some of it.
~ Admiral Ernest J. King[1]

A landing on foreign shore in the face of hostile troops has always been one of the most difficult operations of war. It has now become almost impossible.
~ Sir Basil Liddel Hart[2]

Logistics have always been a governing factor in military operations, as they are the envelope that defines what is possible and what is not. But, there is perhaps no operation where they are more critical than amphibious operations. It is a truism in operations that amphibious operations are some of the most difficult to execute, and that the success or failure of military operations often rests on logistics more than any other function. Accordingly, amphibious and expeditionary logistics are perhaps the most difficult sustainment operations that can be undertaken. In situations where

---

[1] Quoted in Moshe Kress, *Operational Logistics: The Art and Science of Sustaining Military Operations* (Boston, MA: Kluwer Academic Publishers, 2002), viii, https://doi.org/10.1007/978-3-319-22674-3.
[2] Quoted in Jobie Turner, *Feeding Victory: Innovative Military Logistics from Lake George to Khe Sanh* (Lawrence: University Press of Kansas, 2020), 99.

supply lines are contested by an adversary they are even more so. During the course of the twentieth century, the United States military earned a reputation for excellence in amphibious logistics, mostly grounded in the logistical juggernaut that the U.S. military built during the course of the Second World War that sustained simultaneous, large-scale, expeditionary operations in multiple theaters.

Today, the U.S. military is shifting to meet the threat of a near-peer or peer conflict with China or Russia, with a focus on the former. U.S. military leaders expect to face challenges from contested logistics unlike anything the U.S. military has dealt with since the Second World War. In an event with the Center for Strategic and International Studies, a Washington-based think tank, Commandant of the Marine Corps general David H. Berger told the audience, "We have to assume . . . that our supply lines will be contested. We . . . haven't needed to do that in 70 years."[3] In addition, new operating concepts like the Marine Corps' expeditionary advanced base operations (EABO), the Navy's distributed maritime operations (DMO) concept, and the Army's multidomain operations (MDO) will further stress the existing logistics enterprise by distributing units closer to the enemy, which complicates the efficient distribution of supplies and materiel.[4]

This chapter discusses the challenges to U.S. operational logistics in the Pacific and outlines an array of potential solutions in three broad categories: new concepts, new fuels and energy, and new platforms. There are also other innovations in logistics, particularly data analytics and artificial intelligence applications, that will not be discussed. This chapter focuses specifically on the challenges and opportunities for the Marine Corps' new EABO and stand-in forces concepts, but also uses examples from other Services and around the world.[5] At the time of writing, the Marine Corps is in the midst of a major force transformation and redesign that includes how the Corps does logistics and sustainment.

AMPHIBIOUS LOGISTICS IN WORLD WAR II AND BEYOND

The logistical support that enabled U.S. operations in the Pacific theater during the Second World War is unparalleled in history. Logisticians had to package and transport all of the supplies needed to feed, clothe, arm, and supply the millions of U.S. troops spread across the Pacific, and Allied supply lines in the Pacific were at their geographic extreme. The U.S. naval base at Pearl Harbor, Hawaii, is more than 4,000

---

[3] "Maritime Security Dialogue: An Update on the Marine Corps with Commandant Gen. David H. Berger," Center for Strategic and International Studies, 2 September 2021.
[4] Chris Dougherty, *Buying Time: Logistics for A New American Way of War* (Washington, DC: Center for a New American Security, 2023), 10; *Tentative Manual for Expeditionary Advanced Base Operations*, 2d ed. (Washington, DC: Headquarters Marine Corps, 2023); and *The U.S. Army in Multi-Domain Operations 2028*, TRADOC Pamphlet 525-3-1 (Washington, DC: U.S. Army, 2018).
[5] *A Concept for Stand-in Forces*, Marine Corps Doctrinal Paper (Washington, DC: Headquarters Marine Corps, 2021).

kilometers from San Diego, California. From Hawaii, Okinawa is 7,700 kilometers and Manila in the Philippines is nearly 8,900 kilometers. Furthermore, the Imperial Japanese Navy was a very real threat to U.S. maritime supply lines in the Pacific, and many of the battles during the war, especially in the South Pacific, were fought on islands with little to no infrastructure that the Allied forces could rely on, unlike battles fought in Europe where they could use existing roads, railways, ports and other infrastructure.

Once military cargo arrived in the area of operations, it then had to be transferred from ship to shore and distributed to smaller units. This movement over the shore is particularly difficult because it is inherently intermodal and involves shifting supplies from ships to land-based transportation. It also usually involves inter-Service coordination, which has historically been a point of friction in amphibious operations.[6] Over-the-shore logistics are often at their slowest and most vulnerable in predictable locations like landing beaches and ports, making it easier for the enemy to attack them there. Historically, amphibious forces are forced to take an operational pause as they shift combat power over the shore and transition to operations ashore; however, Marine Corps concepts from the 1990s, such as *Operational Maneuver from the Sea* (OMFTS), advocate for planning operations that do not include an operational pause.[7]

During the course of World War II, the U.S. military built a logistics empire capable of sustaining concurrent operations with millions of soldiers, sailors, and Marines on islands large and small, spread across the 60 million square miles of the Pacific Ocean. Allied amphibious operations during the Second World War were enabled by a massive industrial base but also by innovative engineering that enabled the rapid buildup infrastructure like piers, cranes, roads, pipelines, and storage depots. Military historian Jeremy Black has argued that the amphibious campaigns in the Pacific was more a "war of engineers" than anything else, and American excellence in "creating effective infrastructure" was a critical advantage.[8] It was also enabled by new platforms like landing craft with bow ramps and amphibious vehicles like Amtracs and DUKWs that could quickly carry troops and materiel from ships, through the surf, and onto or even past the landing beaches.

Highlighting the growth of the U.S. advantage in logistics was the rapid buildup of U.S. combat power on the South Pacific Island of Guadalcanal in 1942–43, which contrasts with the slow starvation of the Japanese forces on the island.[9] From the Japanese perspective, the Battle of Guadalcanal was really a contest of logistics, and

---

[6] Geoffery Till, *Seapower: A Guide for the Twenty-First Century* (New York: Routledge, 2013), 193.
[7] *Operational Maneuver from the Sea*, Marine Corps Concept Paper 1 (Washington, DC: Headquarters Marine Corps, 1996); and Till, *Seapower*, 272.
[8] Jeremy Black, *Logistics: The Key to Victory* (Havertown, PA: Pen & Sword Books, 2021), 148.
[9] Capt Walker D. Mills, USMC, and Erik Limpaecher, "Sustainment Will Be Contested," U.S. Naval Institute *Proceedings* 146, no. 11 (November 2021).

it was the "toll taken on the convoys headed to Guadalcanal" rather than losses sustained fighting on the island, that were the decisive factor according to World War II historian Phillips P. O'Brien.[10] Naval theorist Milan N. Vego made a similar judgment that the battle was decided by the ability "supply and reinforce ground troops contending ashore for mastery."[11] The Japanese resupply convoys ferrying supplies and reinforcements to Guadalcanal (a.k.a. Tokyo Express) were, according to military logistics historian Jobie Turner, "a makeshift logistics failure that ensured the death of almost two-thirds of the Japanese soldiers on Guadalcanal."[12] The battle was essentially an island siege, and the majority of Japanese casualties came not from combat but starvation, disease, and exposure. The Japanese logistics failure on Guadalcanal came from a combination of hubris and poor planning at a point when Japanese forces were already stretched thin sustaining their forces across the Pacific. It serves as a grim reminder to contemporary forces that logistics in the Pacific define what is possible, wishful thinking notwithstanding.

The Guadalcanal campaign created a logistical gap for the Marine Corps, when supplies delivered by the Navy to Marines ashore were literally washed away by a rising tide because of ineffective coordination for their offloading and a lack of personnel to do the work.[13] Furthermore, U.S. Navy vessels supporting the landing left the area before they had finished unloading their cargo. However, during the next several months the Marines, eventually replaced by the Army, built and insurmountable logistics advantage drawing on the massive U.S. industrial base, but also learning from mistakes and miscalculations earlier in the campaign.

Five months after landing, U.S. forces were well supplied enough to enjoy special meals at Thanksgiving and Christmas, while Japanese forces on the other end of the island were starved and reduced to eating grass and weeds.[14] And by January 1943, Japanese forces on the island were losing an average of 200 soldiers a day to death by starvation.[15] For the Japanese, who assumed that their navy would be able to supply soldiers on remote island outposts or that they would be able to live off the land, starvation became the norm by the end of the war. In the Philippines, as much as 80 percent of the overall Japanese deaths may have been caused by starvation.[16] On other islands like New Guinea, the Japanese military went so far as to authorize cannibalism.[17] Historian Lizzie Collingham estimated that in total, 60 percent of all

---

[10] Phillips Payson O'Brien, *How the War Was Won: Air-Sea Power and Allied Victory in World War II* (Cambridge, UK: Cambridge University Press, 2015), 385.
[11] Milan N. Vego, *Naval Strategy and Operations in Narrow Seas* (New York: Frank Cass, 1999), 119.
[12] Turner, *Feeding Victory*, 146.
[13] Turner, *Feeding Victory*, 111-12.
[14] Turner, *Feeding Victory*, 124-25.
[15] Lizzie Collingham, *The Taste of War: World War Two and the Battle for Food* (London: Penguin Books, 2011), 292.
[16] Collingham, *The Taste of War*, 303.
[17] Collingham, *The Taste of War*, 297-98.

Japanese military deaths between 1941 and 1945, or more than 1 million troops died of starvation and related illness in what was one of the greatest logistical disasters in military history.[18]

But, the culmination of American amphibious logistics would have been the never-executed amphibious landings planned for Japan in November 1945: Operation Olympic in southern Kyushu and then Operation Coronet in Tokyo and the Kanto Plain. The planning for Operation Olympic projected that the operation might land more than a quarter of a million troops on the assault beaches in the first three days of the operation.[19] Backing the amphibious assault were preparations of mammoth scale to sustain their operations, including nearly 150,000 pints of blood for transfusions in specially designed vessels; a shocking number that highlights both the expected casualties and the logistical preparations that planners made to accommodate them.[20]

Since the end of the Second World War, U.S. amphibious and expeditionary logistics have benefited from new platforms and concepts but have nowhere near the capacity that the military enjoyed during World War II. The widespread adoption of the helicopter and the development of Marine Corps and Navy doctrine that incorporated it into amphibious operations added significant logistical capability, but it still does not match the scale of operations during the Second World War or what would be required to fight a major campaign in the Pacific in the twenty-first century.

In recent decades, Marines and other amphibious forces have relied heavily on helicopters to transport both personnel and supplies directly from amphibious ships to objectives ashore. In 2001, U.S. Marines flew from an amphibious ready group (ARG) in the Indian Ocean to seize the airfield that would become Camp Rhino, Afghanistan, hundreds of kilometers inland. Even though the assault force was transported directly from the ship to the objective, the transports had to be refueled en route by Lockheed Martin KC-130 tankers that were flying out of forward operating bases in Pakistan. After the Marines secured Camp Rhino, a detachment of Navy Seabees was required to repair and maintain the runway so that it could receive daily flights from Marine Corps KC-130s and Air Force Boeing C-17 Globemasters. Without established overland supply routes, everything had to be flow in, including thousands of gallons of water each day, an example that shows how much support is required to sustain even a relatively small expeditionary force by air, and the limits of an all-air sustainment approach.[21]

The Marine Corps also invested in prepositioned equipment stored afloat on ships in the Pacific and Indian Oceans that could be quickly offloaded in a crisis and

---

[18] Collingham, *The Taste of War*, 303.
[19] D. M. Giangreco, *Hell to Pay: Operation Downfall and the Invasion of Japan, 1945–47* (Annapolis, MD: Naval Institute Press, 2009), 175.
[20] Giangreco, *Hell to Pay*, 191.
[21] Col Nathan S. Lowrey, USMCR, *U.S. Marines in Afghanistan, 2001–2002: From the Sea*, U.S. Marines in the Global War on Terrorism (Washington, DC: History Division, Headquarters Marine Corps, 2011), 137.

met with personnel flown in from the United States as part of the maritime prepositioning program. And Marines developed innovative concepts like seabasing where major logistical functions are conducted at sea instead of ashore, and *Operational Maneuver from the Sea*, where Marines bypass landing beaches and insert directly on their objectives from helicopters.[22]

These concepts assumed that the U.S. Navy would have assured access to the maritime space adjacent to the area of operations ashore and vessels carrying Marines and their supplies could maneuver unmolested. Since the Second World War, the U.S. military fought major conflicts in Vietnam, Iraq, and Afghanistan with all of the sustainment for those forces arriving by sea, air, or locally procured. Though these conflicts demonstrated that the U.S. military was able to deploy and sustain hundreds of thousands of troops in a war anywhere in the world, the supply chain to those countries was not contested in any serious way and because of that, the United States could rely on commercial transportation and logistics services to supply the troops. In fact, in all three examples, U.S. forces were able to move supplies through intermediate bases that were secure in neighboring countries; and for the wars in Iraq and Vietnam, the U.S. military was able to build up and mass forces relatively unmolested before engaging in major combat operations.

CONTESTED LOGISTICS, A GROWING CONCERN

In recent years, a parade of U.S. military leadership from the Service level down has repeatedly highlighted the difficulty of logistics in a large Pacific conflict. The primary concern is that the U.S. military is overly reliant on large bases, big buildups of material, and secure cargo handling facilities that are all vulnerable to attacks by Chinese long-range missiles and aircraft.[23] Chinese ships and submarines could attack Navy supply ships as they cross the Pacific.[24] The Falklands War offers a modern example that highlights the vulnerability of naval logistics in the missile age, where Argentinian naval aviation crippled the British expeditionary force by sinking several ships, including the SS *Atlantic Conveyor* (1969), which went down with 10 helicopters aboard. This loss severely limited British forces' mobility ashore for the entire campaign and was the primary reason that British units marched across East Falkland from the landing site at San Carlos Bay to Stanley.[25]

In addition to the vulnerability of logistics facilities in theater, defense contrac-

---

[22] *Prepositioning Programs Handbook: Appendix F to Marine Corps Installations & Logistics Roadmap (MCILR)*, (Washington, DC: Headquarters Marine Corps, 2015); and *Operational Maneuver from the Sea*.
[23] Cdr Thomas Shugart, USN, *First Strike: China's Missile Threat to U.S. Bases in Asia* (Washington, DC: Center for New American Security, 2017).
[24] Peter Suciu, "The Really Boring Way China Would Try to Win a War Against America," *Buzz* (blog), *National Interest*, 9 June 2020.
[25] Kenneth L. Privatsky, *Logistics in the Falklands War: A Case Study in Expeditionary Warfare* (Yorkshire, UK: Pen & Sword Books, 2014), 169–71.

tors and factories in the United States might be targets by cyberattacks intended to disrupt the U.S. supply chain in depth.[26] Together, these capabilities would threaten U.S. supply lines in a way that they have not been threatened since the Second World War when the U.S. Merchant Marine had to cross the North Atlantic and brave attacks from German wolf packs and U.S. bases in England and Hawaii could be attacked by German and Japanese planes.

Former Commandant Berger has been one of the most vocal military leaders arguing that the U.S. military needs to modernize its logistical capabilities to operate the way that it wants to in the Pacific.[27] As the deputy commandant for Combat Development and Integration, he wrote in the Marine Corps' functional concept for future installations and logistics development that "in a distributed and contested environment, logistics is the pacing function for the Marine Corps."[28] In his initial *Commandant's Planning Guidance* (2019), he tasked Marines with reimagining their "prepositioning, and expeditionary logistics so they are more survivable, at less risk of catastrophic loss, and agile in their employment."[29]

As Commandant, Berger continued his focus on logistics as the critical challenge for the Corps' future plans. In his 2021 update to *Force Design 2030*, Berger wrote, "We need systemic change in logistics."[30] And argued that "the challenge of providing distribution and sustainment in the context of our emerging concepts makes logistics the pacing function for both modernization and operational planning. Logistics will be contested—in some respects, it is being contested now—by peer and near-peer competitors, along the entire length of the supply chain."[31]

Other Marine leaders have also emphasized the need to update the force's logistical capabilities. Then Assistant Commandant General Eric M. Smith has called contested logistics "a wicked problem" and a "dirty secret" that many leaders would rather avoid discussing.[32] Lieutenant General George W. Smith, commander of the Marine Corps I Marine Expeditionary Force (I MEF), said that he believes the Marine Corps is "not placing enough emphasis on logistics, and particularly logistics in a distributed and contested maritime environment" at an industry conference, and echoed Berger in that "logistics is undoubtedly the pacing function when we talk about operations

---

[26] *Securing Defense-Critical Supply Chains: An Action Plan Developed in Response to President Biden's Executive Order 14017* (Washington, DC: Department of Defense, 2022).
[27] Rich Abott, "Berger Says Marine Corps Must Modernize Logistics Faster," *Defense Daily*, 8 February 2022.
[28] Gen David H. Berger, *Sustaining the Force in the 21st Century: A Functional Concept for Future Installations and Logistics Development* (Washington, DC: Headquarters Marine Corps, 2022), 2.
[29] Gen David H. Berger, *Commandant's Planning Guidance: 38th Commandant of the Marine Corps* (Washington, DC: Headquarters Marine Corps, 2019), 20.
[30] *Force Design 2030: Annual Update* (Washington, DC: Headquarters Marine Corps, 2022), 11.
[31] *Force Design 2030*, 11.
[32] Gen Eric M. Smith, "Lethal and Effective: Marine Corps Force Design 2030 and U.S.–Japan Defense Cooperation," Stimson Center, 15 June 2022; and Parth Satam, "America's 'Dirty Secret': USMC General Admits 'Wicked' Logistics Problems in Western Pacific to Battle China," *EurAsian Times*, 19 June 2022.

in the Pacific. When you look at the vast expanse of the Pacific, and all the attendant challenges, logistics is going to be that pacing function."[33]

Leaders in the other Services have expressed concerns about logistics as well. In 2021, the vice chairman of the Joint Chiefs, Air Force general John E. Hyten told reporters that the Joint Staff had also been focused on contested logistics and what they had seen forced them to change their "entire logistics approach" in thinking about conflict with China or Russia.[34] General Charles Q. Brown, the U.S. Air Force chief of staff, has also made clear that his Service is focused on operational logistics, saying in an interview with *War on the Rocks* that "our aircraft are all static displays without combat support. If you don't have the fuel, you don't have the maintenance, you don't have the airmen then those aircraft will stay parked on the ramp. That combat support is underestimated."[35] U.S. Special Operations Command (SOCOM) is also interested in pursuing novel ways to keep their forces sustained. At an event in May 2022, a SOCOM representative told reporters that "the term 'contested logistics' is at the very top of a lot of our discussions right now" and asked how special operations forces would expect to sustain themselves without regular deliveries or the prestaged stocks that were available in Iraq and Afghanistan.[36] The multi-Service focus on contested logistics is a clear transition from decades of laser-sharp focus on lethality and efficiency when logistics were deprioritized.

The concern about contested logistics extends beyond the Pentagon. At an event hosted by the Center for Strategic and International Studies, Dov Zakheim, a former undersecretary of defense (comptroller), also pointed out the logistical holes in the Marine Corps' *Force Design 2030* plans.[37] Independent analysis from the Center for Budgetary and Strategic Assessments found that "absent dramatic improvements, U.S. sealift forces would face major challenges and may fail to meet Joint Force demands in a major war," a truly damning conclusion.[38] A report from the Center for a New American Security (CNAS) found that "the Department of Defense has systemically underinvested in logistics in terms of money, mental energy, physical assets, and personnel" and argued that in a conflict with Russia or China, both adversaries would focus on degrading and destroying U.S. logistics and sustainment capability, a finding

---

[33] Ricard R. Burgess, "Marine General: Exercises Don't Pressure-Test Logistics for Real-World Operations," *Seapower Magazine*, 17 February 2022.
[34] David Vergun, "DOD Focuses on Aspirational Challenges in Future Warfighting," *DOD News*, 26 July 2021.
[35] Ryan Evans, "A Conversation with Gen. CQ Brown, Chief of Staff of the Air Force," *War on the Rocks*, 25 April 2023.
[36] Stew Magnuson, "SOFIC News: Special Operators Must Learn to Exist without Tethers," *National Defense Magazine*, 16 May 2022.
[37] Mark Cancian et al., "On the Future of the Marine Corps: Assessing Force Design 2030," Center for Strategic and International Studies, 16 May 2022.
[38] Timothy A. Walton, Harrison Schramm, and Ryan Boone, *Sustaining the Fight: Resilient Maritime Logistics for a New Era* (Washington, DC: Center for Strategic and Budgetary Assessments, 2019), 76.

supported by the results of numerous wargames.[39] The Government Accountability Office released similar findings in a 2017 report on U.S. sealift.[40]

The significant Russian military logistics failures during the invasion of the Ukraine have further highlighted the difficulty of contested logistics. During the initial invasion in February 2022, Russian forces struggled to resupply without access to railways in Ukraine.[41] Even before the 2022 invasion, analysts predicted that Russian forces would be "hard pressed" to adequately sustain offensive operations more than 145 kilometers beyond the Russian border and remained heavily reliant on rail transport to sustain their forces.[42] In one now infamous example, a Russian convoy as long as 64 kilometers stalled for days inside Ukraine because of food and gas shortages.[43] Across the front, Russian soldiers who "hadn't brought enough food, water or other supplies for a prolonged campaign" turned to widespread looting to sustain themselves.[44]

Berger highlighted the comparison in testimony to Congress: "As we are witnessing in Ukraine, even a numerically superior force will struggle to sustain itself and protect supply routes against persistent attack and disruption. We cannot allow this occur."[45] Secretary of the Army Christine Wormuth made similar comments. In a speech to the Royal United Services Institute, she said that among the lessons the U.S. Army was drawing from the war in Ukraine, one was "logistics, logistics, logistics."[46] She continued, "Amateurs discuss strategy and experts talk logistics. You can be the best equipped military in the world, but if you can't sustain your forces, it doesn't matter."[47] Watching the first year of open warfare in Ukraine has only reinforced the prioritization of contested logistics in the Marine Corps and the military writ large.

Contested logistics have also become a frequent topic of discussion within the ranks across the Services. Commentary in military and Service-focused publications has also been highly critical of the military's preparedness for contested logistics chal-

---

[39] Dougherty, *Buying Time*, 1, 10.
[40] *Navy Readiness: Actions Needed to Maintain Viable Surge Sealift and Combat Logistics Fleets* (Washington, DC: Government Accountability Office, 2017).
[41] Jack Watling and Nick Reynolds, *Operation Z: The Death Throes of an Imperial Delusion* (London: Royal United Services Institute, 2022), 4.
[42] Alex Vershinin, "Feeding the Bear: A Closer Look at Russian Army Logistics and the Fait Accompli," *War on the Rocks*, 23 November 2021.
[43] Bill Chappell, "Russia's 40-mile Convoy Has Stalled on Its Way to Kyiv, a U.S. Official Says," NPR, 1 March 2022.
[44] Michael Schwirtz et al., "Putin's War," *New York Times*, 16 December 2022.
[45] "Statement of General David Berger Commandant of the Marine Corps as Delivered to Congressional Defense Committees on the Posture of the United States Marine Corps" (congressional testimony, CMC Gen David H. Berger, 9 May 2022), 13.
[46] "Secretary of the Army Christine Wormuth's Royal United Services Institute Ground Forces Symposium (RUSI) Remarks (June 28, 2022) (as Prepared)," Army.mil, 1 July 2022, hereafter Wormuth remarks.
[47] Wormuth remarks.

lenges in a Pacific conflict.[48] In the U.S. Naval Institute *Proceedings*, articles on contested logistics have won prizes and contests three years in a row, and an "Asked and Answered" forum in the April 2022 issue asked the question: "What innovation or asset should the naval services prioritize for future expeditionary warfare?" Different takes on sustainment and logistics were the clear favorite.[49] The March 2023 issue of the *Marine Corps Gazette* had no fewer than 15 articles focused on logistics and sustainment.[50] Commentary in *Defense News* has urged military leaders to seek "new ways of thinking" and make "hard choices . . . that the individual military branches would prefer to avoid" to address logistics challenges in a potential Pacific conflict.[51] It seems as though everyone from junior servicemembers to senior leaders is looking for new and innovative approaches to logistics that can help the Marine Corps and the Joint forces sustain combat operations in a contested environment.

FORCE DESIGN 2030, EABO, AND THE MARINE CORPS

No Service is more preoccupied with the challenges of contested logistics than the United States Marine Corps, perhaps because as a Service, the Marine Corps is the most focused on expeditionary operations and does not have the capability for intertheater logistics, so it is forced to rely on the other Services to supply it. The Marine Corps recently unveiled a new operating concept—expeditionary advanced base operations (EABO)—that envisions deploying Marine units distributed on islands in the Pacific that can contribute to a larger maritime or Joint campaign through reconnaissance, fires, and other means.[52] Importantly, these units, called stand-in forces, will be based within reach of adversary weapons like long-range missiles and land-based aircraft, putting not just them at risk but also any units or platforms attempting to resupply or sustain them logistically.[53] The long range and lethality of these adversary weapons means that the Marines and the Navy will likely not be able to bring large amphibious or logistics vessels close to shore to resupply Marine forces and they will have to stay out of reach of existing ship-to-shore connectors. At the same time, distributed operations will further stretch logistics as units cannot be centrally resupplied. While the Marine Corps is in the middle of *Force Design 2030* that will allow the force to operationalized EABO, these logistical challenges remain unsolved.[54] Howev-

---

[48] Mills and Limpaecher, "Sustainment Will Be Contested."
[49] Mills and Limpaecher, "Sustainment Will Be Contested"; Maj Dustin Nicholson, USMC, "Marines Need Regenerative Logistics," U.S. Naval Institute *Proceedings* 148, no. 11 (November 2022); LtCol Brian Donlon, USMC, "Logistics 20203: Foraging Is Not Going to Cut It," U.S. Naval Institute *Proceedings* 149, no. 11 (November 2023); and "Asked and Answered," U.S. Naval Institute *Proceedings* 148, no. 4 (April 2022).
[50] *Marine Corps Gazette* 107, no. 3 (March 2023).
[51] K. Bremer Maximillian and Kelly Grieco, "The Pentagon Needs Fresh Ideas for Evading Taiwan Logistics Pitfalls," *Defense News*, 4 December 2023.
[52] *Tentative Manual for Expeditionary Advanced Base Operations*.
[53] *A Concept for Stand-in Forces*.
[54] *Force Design 2030* (Washington, DC: Headquarters Marine Corps, 2020).

er, there are a range of platforms, technologies, and concepts that could contribute to helping EABO and the Marine Corps overcome the challenges of contested logistics in a Pacific scenario, and Marine Corps leaders have made clear that they believe overcoming the logistical challenges of EABO is a top priority.

As part of the Marine Corps' *Force Design 2030* effort, the Corps released two key documents in early 2023 that map how the Service is thinking about logistics. First, in February 2023, came *Installations and Logistics 2030*, which "chart[ed] the way ahead for [the] Marine Corps Installations and Logistics Enterprise" in the mold of earlier *Force Design 2030* reports on *Talent Management 2030* and *Training and Education 2030*.[55] Signed by Commandant Berger, the report was both a roadmap for where the Marine Corps wants to go with its installations and logistics enterprise and an compilation of actual tasks for specific suborganizations. Organizationally, the Marine Corps has a deputy commandant for installations and logistics as a single advocate for both areas.

The report identified five key objectives that the Marine Corps is pursuing to reorient its logistics enterprise for contested logistics in a Pacific conflict with a peer adversary. First is an effort to "improve logistics awareness," that will increase real-time information sharing on where things are and what is needed by units.[56] The second and third focus on "improving sustainment" and "diversify distribution" to ensure the platforms and services used by the Marine Corps are ready to supply stand-in forces.[57] And the last two objectives concern installations and talent management—both areas that the Corps recognizes are foundational to the logistics enterprise. With a new Commandant expected to replace General Berger in summer 2023, it remains to be seen how closely his successor will hew to the specific objectives and tasks in *Installations and Logistics 2030*.[58]

In March 2023, the Marine Corps released a revised version of *Logistics*, Marine Corps Doctrinal Publication 4 (MCDP 4). It was the first time the doctrinal publication was revised since 1997, and it was rewritten in the style of *Warfighting* (MCDP 1), which famously explains how Marines think about war and conflict. Similarly, *Logistics* explains how Marines think about logistics, and what logistics are; it is not an instructional manual that explains how to "do" logistics. As the publication puts it, the manual "describes the theory and philosophy of military logistics as practiced by the United States Marine Corps."[59] The manual includes both historical examples of logistics and fictional vignettes that has Marines fighting a war against an unnamed adversary in the Pacific and deploying future technology like unmanned resupply drones and bladders of fuel anchored to the seafloor. It emphasizes that Marines need to work on both sides of the logistics equation, by reducing demand

---

[55] *Installations and Logistics 2030* (Washington, DC: Headquarters Marine Corps, 2023), 1.
[56] *Installations and Logistics 2030*, 1.
[57] *Installations and Logistics 2030*, 1.
[58] Malory Shelbourne, "Senate Confirms Eric Smith as New Marine Corps Commandant," *USNI News*, 21 September 2023.
[59] *Logistics*, Marine Corps Doctrinal Publication 4 (Washington, DC: Headquarters Marine Corps, 2023).

and increasing self-sufficiency as well as by leveraging new technology to push more supplies to forward units.

The Marine Corps is in a period of rapid transformation that includes how the Service executes and conceptualizes logistics, but it is clear that the transformation is a work in progress. Most of the tasks and objectives that the Commandant has laid out for the Service have not yet been completed and, as the *Force Design 2030* name suggests, they are not expected to be completed for several more years. It is also clear that within the U.S. military, the Marine Corps is out in front of the other Services on rethinking how it will do logistics in a future conflict. Senior Marine Corps leaders have consistently been the most vocal about the future of contested logistics, and the Marine Corps is the only one of the Services to have released new, unclassified documents like *Installations and Logistics 2030* or revamp logistics doctrinal manuals like *Logistics*. This makes sense because the Marine Corps concept for stand-in forces will require a transformation of logistics capability to make it feasible, and the Marine Corps has a history of leaning into new concepts and technology like amphibious warfare and helicopter operations.[60]

NEW LOGISTICS CONCEPTS

New ways of thinking about logistics and new logistics concepts have been developed and are percolating through the defense establishment. The number of different ideas is proof of both how seriously leaders in the military and defense establishment view the problem of contested logistics but also evidence that there is no clear solution to the problem yet or consensus on what one might be.

A CNAS report on contested logistics by Chris Dougherty discusses "adaptive logistics," which is "a temporary, conditions-based concept for contested and degraded environments."[61] He explains that "an adaptive joint logistics enterprise would be capable of switching from efficient methods to resilient methods depending on threats, the character of U.S. operations, or the status of U.S. logistical networks."[62]

In professional journals like *Proceedings* and the *Marine Corps Gazette*, officers have put forth a range of award-winning ideas for logistics frameworks and concepts. "Regenerative logistics" is one idea where Marine units should have logistics akin to "a lizard that can discard its tail to save its life—and then go on to grow another life-saving tail."[63] Marines will leverage future and emerging technologies so that stand-in forces can "produce, consume, reproduce, and reconsume organically with limited outside support" in a "closed system" to the greatest extend possible.[64] Clandestine forward caching, or "sleeper cell logistics," is another way that the Corps could try

---

[60] B. J. Armstrong, "The Answer to the Amphibious Prayer: Helicopters, the Marine Corps, and Defense Innovation," *War on the Rocks*, 17 December 2014.
[61] Dougherty, *Buying Time*, 11.
[62] Dougherty, *Buying Time*, 11.
[63] Nicholson, "Marines Need Regenerative Logistics."
[64] Nicholson, "Marines Need Regenerative Logistics."

to overcome logistical challenges. Instead of prepositioning large equipment sets on prepositioning ships, logisticians could hide or cache critical components forward with or without the knowledge of the host country so that it would be immediately available in a conflict.[65]

Twenty-first century foraging refers to an idea that has been introduced by Marine leaders and pitched as a way to help solve some of the logistical challenges inherent in EABO, but the origins of foraging as a logistics concept are as old as war. Simply put, to sustain an army, the army draws on the available supplies of the local population, usually in recently captured territory, and the army has to keep moving so as not to exhaust the local supply base.[66] In the West, it was not until the end of the Thirty Years' War (1618–48) that armies shifted away from a reliance on foraging for their basic needs and toward other systems of supply.[67]

Twenty-first century foraging does not have a formal, doctrinal explanation, but it is a combination of reducing demand for consumable commodities, local contracting, and scavenging for locally available resources like food and water. In 2021, Assistant Commandant General Eric Smith explained the idea at an industry event:

> *The first thing about being able to handle a logistics enterprise support you in a distributed environment is need less. . . . Why would I move water to the South China Sea? That's insane, why would I move food? It's called expeditionary foraging.*[68]

Even though it may not be fully fleshed out, Marines have already begun to experiment with the concept in exercises.[69] It has also been incorporated into training. The Basic School in Quantico, Virginia, where the Marine Corps trains its entry-level officers, recently added lessons on foraging for food and butchering animals so that the students could "consider augmenting their resupply with local resources in order to sustain their force," according to an instructor from the course.[70]

While twenty-first century foraging is a promising concept that could reduce the demand for supply by Marine units, the Corps needs to be careful that the emphasis falls more on local contracting and less on hunting and preparing game at the unit level. The Japanese experience on Guadalcanal and at other islands in the Pacific where units were left to "wither on the vine," demonstrates the risk associated with

---

[65] Capt Michael Sweeney, "Sleeper Cell Logistics: Sustaining New Warfighting Concepts," *Marine Corps Gazette* 105, no. 1 (January 2021): 64–66.
[66] Martin van Creveld, *Supplying War: Logistics from Wallenstein to Patton*, 2d ed. (Cambridge, UK: Cambridge University Press, 2004), 12.
[67] van Creveld, *Supplying War*, 17.
[68] Philip Athey, "Is Expeditionary Foraging in the Corps' Future?," *Defense News*, 6 August 2021.
[69] Philip Athey, "31st MEU Put Corps' Littoral Tactics, '21st Century Foraging' to the Test," *Marine Corps Times*, 21 January 2020.
[70] Philip Athey, "Marine-style Barbecue?: Marines Add Foraging Class to The Basic School," *Marine Corps Times*, 3 December 2021.

planning that assumes units can adequately supply themselves locally.[71] Logistics challenges cannot be overcome solely at the tactical level or reduced to an oversimplified problem of moving "pelican cases and seabags."[72] It is also important to remember that since the early twentieth century the amount of subsistence required by military units in combat has been relatively small as a percentage of the total logistical requirement, most of the it is ammunition and fuel. Van Creveld notes that by the end of the Second World War, "subsistence accounted for only eight to 12 percent of all supplies," and since then the amount of fuel used per soldier has increased dramatically, with U.S. forces in Afghanistan using as much as 22 gallons of fuel a day per deployed soldier.[73]

Other innovative concepts might focus on the production or fabrication of supplies at or near the battlefield. Additive manufacturing, often called 3D printing, is an idea that the Marine Corps is already experimenting with; in 2020, it released a Marine Corps order on additive manufacturing that details "who can print what, where, part approval process, training and education, and it also covers legal implications."[74] Champions of the technology have called it a "game changer" and asserted that with additive manufacturing the Corps "can construct essential components right on the battlefield, making us nimbler and more responsive in any combat scenario."[75] So far, the Corps is focusing on using 3D printing to fabricate specific parts and tools that are otherwise unavailable rather than mass producing things like weapons or munitions.[76] In 2022, a group at the University of Maine demonstrated the ability to 3D print two boats capable of carrying a Marine Rifle Squad and their gear in only three days, but the equipment to do so is so far only available at the university.[77] A more tactical variant of victory gardens is another idea that has been pitched by a Marine officer as a way to produce food closer to the battlefield.[78] Moving forward, it will be critical for the U.S. military and militaries around the world to look at innovative solutions for their logistical challenges. This will require a degree of humility and outside-the-box thinking for a defense bureaucracy accustomed to being a world leader in logistics.

Insurgents and traditional adversaries may also offer examples of logistics net-

---

[71] Collingham, *The Taste of War*, 298.
[72] Donlon, "Logistics 20203: Foraging Is Not Going to Cut It."
[73] van Creveld, *Supplying War*, 233; and Noah Shachtman, "Afghanistan's Oil Binge: 22 Gallons of Fuel Per Soldier Per Day," *Wired*, 11 November 2009.
[74] Gidget Fuentes, "Marine Corps Wants a Digital Blueprint Locker for Access to 3D Printing Plans Anywhere," *USNI News*, 5 July 2021.
[75] Johannes Schmidt, "Forging the Future: How Advanced Manufacturing Is Revolutionizing Marine Corps Logistics," Marine Corps Systems Command, 4 October 2023.
[76] Fuentes, "Marine Corps Wants a Digital Blueprint Locker for Access to 3D Printing Plans Anywhere."
[77] "UMaine Advanced Structures and Composites Center Produces World's Largest 3D-printed Logistics Vessel for U.S. Department of Defense," UMaine News, 25 February 2022.
[78] Ben Cohen and Leo Blanken, "Reviving the Victory Garden: The Military Benefits of Sustainable Farming," *War on the Rocks*, 20 January 2022.

works in contested environments. For example, are logisticians studying the network that supported Taliban fighters in their routing of the Afghan National Army in 2021?[79] What can the U.S. military learn about logistics from cocaine trafficking networks?[80]

Both old and new concepts can help the Marine Corps overcome some of the logistical challenges associated with EABO, but concepts alone are likely not enough. They may also need to be supported by new technologies and logistics platforms to truly adapt the way the Marine Corps does logistics for EABO.

NEW ENERGY TECHNOLOGIES

Some of the most promising technologies for contested logistics are technologies that might reduce or replace entirely, the military's reliance on petroleum-based fuels. In 2019, Marine Corps deputy commandant for installations and logistics, Lieutenant General Charles G. Chiarotti, told the 24th Annual Expeditionary Warfare Conference in Annapolis that "fuel is the pacing commodity" for Marine Corps operations.[81] Fuel is the single most important commodity for modern operations and often up to 50 percent by volume of the supplies needed to sustain an operational unit. Historically, the military has incurred significant risk and cost transporting that fuel to the battlefield. An Army study found that in Afghanistan between 2003 and 2007, U.S. forces suffered one casualty for every 24 fuel supply convoys, and that between Iraq and Afghanistan as many as 18 percent of all casualties occurred during resupply operations.[82] Transporting bulk fuel across contested sea lines of communication may prove even more dangerous than over land, during the Second World War the U.S. Merchant Marine suffered a casualty rate of approximately four percent, the highest casualty rate of any branch of Service.[83]

Electric vehicles have been repeatedly pitched as one way to help cut the military's tether to fossil fuels.[84] Both the Army and the Navy have committed to acquiring electric vehicles for tactical and nontactical uses in the future. The Department of the Navy has committed to acquiring 100 percent electric vehicles by 2035, and the Army has committed to developing "hybrid-drive tactical vehicles" by 2035 and "fully

---

[79] Jonathan Schroden, "Lessons from the Collapse of Afghanistan's Security Forces," *CTC Sentinel* 14, no. 8 (October 2021).

[80] Capt Walker D. Mills, "Contested Logistics: Look to the Drug Trade," U.S. Naval Institute *Proceedings* (August 2021).

[81] Todd South, "Not Just Riflemen Anymore: Marines Must Self Sustain in the High End Fight," *Marine Corps Times*, 8 November 2019.

[82] David S. Eady et al., *Sustain the Mission Project: Casualty Factors for Fuel and Water Re-supply*, Final Technical Report (Johnstown, PA: Concurrent Technologies, 2009), 2–6.

[83] "Supplying Victory: The History of the Merchant Marine in World War II," National WWII Museum, 7 February 2022.

[84] Cdr Michael Knickerbocker, "Military EVs Are a Necessary Awakening–Not 'Wokeness'," *Hill*, 27 April 2022.

electric tactical vehicles" by 2050.[85] Oshkosh Defense has already developed a hybrid version of its Joint Light Tactical Vehicle and there are electric versions of smaller vehicles as well.[86] Hybrid vehicles, while not able to cut their reliance on petroleum fuels, offer clear savings in efficiency over legacy models and would be a relatively easy way for the military to reduce petroleum consumption reduce some strain on logistics.[87] The Air Force has also acquired an "electric passenger aircraft capable of taking off and landing vertically," marketed as an "air taxi" that it plans to use for testing and experimentation.[88]

However, there are serious questions about the feasibility of all-electric tactical vehicles with existing technology, though the Services are collaborating to develop better lithium-ion battery technology to support the development of future vehicles. It is not clear how expeditionary forces would charge high numbers of electric vehicles without relying on large generators running on petroleum fuel that would only add to the logistics burden. There are also valid concerns about the safety of lithium-ion batteries aboard ships, especially after the car transport ship *Felicity Ace* (2005) burned out of control in 2022 because of a fire in one of the electric vehicles it was carrying.[89] But the rapid pace of electric vehicle development in the private sector, including for aircraft, may lead to technological breakthroughs or impressive gains in performance that make eclectic vehicles more attractive for expeditionary operations.[90]

Advances in the production of hydrogen have made it possible to produce hydrogen from aluminum feedstock at the tactical edge of the battlefield.[91] This breakthrough, combined with the increasing interest in hydrogen in the commercial sector, has the potential to make hydrogen attractive for military applications.[92] Tactical platforms running off of hydrogen fuel cells would also have significant tactical benefits over legacy platforms running on internal combustion engines, much like electric and hybrid vehicles. Fuel cell-powered platforms would be much quieter, have a

---

[85] *U.S. Army Climate Strategy* (Washington, DC: Department of the Army, 2022), 10; and *Climate Action 2030* (Washington, DC: Department of the Navy, 2022), 13.
[86] Caleb Larson, "Could the U.S. Army and Marine Corps Get Hybrid Vehicles?," *Buzz* (blog), *National Interest*, 27 January 2022.
[87] Marcus Weisgerber, "Hybrid-Electric Troop Transports Are Moving Toward the Battlefield," *Defense One*, 14 October 2022.
[88] Niraj Chokshi, "Air Force Receives It's First Electric Air Taxi," *New York Times*, 25 September 2023.
[89] "Lithium-ion Batteries 'Keep the Fire Alive' on Burning Cargo Ship Carrying Luxury Cars," *ABC News*, 21 February 2022.
[90] Niraj Chokshi, "Electric Planes, Once a Fantasy, Start to Take to the Skies," *New York Times*, 3 November 2023.
[91] Jonathan Thurston Slocum, "Characterization and Science of an Aluminum Fuel Treatment Process" (diss., Massachusetts Institute of Technology, February 2018).
[92] Walker Mills and Erik Limpaecher, "The Promise of Hydrogen: An Alternative Fuel at the Intersection of Climate Policy and Lethality," Modern War Institute at West Point, 27 December 2021.

lower thermal signature, and longer range.[93] General Atomics proposed a hydrogen-powered version of its MQ-1C Gray Eagle unmanned aircraft, and the U.S. Army has expressed interest in the ZH2, a hydrogen-powered Chevrolet Colorado.[94] A major shift to hydrogen-powered vehicles in the Department of Defense would likely take decades, but units within the Marine Corps like the Marine Littoral Regiment could make the switch much faster and reap the tactical benefits and operational benefits of being freed from the tether to petroleum fuels.

Synthetic fuels are another technology that could help cut or shift reliance on fossil fuels. The U.S. Air Force is pursuing synthetic fuels like the Fischer-Tropsch process fuel as a way to cut its reliance on petroleum fuels.[95] Developed in the 1920s, the Fischer-Tropsch process fabricates synthetic fuel, usually using coal, natural gas or hydrogen. Today, aircraft make up the bulk of petroleum consumption in the military and are more difficult to transition to electric, hybrid, or hydrogen.[96] However, synthetic fuels can in most case be used as drop-in replacements for petroleum that provide more flexibility to logisticians because they can be manufactured on demand and closer to the point of use, and in some cases even produced out of "thin air."[97]

There is also a long history of effective synthetic fuel production and use at industrial scale. During the Second World War, Germany was heavily reliant on synthetic fuel. This was made possible by major investments in synthetic production by the German government and commercial industry in the 1930s, despite the widespread availability of cheaper, imported fuel.[98] Between 1939 and 1945, almost one-half of the fuel used in Germany and by its military was synthetic fuel produced from coal. British military officer and historian J. F. C. Fuller went so far as to argue that without synthetic fuel the Germans "could not have declared war, let alone waged it."[99]

The Air Force has used some synthetic fuel mixtures since 2008, and in 2012, it completed certifications for all of its aircraft to fly on a blend of 50-50 petroleum fuel and Fischer-Tropsch synthetic fuel.[100] In 2020, the Air Force partnered with a

---

[93] Capt Walker D. Mills, Maj Jacob Clayton, and Erik R. Limpaecher, "Powering EABO: Aluminum Fuel for the Future Fight," *Marine Corps Gazette* (August 2022), 82–85.
[94] David Vergun, "Army Showcases Stealthy, Hydrogen Fuel Cell Vehicle," U.S. Army, 30 January 2017.
[95] Corrie Poland, "The Air Force Partners with Twelve, Proves It's Possible to Make Jet Fuel Out of Thin Air," U.S. Air Force Reserve Command, 22 October 2021.
[96] Neta C. Crawford, *Pentagon Fuel Use, Climate Change and the Costs of War* (Providence, RI: Watson Institute, Brown University, 2019).
[97] Poland, "The Air Force Partners with Twelve, Proves It's Possible to Make Jet Fuel Out of Thin Air."
[98] Adam Tooze, *The Wages of Destruction: The Making and Breaking of the Nazi Economy* (New York: Penguin, 2006), 116–18.
[99] Robert Gorlaski and Russel W. Freeburg, *Oil & War: How the Deadly Struggle for Fuel in WWII Meant Victory or Defeat* (New York: William Morrow and Company, 1987; Quantico, VA: Marine Corps University Press, 2022 reprint), 26, 278, https://doi.org/10.56686/9780160953613.
[100] "USAF Completes Fleetwide Certification of Fischer-Tropsch Alternate Fuel," *Inside Defense*, 26 April 2012.

company that produces synthetic fuel from captured carbon dioxide from "thin air."[101] Since then, testing has confirmed that the synthetic fuel made from captured carbon dioxide "matches the properties and performance of Jet A-1 [kerosene-based fuel], and contains all necessary components of jet fuel, including aromatics."[102]

In the United Kingdom, the Royal Air Force has become a leader in synthetic fuels, flying the first aircraft run on 100 percent synthetic fuel in 2021.[103] The chief of the Royal Air Force Sir Mike Wigston believes it could be a logistics game changer along with other technology:

> *Renewable power generation, like solar or small hydrogen power units, removes the requirement for a massive fuel and logistics supply tail, and the vulnerability and headaches that attracts. And taking it one step further, just imagine if the synthetic fuel plant . . . could be deployable too, and we were able to make our own jet fuel at a deployed operating base or at sea.*[104]

Synthetic fuels are also pitched as a way to help the United Kingdom's Ministry of Defense meet its net-zero climate goals, and could be a way to help U.S. forces in Europe cut their reliance on petroleum fuels sourced from Russia.[105]

The U.S. Army and Air Force are both pursuing different micronuclear reactor projects to generate power for austere bases and reduce their consumption of petroleum fuel. The Army's Project Pele will demonstrate a "mobile microreactor" and the Air Force plans to operate a microreactor at Eielson Air Force Base in Fairbanks, Alaska, by 2027.[106] Though these systems may not be small enough to be deployed to expeditionary advanced bases, they are projected to supply between one and five megawatts of power each, more than enough to power a forward operating base or a base in an austere location.[107] These systems could also provide enough power to make charging fleets of electric tactical vehicles more realistic, but there are concerns about how they would handle missile or bomb strikes.

There are several technologies that already exist, such as hybrid, electric, and hydrogen fuel-cell propulsion, that are under development like microreactors that

---

[101] Poland, "The Air Force Partners with Twelve, Proves It's Possible to Make Jet Fuel out of Thin Air."
[102] Maj Nicole Pearl, Paul Wrzeninski, and Capt Kaleb Mitchell, "Project FIERCE Fuels the Future of Synthetic Jet Fuel Generation," Air Force Research Laboratory, 8 November 2022.
[103] Andrew Chuter, "British Air Force Chief Envisions Synthetic Fuel Produced on Deployments," *Defense News*, 24 November 2021.
[104] Chuter, "British Air Force Chief Envisions Synthetic Fuel Produced on Deployments."
[105] "Mapping U.S. Military Dependence on Russian Fossil Fuels," Climate Solutions Lab, Watson Institute, Brown University, 28 April 2022.
[106] Department of Defense, "DOD to Build Project Pele Mobile Microreactor and Perform Demonstration at Idaho National Laboratory," press release, 13 April 2023; and Secretary of the Air Force Public Affairs, "Request for Proposal Released for Eielson Air Force Base Micro-reactor Pilot Program," press release, 26 September 2022.
[107] "Project Pele Mobile Microreactor to Go Ahead," World Nuclear News, 14 April 2022.

could dramatically upend how the military gets its operational energy.[108] The rapid advancement of renewable and alternative energy technology in the commercial sector also makes it quite possible that the next breakthrough that will change military energy usage and generation is imminent.

A growing awareness of climate change has also created new reasons for the Department of Defense to reimagine how it manages operational energy. All of the Services published climate action plans in 2022 that promise shifts to electric and hybrid tactical vehicles to improve resilience to climate change, but meeting those promises will take significant investment and effort.[109] Petroleum fuel use is entrenched in not just the platforms the military uses but also the infrastructure that transports and stores fuel, and widespread change would take years if not decades and face significant headwinds.

It is also possible that public and political pressure will push the U.S. military to invest in renewable and alternative energy technologies to limit the military's contribution to greenhouse gas emissions faster than it already is, especially because the Department of Defense is the world's single largest institutional contributor of emissions.[110] This has already happened in the United Kingdom, and the Ministry of Defense has committed to being net-zero by 2040.[111]

New energy technology could fundamentally reshape operational logistics in a way not seen since the mechanization of military formations in the first half of the twentieth century in unpredictable ways. Increasing electrification of military platforms is already being promised and with that will come requirements for electrical energy storage solutions, like tactical battery banks, and a more diverse set of options for tactical power generation. These developments may reduce the requirement for petroleum fuels but it will also complicate tactical logistics by requiring other ways to source electricity for vehicle fleets.

NEW PLATFORMS

The U.S. military has a long history of creating new platforms to meet changing operational needs. The development of landing ship, tanks (LSTs) and other amphibious vehicles are examples of how new platforms were adapted or designed to meet the challenges of amphibious operations.[112] Today, there are several platforms that could potentially help the U.S. military and the Marine Corps meet the challenges of con-

---

[108] Paul J. Kern et al., "An Albatross Around the US Military's Neck: The Single Fuel Concept and the Future of Expeditionary Energy," Modern Warfare Institute at West Point, 29 June 2021.
[109] *Army Climate Strategy: Implementation Plan, Fiscal Years 2023–2027* (Washington, DC: Department of the Army, 2022); *Department of the Air Force Climate Action Plan* (Washington, DC: Department of the Air Force, 2022); and *Climate Action 2023* (Washington, DC: Department of the Navy, 2022).
[110] Crawford, *Pentagon Fuel Use, Climate Change and the Costs of War*.
[111] *Ministry of Defense Climate Change and Sustainability Strategic Approach* (London: UK Ministry of Defense, 2021).
[112] William L. McGee, *The Amphibians Are Coming!: Emergence of the 'Gator Navy and Its Revolutionary Landing Craft* (Napa, CA: BMC Publications, 2000).

tested logistics. Most of them are specifically focused on delivering cargo the last tactical mile or to the end user on the battlefield, the segment of the supply chain that is often the most difficult and dangerous.

The Marine Corps believes that new amphibious platforms will be key to operationalizing the EABO concept. It wants to acquire up to 35 of a new class of ship, the Light Amphibious Warship (LAW) also called the Landing Ship Medium (LSM), to help support the logistical requirements of its EABO concept. The LAW is intended to be much smaller than existing amphibious vessels but bigger than ship-to-shore connectors. It will be capable of carrying a platoon or company of Marines with vehicles and equipment and delivering them to a beach or pier.[113] These vessels are intended to support interisland movement and bring in supplies to Marine units. The Marine Corps has made the program a priority, though it is unclear if the Navy feels the same, and it is unknown when the Corps will receive their new vessels. The earliest the Corps could see them is 2025, though that may get pushed back.[114] However, the Marines may be able to use similar vessels from the Army watercraft fleet for experimentation in the meantime.[115]

The Army is also recapitalizing its watercraft fleet with the acquisition of 36 Maneuver Support Vessel-Light (MSVL) intended to replace Vietnam-era landing craft, mechanized (a.k.a. LCM-8 or Mike Boat) that carry heavy vehicles and equipment from larger ships to shore or that could be used to transport troops and equipment between islands.[116] These vessels are too small for what the Marine Corps needs, but they will still be useful in experimentation and concept refinement. In 2023, the Army established a cross-functional team focused on contested logistics that will initially prioritize further watercraft recapitalization, including replacing the Maneuver Support Vessel-Heavy (MSVH), which is used for intertheater lift of supplies and heavy equipment.[117]

Unmanned aerial vehicle (UAV) systems have been repeatedly pitched as a solution to delivering supplies to units in contested environments. David Beaumont, an Australian military logistics expert, argued that "automation offers military logisticians tremendous advantage and has to be part of their future," and there are reports that British-supplied Malloy T400 UAVs have been used for tactical resupply in the

---

[113] Ronald O'Rourke, *Navy Light Amphibious Warship (LAW) Program: Background and Issues for Congress* (Washington, DC: Congressional Research Service, 2022).
[114] Todd South, "Marines Will Have to Wait at Least until 2025 for Light Amphibious Warship," *Marine Corps Times*, 28 March 2022.
[115] Capt Walker D. Mills and Lt Joseph Hanacek, "The US Navy and Marine Corps Should Acquire Army Watercraft," *Defense News*, 22 June 2020.
[116] Joseph Trevithick, "The US Army Is Buying New Boats to Replace Vietnam-Era Landing Craft," *Drive*, 29 June 2019.
[117] Jen Judson, "US Army Official Reveals Watercraft, Networks as Logistics Focus Areas," *Defense News*, 11 April 2023.

Ukraine conflict.[118] There is growing interest in unmanned aircraft, either remotely piloted or fully autonomous, for use in a logistics role and these platforms are receiving significant investment from both the military and the private sector. The Marine Corps successfully flew a modified, unmanned Kaman K-MAX helicopter in Afghanistan in 2011 and was pleased with the results, but the program was not continued.[119] Marines have also been experimenting with smaller UAVs like the tactical resupply vehicle TRV-150C to deliver supplies at the tactical edge of the battlefield and plans to establish a new a military occupational specialty for operators called "Small Unmanned Logistics System–Air Specialist."[120] The TRV-150C has been used in exercises with foreign partners like Balikatan in the Philippines, and has a purported useful range of approximately 14 kilometers with a 150-pound payload.[121]

Various private companies have also been experimenting with custom built unmanned aircraft of different sizes to market to the military, but limitations on weight and range restrict their utility.[122] The opportunity for commercial drone-based delivery services in the United States will likely continue to drive innovation with unmanned systems, but over-hyped programs like Amazon Prime Air have so far delivered less than promised. According to the *New York Times*, Prime Air "as it currently exists is so underwhelming that Amazon can keep the drones in the air only by giving stuff away," and it is limited to delivering a handful of products like canned soup and breath mints.[123] Ideally, UAVs would be a cheap and potentially disposable option for delivering small amounts of cargo rapidly and in any type of terrain. An experimental unmanned glider that can be dropped from transport aircraft and flown to "within 30 meters of its intended target" that was tested by the Army is an example of this approach, where payloads are delivered by single-use, relatively cheap means.[124] Both sides in the ongoing conflict in Ukraine are also pushing the boundaries of what UAS are capable of and it should come as no surprise if tactically useful UAS resupply comes out of wartime innovation.

A more extreme version of an unmanned resupply is the Air Force's interest in rocket-delivered cargo that would fly through space and be deliverable worldwide in minutes; but it is unclear if the Service will move forward with the concept, and it

---

[118] David Beaumont, "Sustaining Machines: Logistics and Autonomous Systems," Defense.info, 21 April 2021; and "Ukraine to Get a New Batch of Malloy Drones, and in This Case the Size Matters," *Defense Express*, 22 July 2023.
[119] Alex Davies, "The Marines' Self-Flying Chopper Survives a Three-Year Tour," *Wired*, 30 July 2014.
[120] LCpl David Brandes, "Tactical Resupply Unmanned Aircraft System Demonstration," Marines.mil, 11 April 2023.
[121] Cpl Tyler Andrews, "3d LLB Tests Capabilities of TRV-50 TRUAS," DVIDS, 22 April 2023; and Kelsey D. Atherton, "The Marines Are Getting Supersized Drones for Battlefield Resupply," *Popular Science*, 27 April 2023.
[122] David Hambling, "U.S. Army Pushes Ahead with Battlefield Resupply Drones," *Forbes*, 16 March 2021.
[123] David Streitfeld, "Look, Up in the Sky! It's a Can of Soup!," *New York Times*, 4 November 2023.
[124] Jared Keller, "Green Berets Are Testing a Prototype Glider Drone for Speedy Resupply," *Task & Purpose*, 31 March 2023.

raises obvious questions about cost and limits on the amount of cargo that can be delivered.[125] Rocket-delivered logistics would in some ways contradict what most leaders are calling for because of the high price tag and low numbers available. Military innovation with unmanned systems will likely continue to focus on sensing and strike roles, with increasingly large and complex systems fielded at the tactical level. Innovation with unmanned systems for carrying cargo is more likely to be driven by the private sector where there will be major market advantages for the first companies to make "drone delivery" efficient and low-cost. Military advances in unmanned cargo capacity will likely follow the commercial sector and feature most prominently at the tactical edge, with larger, higher-capacity drones as the technology improves, though multiple Services are experimenting with different capabilities and missions.[126]

Unmanned submersibles or semisubmersibles have also been suggested as a covert and long-range option for resupplying Marines on islands.[127] These vessels would move slowly but have a 1,600-kilometer or more range and be difficult to detect except with advanced sensors for hunting submarines. This idea was inspired by cocaine traffickers in the Caribbean who have been using semisubmersibles and low-profile vessels since the early 1990s to stealthily move multiton shipments of cocaine through the Caribbean.[128]

But unmanned systems and the associated technology are not and will not be a logistics panacea. These systems are key to improvements in military logistics but cannot address of the challenges presented by contested logistics. Further, to be effective, unmanned systems need to be integrated into processes and systems that leverage their unique advantages and are employed at the organizational level.[129] Also, as the Marine Corps recognizes in the revised *Logistics*, MCDP-4, human beings are at the center of the logistics enterprise: "Logistics is about how people interface with machines."[130] Even while we look to the promise of unmanned systems, they are not an end themselves, but rather a new set of tools for the logistician to employ as part of an overarching concept or framework.

Amphibious aircraft and seaplanes have also been heralded as an answer to pieces of the contested logistics puzzle.[131] Seaplane advocates argue that in any conflict with China, one of the first targets for Chinese air and missile strikes would be the

---

[125] Kyle Mizokami, "The Air Force Wants to Drop 100 Tons of Cargo from Space," *Popular Mechanics*, 4 June 2021.
[126] Dan Parsons, "Navy Considering Drone Delivery for Essential Parts at Sea," *USNI News*, 5 August 2021.
[127] Walker D. Mills, Dylan Phillips-Levine, and Collin Fox, "Cocaine Logistics for the Marine Corps," *War on the Rocks*, 22 July 2020.
[128] Byron Ramirez and Robert J. Bunker, *Narco-Submarines: Specially Fabricated Vessels Used for Drug Smuggling Purposes* (Fort Leavenworth, KS: U.S. Army Foreign Military Studies Office, 2015).
[129] Robbin Laird, "Shaping the Eco-System for Logistics Innovation: The Impact of Automation and Autonomous Systems," Defense.info, 23 March 2021.
[130] *Logistics*, 4-7.
[131] Capt Walker D. Mills, USMC, and LCdr Dylan Phillips-Levine, USN, "Give Amphibians a Second Look," U.S. Naval Institute *Proceedings* 146, no. 12 (December 2020).

runways that U.S. aircraft rely on; but they acknowledge that seaplanes would be unaffected by the strikes and able to continue operating across the Pacific moving personnel and supplies where needed.[132] Designed for long-range travel, amphibious aircraft could "be a logistical enabler across the Pacific" and help defeat the "tyranny of distance."[133] Other advocates highlight the major contribution that Allied seaplanes made to the war effort during the Pacific campaign in Second World War, where they served in scouting and reconnaissance, search and rescue, and even bombing roles.[134] Seaplanes have also been put forward in a tanking role, where they would be able to provide fuel for land and carrier-based aircraft.[135]

Other countries in the Pacific region already use seaplanes. China has developed a large AVIC AG600 Kunlong seaplane, the largest flown since the famous *Spruce Goose* was flown in 1947.[136] The Japanese Self-Defense Force flies several ShinMaywa US-2 short-takeoff and landing planes for maritime search and rescue missions, and it has drawn significant interest from the U.S. Air Force, and there are also Russian seaplane models in service.[137] Within the U.S. military, Special Operations Command is also exploring the idea of an amphibious version of the venerable C-130 aircraft, called the MC-130J amphibious capability (MAC) that would likely be used to transport troops and supplies within the Pacific.[138]

A subset of amphibious aircraft are wing-in-ground (WIG) effect aircraft. These aircraft are designed to fly close to the surface of the water to take advantage of the WIG effect and have significant gains in efficiency and carrying capacity when they do. WIG aircraft would be an ideal candidate for a logistics aircraft because of their large carrying capacity.[139] The Defense Advanced Research Projects Agency (DARPA) is building a prototype WIG aircraft for theater logistics called the Liberty Lifter.[140] The Marine Corps has also expressed interest in a niche class of aircraft called "sea gliders" that operate on a combination of hydrofoil and WIG capability.[141]

But even the deployment of significant numbers of amphibious aircraft would

---

[132] David Alman, "Bring Back the Seaplane," *War on the Rocks*, 1 July 2020.
[133] Christopher D. Booth, "Overcome the Tyranny of Distance," U.S. Naval Institute *Proceedings* 146, no. 12 (December 2020).
[134] David Alman, "Seaplanes Go to War," *Naval History Magazine* 35, no. 4 (August 2021).
[135] David Alman, "Extend Air Wing Range with Seaplane Tankers," U.S. Naval Institute *Proceedings* 147, no. 5 (May 2021).
[136] Bryan Hood, "China Just Flew the Largest Seaplane Since the *Spruce Goose*," Robb Report, 28 July 2020.
[137] Thomas Newdick, "U.S. Air Forces Trains with Japan's US-2 Flying Boat as It Looks Forward to Its Own Amphibious Plane," *Drive*, 23 February 2022.
[138] Peter Ong, "USSOCOM Update on MC-130J Amphibious Capability or MAC," Naval News, 17 July 2022.
[139] Walker D. Mills, Joshua Taylor, and Dylan Phillips-Levine, "Modern Sea Monsters: Revisiting Wing-in-Ground Effect Aircraft for the Next Fight," U.S. Naval Institute *Proceedings* (September 2020).
[140] Peter Ong, "DARPA Responds on 'Liberty Lifter' Plane," Naval News, 16 June 2022.
[141] Hope Hodge Seck, "Marine Corps Looks at Ocean Glider for Rapid Resupply to Fight China," *Marine Corps Times*, 30 November 2023.

not solve the logistics problems, as aircraft would struggle to deliver the volume of supplies needed to sustain even relatively small forces like a Marine Corps' littoral regiment or an Army multidomain task force. During the initial invasion of Afghanistan, Marines from Task Force 58 were flown into Kandahar to establish a forward operating base. Almost immediately, they received nightly deliveries from both KC-130 and C-17 aircraft to sustain operations.[142] The initial assault force could be delivered by helicopter, but the force could not be sustained organically. In addition to aircraft and unmanned systems, the Marine Corps will need to be creative and look for lower technology platforms to augment logistics capabilities like using clandestine vessels that are outwardly civilian appearing, or they could turn to pack animals for land-based transportation to cut the requirements for fuel and spare parts.[143]

The rapid improvement of logistics technology, especially with regard to unmanned systems is an opportunity for the U.S. military and the Marine Corps but not an end state. The technology is only going to be as effective as the way it is employed and the servicemembers who are employing it. Unmanned technology also presents new challenges to logisticians who will need to figure out how these platforms are managed, maintained, refueled, and employed if they do not have crews on board. Further, any new platforms or systems need to be integrated into logistics concepts and tactics for their benefits to be realized.

CONCLUSION

It is clear from studying the problem that the Marine Corps and the U.S. military are in desperate need of new ways to sustain forces in a contested environment. However, there are already a wide range of different options for meeting logistical needs ranging from new concepts like twenty-first century foraging and regenerative logistics to narco-inspired semisubmersibles and cargo rockets. The challenge for the military is three-fold. First, the Services need to prioritize acquisition focused on logistics technologies and decide which technologies and platforms have true revolutionary potential and which are no better than snake oil. Second, the Services need to integrate these technologies and platforms at scale into new concepts that can maximize their benefits and effectively organize logistics efforts. And third, the Services need to coordinate with each other to ensure that their efforts are complimentary, and their concepts can be integrated in a conflict. Any true solution will be a marriage of new platforms and technology with updated or innovative operational concepts that can best leverage the capability of new and existing platforms. Then these collaborations will have to be wargamed and tested to refine and validate their effectiveness. Any effective solution will also be a combination of different technologies and platforms

---

[142] Arthur P. Brill Jr., "Afghanistan Diary: Corps Considerations: Lessons Learned in Phase One," *Seapower Magazine*, April 2002.
[143] Christopher D. Booth, "The Modern Shetland Bus: The Lure of Covert Maritime Vessels for Great–Power Competition," *War on the Rocks*, 29 December 2020; and Capt Walker D. Mills and Christopher D. Booth, "Marines Need a Few Good Mules," U.S. Naval Institute *Proceedings* 148, no. 4 (April 2022).

rather than any single perfect solution. Beyond that, the logistical enterprise needs to backed by the industrial might of the U.S. economy, because even the most well-designed platforms will experience losses in a contested environment. The ongoing conflict in Ukraine has proven that logistics and sustainment start at the factory, but a discussion of the defense industrial base is beyond the scope of this chapter.

Within the Marine Corps, change in the logistics enterprise will require significant reorganization of the support units and the requisite experimentation to validate and refine those changes.[144] The revised version of *Logistics*, MCDP-4, has laid the doctrinal foundation for future changes and *Installations and Logistics 2030* has set the initial guidance for a transformation of the Marine Corps' logistics enterprise. An additional challenge for the Marine Corps is that it is reimagining logistics at a time of overall force redesign, so the logistics enterprise is in competition with other functions for resources and focus.

Anything less than major changes in how the Marine Corps and the military are ready to sustain their forces will result in disaster or may even preclude involvement in a major Pacific conflict altogether. Adversaries like China and Russia have made clear that they would target U.S. sustainment capabilities like tanker aircraft, logistics ships, critical infrastructure, and propositioned supplies at the outset of any conflict. These targets are all vulnerable and at present not easy to replace. This would leave U.S. forces in a precarious position and without the support they expect and require as they fought in the most intense conflict since the Second World War. It is not just that U.S. forces would struggle to sustain themselves, in many cases they would never be able to deploy in the first place. Functional logistics are a precondition for military operations. An inability to sustain combat forces in a contested environment will limit the options for commanders and could tie the hands of political leadership. Fortunately, leaders inside and outside of the military recognize the challenges of contested logistics; and if prudent investments and innovative thinking follow, the military and the Marine Corps will adapt to meet the challenge. For the Marine Corps and the U.S. military, amphibious and expeditionary logistics in a contested environment marks a return to the past. And in the past, the military was able to rapidly adapt and build an unmatched logistics organization capable of projecting air, sea, and land power thousands of kilometers across the Pacific Ocean into the heart of Imperial Japan.

---

[144] Paul S. Panicacci, "How to Do Logistics in EABO: It's a MAGTF, Not a MAGLTF," *Marine Corps Gazette* 104, no. 12 (December 2020).

CHAPTER FIVE

Amphibious Juggernaut

How the Landing Ship, Tank, and Landing Vehicle, Tracked, Created the Most Powerful Amphibious Assault System of World War II

Douglas E. Nash Sr.

In the modern era, the pace of technological advances has always accelerated during wartime. The development of the telegraph, railroad, wireless, submarine, and aircraft leapt ahead when put to use on the battlefield, often vaulting over a process that would normally take decades during peacetime. Even more influential has been the multiplying or synergistic effect that takes place when new technologies supplement or complement other technologies, achieving an effect far greater than had they occurred in isolation. An excellent example of this synergy of technologies was the combination of radios with aircraft, enabling reconnaissance flights to gather and relay current information to ground headquarters that can materially affect the outcome of a battle.

Another example, one from World War II, involves the mutually complementary synergistic effect that occurred when the Allies' Landing Ship, Tank (LST) was joined with the Landing Vehicle, Tracked (LVT) in the Pacific theater of operations. The resulting combination of two completely different systems—each developed for a specific, limited military purpose—resulted in a completely new method of conducting amphibious assault against a defended beachhead, a synergy that dramatically reduced casualties and allowed the Marine Corps and Navy's amphibious force to rapidly build up combat power ashore. This chapter focuses on how these platforms were developed separately by the Navy and Marine Corps, and how, almost by happenstance, they were combined to create a new tactical system for conducting

FIGURE 1
Adm Edward C. Kalbfus, commander battleships, U.S. Navy. He was the first to spot the October 1937 *Life* magazine article about Roebling's Alligator when he and MajGen Little were sharing a drink at his quarters in Norfolk, VA.
Source: official U.S. Navy photo NH48682

FIGURE 2
MajGen Louis M. Little, commanding general, Fleet Marine Force, Atlantic. He realized the significance of Roebling's invention and spurred the Marine Corps to investigate its potential for use in landing operations.
Source: official U.S. Marine Corps photo

amphibious assault that enabled the realization of the amphibious warfare theories espoused by the Marine Corps in the 1930s.

The LVT was first developed in 1935 by Donald Roebling, an inventor and manufacturer, at his workshop in Clearwater, Florida. Originally intended as a rescue vehicle designed to operate in swampy terrain as well as on water, the fully tracked vehicle, known unofficially by Roebling as the "Alligator," attracted the Navy and Marine Corps' attention in October 1937 when a *Life* magazine article was seen by Admiral Edward C. Kalbfus and Major General Louis McCarty Little, commanding general of the newly created Fleet Marine Force.[1] Both men quickly realized the vehicle's potential as an adjunct to the amphibious fleet.[2] Major General Little brought the Alligator to the attention of Major General John H. Russell Jr., Commandant of the Marine Corps, who quickly forwarded the information to the Marine Corps

---

[1] "Roebling's 'Alligator' for Florida Rescues," *Life*, 4 October 1937, 94–95.
[2] Maj Alfred D. Bailey, USMC (Ret), *Alligators, Buffaloes and Bushmasters: The History of the Development of the LVT through World War II* (Washington, DC: History and Museums Division, Headquarters Marine Corps, 1986), 34.

FIGURE 3
Donald Roebling. The eccentric Florida businessman and inventor who developed the Alligator as a fully tracked swamp rescue vehicle that later evolved into the Landing Vehicle, Tracked (LVT).
Source: official U.S. Marine Corps photo

Equipment Board, which sent a representative to Florida to evaluate the vehicle four months later. After viewing the Alligator in action, the evaluator, Major John W. Kaluf, was impressed enough to endorse the project by stating that it "has possibilities for use in landing troops and supplies at points not accessible to other types of small boats."[3] Thus began a close working relationship between the Marine Corps and Donald Roebling that would last throughout World War II.

After several years trying to convince the Navy that it should spend its Bureau of Ships design and procurement funds on an "experimental" oddity, the Marine Corps finally succeeded in October 1940, when the first prototype Alligator built to military specifications was delivered. This initial LVT was successfully demonstrated to the Commandant of the Marine Corps and several other high-ranking Army and Navy officers in Quantico, Virginia, later that month. The Navy, however, insisted on modifications to the prototype, such as requiring that its hull be constructed from steel instead of aluminum to increase its durability. Less than a week later, the Navy awarded a contract to Roebling to build 100 in cooperation with the Food Machinery Corporation at its factory in Dunedin, Florida, which would be known thereafter as Landing Vehicle, Tracked Model 1 (LVT-1). A small test detachment was formed in May 1941 at Dunedin to train and familiarize Marines with the new vehicle. After nearly a year of additional testing and evaluation, the 1st Amphibious Tractor Battalion was formed by 16 February 1942 and assigned to the 1st Marine Division.[4]

The initial production run of LVT-1s were all-steel construction, weighing in at 17,500 pounds empty and 22,000 pounds when fully loaded with fuel, crew, and cargo. The first amtracks—slang for amphibious tractor—as they were quickly nicknamed, were 21 feet long, 9 feet, 10 inches wide and 7 feet, 8 inches high. Powered by a 150-horsepower V-8 Hercules engine, it was capable of 19 kilometers per hour on land and up to 11 kilometers per hour in the water. Steered manually by dual lateral controls, it could turn in the water in its own length, an important feature when

---

[3] Bailey, *Alligators, Buffaloes and Bushmasters*, 34.
[4] Bailey, *Alligators, Buffaloes and Bushmasters*, 40–42.

FIGURE 4
The earliest version of the Marine Corps' amphibious tractor or amtrac, the LVT-1. Shown here in use as a logistics vehicle during the landing operation at Guadalcanal, 7 August 1942, with the attack transport USS *President Hayes* (AP 39) at anchor in the distance.
Source: *official U.S. Navy photo NH97749*

conducting water operations in confined seaways. Its two-person crew consisted of a driver and assistant driver, both of whom sat in a small crew compartment in the front of the vehicle.[5] It did not have a rear ramp or access doors, requiring anyone entering the vehicle to climb up and over the side using scalloped handholds located in the flotation sponsons (hollow box-like structures built into both sides of the hull) on either side. The vehicle's gasoline tank could hold up to 50 gallons, giving it a limited land cruising range of 193 kilometers. One disadvantage though was that its engine and drivetrain had a life expectancy of only 200 hours, but the advantages that the Alligator provided the Fleet Amphibious Force far outweighed its deficiencies.[6]

Viewed by the Marine Corps as primarily a logistics support vehicle, the LVT-1 was capable of transporting up to 4,500 pounds of supplies and troops from ship to shore, though it quickly proved equally able to negotiate swampy or marshy terrain beyond the beachhead. Few considered it a combat vehicle, because of its low speed in the water, lack of armor, and general lack of mechanical robustness that would make it unreliable in battle. Still, in 1941, the Marine Corps was satisfied with its purchase

---

[5] Bailey, *Alligators, Buffaloes and Bushmasters*, 43.
[6] Bailey, *Alligators, Buffaloes and Bushmasters*, 62, 97.

and, in the wake of the country's entry into World War II, began to raise an additional battalion, intending to provide each of the two existing divisions—the 1st and 2d Marine Divisions—an amphibious tractor battalion for support of logistical operations.[7]

Some senior Marines, though, had other ideas, believing that the LVT-1 could potentially perform more types of missions than the mundane task of ferrying supplies ashore. Some futuristic thinkers had already been working behind the scenes to bring about the necessary doctrinal changes that would allow expanded usage of the vehicle. Already, change 1 to the *Landing Operations Doctrine*, Fleet Training Publication 167, the Navy and Marine Corps' manual for the conduct of amphibious warfare, issued on 2 May 1941, described how best to employ LVTs in a logistics role during an amphibious operation.[8]

One of these visionaries was Major General Holland M. Smith, who had taken over command of the newly activated headquarters, Marine Amphibious Force, Atlantic Fleet, on 13 June 1941. Based in Quantico, Smith, who was in charge of training the new 1st Marine Division and the Army's 9th Infantry Division, was a passionate advocate of amphibious warfare and the Marine Corps' position as the nation's leading specialists in amphibious operations. During the late 1930s, Smith pioneered many amphibious tactics, techniques, and procedures and had been able to translate the new *Landing Operations Doctrine* from doctrine into practice through a series of realistic amphibious exercises in the Caribbean.[9]

When he was appointed commander of the 1st Marine Amphibious Brigade in September 1939, which was expanded into a division two years later, he oversaw several large-scale landing exercises at Guantánamo Bay, Culebra, and Vieques Island. By the time he had been appointed to command Amphibious Forces, Atlantic Fleet, in the early summer of 1941, Smith had become the nation's foremost expert on the practice of amphibious warfare. He quickly set about preparing his new command for the war that he knew was to come. Though hampered by a shortage of nearly everything, especially landing craft and troop transports, Smith put his troops through a rigorous training regimen that would serve them in good stead when committed to battle at Guadalcanal a year later.

Never content to appear complacent when newer and more promising ways beckoned, Smith recommended in a letter on 21 March 1942 to the commander of the U.S. Army's ground forces, Lieutenant General Lesley J. McNair, that an amtrac battalion be assigned to each Army and Marine division for beach assault, stating that "the use of the amphibian tractor permits a wider selection of landing places and more freedom of maneuver for the attacker." He followed up two weeks later with a similar letter to Admiral Ernest King, commander in chief, U.S. Fleet, stating that

---

[7] Bailey, *Alligators, Buffaloes and Bushmasters*, 43.
[8] *Landing Operations Doctrine*, FTP-167, change 1 (Washington, DC: Office of Naval Operations, Division of Fleet Training, U.S. Navy, 1942).
[9] Holland M. Smith and Percy Finch, *Coral and Brass* (New York: Charles Scribner's Sons, 1949), 83–85.

FIGURE 5
LtGen Holland M. Smith (right) pictured in Saipan with Adm Raymond A. Spruance (left), ca. 1944. Smith was the staunchest advocate of Marine Corps amphibious warfare doctrine and equipment in the prewar era.
Source: *official U.S. Navy photo NH80-G-287225*

"these machines . . . will be of inestimable value for direct ship-to-shore movement of supplies and transportation of tactical units ashore through hydrography or topography which will not permit the use of conventional boats or motor transport."[10] But with initial production proceeding slowly (only 72 were produced in 1941), Smith's amphibious dreams would have to wait until the nation's industrial capacity geared up to full production.[11] Unfortunately, that would not happen until the attack on Pearl Harbor on 7 December 1941 awoke the United States from its long slumber.

---

[10] MajGen Holland M. Smith, Letters, 13 and 31 March 1942, in Holland M. Smith: A Register of His Papers in the Marine Corps Archives and Special Collections, Box 1, Series 1.1, Folder 3, Marine Corps Archives and Special Collections Branch Library of the Marine Corps Gray Research Center Quantico, VA, 7.
[11] U.S. Civilian Production Administration, *Official Munitions Production of the United States by Months, July 1, 1940–August 31, 1945* (Washington, DC: War Department Production Board, 1947), 102.

Smith did not idly stand by and complain however. Due to his insistence and similar urging from other like-minded Marines, change 2 to *Landing Operations Doctrine* was published on 1 August 1942. This doctrinal change, which appeared the same month that the invasion of Guadalcanal took place, provided more detailed guidance concerning the possible employment of LVTs. It stated:

Landing vehicles, track, will be useful and should be available for the following employment:
- a. *Crossing water too shoal for regular landing boats.*
- b. *Crossing coral reefs.*
- c. *Negotiating obstacles both under water and on land.*
- d. *Crossing swampy or marshy areas.*
- e. *Movement of personnel, equipment, and supplies from transports to locations inland without unloading at the beach.*
- f. *In lieu of tractors and trailers in the early phases of an operation before motor transport has been landed.*[12]

These significant additions to landing operations doctrine, particularly the subparagraph pertaining to the "movement of personnel, equipment, and supplies from transports to locations inland without unloading at the beach," opened the door for Marines, such as Major General Smith, to consider the employment of LVTs in an amphibious assault role. However, the LVT-1 in use at the time was poorly suited for this purpose, as it was considered too fragile and unreliable to entrust the lives of Marines let alone to serve as an assault platform. Though some Marines, such as Smith, sensed the vehicle's potential, little testing or experimentation was carried out; the few vehicles then available were used primarily for training and familiarization.[13]

For the invasion of Guadalcanal and Tulagi islands on 7 August 1942, 13 old troop transports, 6 cargo ships, and 4 small high-speed transports would carry 19,000 troops of the 1st Marine Division to their objectives.[14] Landing Craft, Vehicle, Personnel (LCVPs or Higgins Boats for inventor Andrew Higgins) were used to bring the assault troops ashore, a long and laborious process that usually took up to four hours to complete before the initial assault wave was formed up to begin the landing, thus spoiling the element of surprise. This process meant that troopships had to lower each LCVP into the water using shipboard booms, because older ships lacked the new Welin davits, which could carry and launch up to three landing craft each.[15]

Launching was then followed by the assault troops having to climb aboard the

---

[12] *Landing Operations Doctrine*, change 2 (Washington, DC: Office of Naval Operations, Division of Fleet Training, U.S. Navy, 1942), 61, sect. 401.
[13] *Landing Operations Doctrine*, change 2, sect. 401, para. 3(e), 61.
[14] Maj John L. Zimmerman, USMCR, *The Guadalcanal Campaign* (Washington, DC: Historical Division, Headquarters Marine Corps, 1949), 24.
[15] Mike Whaley, "The Higgins Boat," Stanford University Department of Engineering, accessed 28 September 2023.

36-foot craft bobbing alongside via the tried-and-true method of cargo nets laid along the side of the troopships. After sailing in a circular pattern until all boats were loaded, the LCVPs would then form up into assault echelons that would then run in to shore, a process that could take as long as an hour, even with their top speed of 12 knots. Once ashore, the troops would immediately disembark and begin their assault, while the LCVP's coxswain would back the craft off of the beach and return to the troopship for another load of troops or supplies. Though it signified a tremendous step forward for the Marine Corps' amphibious assault capability, the LVCP's use was limited to the water's edge.[16]

While a number of LVT-1s assigned to the 1st Amphibian Tractor Battalion participated in the landing at Guadalcanal, they were used primarily for resupply and other mundane missions, and not to carry the assault wave of troops ashore.[17] Heavy and ungainly when aboard ship, the LVTs still had to be hoisted over the side of cargo ships using shipboard material handling equipment. Though LCVPs and LVT-1s were new and particularly useful additions to the amphibious force, practically everything else about the landing operation at Guadalcanal from the standpoint of the shipping and equipment involved was carried out in virtually the same fashion as it had been 44 years earlier during the landings at Guantánamo Bay during the Spanish-American War in 1898. Fortunately, the landing beach at Guadalcanal was undefended, with the Japanese garrison fleeing into the jungle during the initial naval bombardment. Throughout the rest of the campaign, the 1st Marine Division's LVT-1 battalion provided yeoman service in a variety of logistics-related roles, from carrying supplies from the beachhead to forwards units, serving as foundations for a mobile pontoon bridge, and for the evacuation of wounded troops from the jungle. Roebling's amphibious tractor had indeed lived up to its optimistic expectations.[18]

The 1st Marine Division's after action report for the Guadalcanal campaign, completed on 19 January 1943, several months after the initial landings, was not as sanguine about the nonlogistic employment of the LVT in amphibious assault role, despite General Smith's belief. The 1st Marine Division's after-action report stated unequivocally that LVTs should be used strictly for their intended logistics purpose.[19] According to the report's author, "In the past, the uses of this distinctive vehicle have been misunderstood in many quarters. The vehicle is definitely a supply unit. . . . Indiscreet publicity and an inefficient investigation of its potentialities have handi-

---

[16] The tactics, techniques, and procedures for conducting an amphibious assault during that period of the war is described in detail in Gordon L. Rottman, *U.S. World War II Amphibious Tactics: Army and Marine Corps, Pacific Theater* (New York: Osprey, 2004), 49–53.
[17] Zimmerman, *The Guadalcanal Campaign*, 84.
[18] Bailey, *Alligators, Buffaloes, and Bushmasters*, 51, 53; and Victor J. Croizat, *Across the Reef: The Amphibious Tracked Vehicle at War* (Quantico, VA: Marine Corps Association, 1989), 46–47
[19] Guadalcanal Operation after action report, "Employment of the Amphibian Tractor Battalion in the Solomons," 19 January 1943, para. 1, U.S. Marine Corps First Division, "Final Report on Guadalcanal Operation," vol. 5 (Norfolk, VA: Library, Armed Forces Staff College), 102, hereafter Guadalcanal after action report.

FIGURE 6
The old way of unloading. Prior to the introduction of the LST as an amphibious warfare platform, LVTs were carried as deck cargo or in the lower holds of attack cargo ships and lowered over the side using the ship's booms. This method was slow and laborious, leading to the Marine Corps and Navy's quest for a better method.
Source: official U.S. Marine Corps photo, Archives Branch, History Division

capped its use."[20] However, the report concluded by stating that "it might be assumed that employed judiciously, the amphibian tractor has a definite and valuable place in the present scheme of war, particularly so in tropical areas."[21] The evaluation of the amtrac would be markedly different 10 months later after the completion of Operation Galvanic (November 1943), the invasion of the Gilbert Islands that culminated in the amphibious assault at Tarawa.

The route to Tarawa did not follow a straight line, however, especially with the use of the LST in an amphibious assault role. This ship, which originated as a British-inspired design in 1941, featured a flat bottom, floodable compartments, and large bow clamshell doors that would enable it to beach on the objective after ballasting

---

[20] Guadalcanal after action report.
[21] Guadalcanal after action report, para. 8.

down and disgorge its cargo directly onto the shore over a retractable ramp, thus eliminating the immediate need for piers and loading docks. Intended to carry out this task after a landing beach had been taken, the potential for other uses of the LST was readily apparent. After the Navy's Bureau of Ships modified the British design in early 1942 for American shipyards, the keel of the first LST was laid that summer in the United States, with the first production model being launched in September. Though the first dozen LSTs were given to Britain under the provisions of the Lend-Lease Act (1941), the Navy accepted its first ship, the USS *LST-383*, on 28 October 1942.[22]

Exhaustive testing quickly followed. During December 1942, a series of tests, codenamed "Goldrush," were carried out by the Amphibious Force, U.S. Atlantic Fleet, at the Norfolk, Virginia, Navy base in conjunction with the Army, which provided troops, equipment, and materials to be loaded on the newly commissioned *LST-387*. The purpose was to determine how much cargo an LST could carry, how best to approach a shore for beaching the ship, how to discharge cargo, how best to prepare the beach to receive the ship, and other related tasks. Judged a success, the results of the tests were widely disseminated to all of the Services, including the Marine Corps, which still had forces engaged in combat in Guadalcanal, but was planning follow-on amphibious operations in the northern Solomons.[23] At that time, the possibility of combining LSTs with LVTs was not yet appreciated, and the increasing demand for these ships ensured that they would be pressed into service immediately after being commissioned for their intended purpose of delivering tanks, other vehicles, and equipment to the various invasion beaches for the campaigns then being contemplated, such as New Georgia in the southwest Pacific (June 1943), Sicily (July 1943), and Salerno in southern Italy (September 1943).[24]

So, the question arises, what were the origins and who were the originators of the idea of using LSTs as an assault platform for launching LVTs? None of the contemporary or postwar accounts describe how this pairing of two such seemingly noncomplementary conveyances came about; it seems to have been accepted as a matter of course or as something so obvious that it bears no further comment or mention in the

---

[22] Brandon C. Montanye, "Analysis of the Landing Ship Tank (LST) and Its Influence on Amphibious Warfare During World War Two" (thesis, U.S. Army Command and General Staff College, 2013), 24, 27. Launched and acceptance are two different dates. Launched means when it departs the dry dock where it was built, with a period of fitting out that takes several weeks then follows. A ship is not accepted until it is fully ready to sail with crew. In this particular instance, the acceptance date is more important than the launch date, which was September.

[23] "Subject: Goldrush Project-Test Debarking of Type Equipment form Tank Landing Ship, 17 December 1942," Op-30-B6-ISK, Ser. 0327750, Navy Department, Office of the Chief of Naval Operations, Washington, DC.

[24] VAdm George C. Dyer, *The Amphibians Came to Conquer: The Story Admiral Richmond K. Turner*, FMFRP 12-109-I (Washington, DC: Headquarters Marine Corps, 1991), 481; and Gen Holland M. Smith, "The Development of Amphibious Tactics in the U.S. Navy, Part IV," *Marine Corps Gazette* 30, no. 9 (November 1946): 39.

FIGURE 7
BGen David R. Nimmer serves as the senior Marine Corps planner as a colonel with Joint planning staff for the Joint Chiefs of Staff in Washington, DC, ca. 1943. The Battle of Guadalcanal veteran was quick to see the potential of the LVT as an amphibious assault vehicle and insisted that its inclusion in the invasion of the Marshall Islands plan was clearly spelled out as a requirement in the wake of the Trident Conference, May 1943.
Source: official U.S. Marine Corps photo #113182

official records. However, after studying how new systems, techniques, or tactics were introduced into the U.S. armed forces during World War II, one comes away with the overall impression that much study, testing, and analyses were conducted before any such novel items or ways of fighting were introduced to the troops, air wings, or fleets operating in the various theaters of war.

For example, when the Army's amphibious 2.5-ton truck, the duplex-drive DUKW (nicknamed the "Duck") was introduced by the General Motors Corporation in late 1942, it was subjected to exhaustive testing and evaluation on land and sea in December 1942 by both the Navy and Army at their test facility in Norfolk.[25] Accepted by the equipment evaluation board, it quickly became the Army's preferred ship-to-shore logistics vehicle, seen as more versatile, reliable, and effective than the LVT. The DUKW, however, was a wheeled vehicle and, as such, not suited for traversing soft landing surfaces, such as a sandy beach. Its lack of armor ensured that it would never be used to carry out amphibious assaults.

Despite a lack of evidence concerning its origins, overwhelming circumstantial evidence points toward the one person who would become the catalyst for bringing the LST and LVT together, and that was Marine Corps colonel David R. Nimmer.

While serving as brigade and division G-3 until Smith was promoted to major general and transferred to command the Amphibious Force, U.S. Atlantic Fleet, in June 1941, Colonel Nimmer was in daily, even hourly contact with the mercurial Smith, who mercilessly drove his brigade and then division through a series of increasingly complex amphibious training exercises in the Caribbean. Nimmer would have been present when amtracs were first introduced in the 1st Marine Brigade in

---

[25] Commander, Amphibious Force, "U.S. Atlantic Fleet: Tests of 2 1/2 Ton Amphibian Cargo Truck (DUKW)," 11 December 1942, Archives Collections Branch, Naval History and Heritage Command, Washington, DC.

1940 and would have been there whenever Smith spoke passionately about his ideas on amphibious warfare to his staff as well as with Navy officers involved in the landing exercises; it would have been nearly impossible for an officer as intelligent and experienced as Nimmer to not have been impressed with Smith's ideas on how the Marine Corps should be prepared to fight the impending war.[26]

In the spring of 1942, much to his disappointment, Nimmer was transferred out of the 1st Marine Division, which was sent to fight in the South Pacific, and was instead given command of the Marine barracks at Guantánamo Bay, Cuba.[27] Here, he was tasked with the establishment and organization of the new 9th Defense Battalion beginning on 1 October 1942 and preparing it for overseas deployment. In November 1942, his battalion was finally transferred to Guadalcanal, where it joined his old 1st Marine Division, which was still involved in heavy fighting against Japanese defenders. During the time he commanded the battalion until his departure in April 1943, Nimmer gained an enormous amount of combat experience as well as an appreciation of the capabilities of the LVT-1, which was the workhorse of the Marines' logistics effort ashore. Nimmer would also have learned first-hand how vulnerable amphibious forces are once ashore and deprived of the necessary naval support required to keep and expand the beachhead. Though he and the 9th Defense Battalion did not participate in the initial landings the previous August, Nimmer would still have acquired a healthy appreciation of the conditions existing there and what was required to wage amphibious warfare in the Pacific.[28]

Much to his surprise, Colonel Nimmer was relieved of command of the 9th Defense Battalion on 17 April 1943, and he was transferred from Guadalcanal to Marine Corps Headquarters in Washington, DC, where he was assigned to the Joint Chiefs of Staff's Joint Planning Staff. Here, he would serve in the newly constructed Pentagon building for a year and a half as the senior officer of the planning group charged with the responsibility for crafting war plans for the Pacific theater of operations. Based on his own observation, he quickly ascertained that he had more experience with actual landing operations than anyone else in his group, including his Navy colleagues, and had the formal professional military education to back it up.[29]

One of the first tasks Nimmer faced with his fellow planners was the need to flesh out the details of the general plan for waging the war against Japan. The central element of this plan, intended to begin by the end of 1943, was a two-pronged offensive designed to bring the war to the enemy's home islands via the Southwest Pacific, which would be led by General Douglas MacArthur, and via the Central Pacific, led by Admiral Chester W. Nimitz. Both of these offensive prongs would require that a

---

[26] *MajGen David R. Nimmer Oral History*, vol. 3 (Washington, DC: History and Museums Division, Headquarters Marine Corps, 1970), 1–3, hereafter *Nimmer Oral History*.
[27] The brigade was upgraded to a division and on 1 February 1941; so by the time Nimmer was shipped to Guantánamo, it had been officially a division for a year.
[28] *Nimmer Oral History*, 14–15.
[29] *Nimmer Oral History*, 107.

number of amphibious operations be conducted in close partnership by the Army, Army Air Forces, Navy, and Marine Corps. The beaches to be seized by MacArthur's forces included those in the Solomons Islands, New Guinea, and the Philippines. These beaches were generally of the sandy variety, with no surrounding coral reefs to contend with. Though many of them faced the jungle a few yards off the landing site, conventional landing craft, including LSTs and LCVPs, could land with little difficulty.[30]

The landing beaches in the Central Pacific, which included the Gilbert, Marshall, Caroline, Mariana, and Bonin Islands, were altogether different. Many of these islands were volcanic in origin and others were little more than coral atolls, surrounded by reefs that would allow most conventional landing craft to pass through only at high tide. Though some of these islands had dredged shipping channels that allowed the approach of large vessels without grounding, many did not, a fact that posed particular problems to anyone contemplating an amphibious assault.[31]

For example, there were doubts that the LCVP could pass over a coral reef at low tide; although it had a draft (clearance) of three feet, this was thought to be insufficient should the reef be exposed at low tide, forming an unsurpassable obstacle that would require the embarked assault troops to be landed at the reef and then wading through several hundred yards of surf before they reached the shore. Should the enemy survive the preinvasion bombardment, such troops would be exposed to a withering fire all the way to the beach. There was a general appreciation by the Joint Staff Planners, particularly anyone with amphibious warfare experience, that something besides the LCVP was needed if these islands were to be assaulted successfully.[32] But what?

Fortunately, Colonel Nimmer remembered his experience using LVTs at Guadalcanal as well as the exhortations of his former commander, Major General Holland Smith, that LVTs could potentially be used as an amphibious assault vehicle that was capable of crossing a coral reef. Additionally, the *Landing Operations Doctrine* with change 2, which had been distributed the previous August, recommended their possible use in such cases. But there was one problem: no one actually knew whether this could be done, since it had never been tested under real conditions. In late April 1943, shortly after Nimmer's arrival and before the Joint Planning Staff began work on the concept plan for the upcoming Central Pacific drive, he proposed through Marine Corps command channels that tests be secretly conducted in the Pacific using LVTs. Within days, a message transmitted through the office of the Commandant Lieutenant General Thomas Holcomb reached the desk of the commander of the I Marine Amphibious Corps, Major General Clayton B. Vogel, then commanding all Ma-

---

[30] *Nimmer Oral History*, pt. 3, 143–44, 163–65.
[31] Based on author's study of numerous area maps, Goode's *World Atlas*, U.S. CIA Country Studies, and analysis of the area using a variety of open sources.
[32] Bailey, *Alligators, Buffaloes, and Bushmasters*, 82–83.

rine Corps units in the South Pacific.³³

On 24 April 1943, within days of Nimmer's request, Lieutenant Colonel Victor H. Krulak, commander of the Marine 2d Parachute Battalion who had previous experience operating LVTs while assigned to the 1st Marine Division in 1941, was ordered by Vogel to conduct the test using four LVT-1s. During the next several days, Krulak and his handpicked team of LVT crew put the vehicles through the exhaustive tests, driving them over coral reefs ringing the island of New Caledonia in a variety of conditions and configurations. Both the vehicles and operators were beat up, but they had proved that the LVT could be used to cross coral reefs at high and low tides while loaded with troops or equipment.³⁴

Classified as top secret, the test results were back in the hands of General Holcomb by 5 May 1943. Nimmer, as Holcomb's representative on the Joint Planning Staff, would have received the same message that day or shortly thereafter.³⁵ A month later, the results were also shared with the commanding generals of Camp Pendleton, California; Camp Lejeune, North Carolina; the 4th Marine Division; Marine Corps Schools in Quantico; and the Amphibious Tractor Detachment

FIGURE 8
Capt Clifford G. Richardson, while assigned to the administrative command, Amphibious Forces, U.S. Atlantic Fleet in Norfolk. Richardson supervised a series of experiments (Operation Goldrush) with the newly introduced LST between December 1942 and April 1943 that included launching DUKWs from LSTs while underway. Richardson, who had been one of the original authors of the *Tentative Landing Manual* while assigned as an amphibious warfare instructor at Marine Corps Schools in Quantico, was a visionary and a persistent advocate of the use of the LST and LVT as an amphibious assault combination to carry out amphibious assaults more effectively in the Pacific.
Source: *official U.S. Navy photo #NH 84435*

---

³³ I Marine Amphibious Corps was renamed III Amphibious Corps on 15 April 1944. *Nimmer Oral History*, vol. 2, 116; and "Report for Commander, First Amphibious Corps: Tests of Amphibian Tractor under Surf and Coral Conditions," 3 May 1943, Historical Amphibious File (HAF) 750, Archives Branch, Marine Corps History Division, Quantico, VA, 2.
³⁴ "Report for Commander, First Amphibious Corps: Tests of Amphibian Tractor under Surf and Coral Conditions," 3 May 1943, 5–10.
³⁵ LtCol Victor H. Krulak, I Amphibious Corps Report, "Test of Amphibian Tractor under Surf and Coral Conditions," 9 June 1943, Historical Amphibious File, Box 42, HAF 750, History Division, Marine Corps University, Quantico, VA.

in Dunedin, Florida. For some unknown reason (possibly on account of security), the results were not shared with the Marine forces assigned to the Pacific, who would be the ones carrying out the upcoming amphibious assaults. Armed with the results of Krulak's test, Nimmer was now ready to move on to the second part of his investigation: What was the best kind of ship to deliver LVTs to the objective? The time-tested method of lowering them from cargo ships into the sea using ship's gear was too slow and cumbersome. There had to be a better way.

While the Marine Corps was continuing to broaden its base of knowledge about LVTs, the Navy had not stood idly by either, especially regarding their new equipment, the LST. Encouraged by the result of their Goldrush project tests the previous December, Admiral Alan G. Kirk, the commander of the administrative command, Amphibious Forces, Atlantic Fleet, directed that the LST undergo additional tests as specified in an order dated 25 January 1943.[36] Secure in the knowledge that the LST could perform the minimum expected tasks (i.e., beaching onto a shore and discharging cargo), the next series of tests, which were to run until 17 March, would be analyzed by another special investigative board convened on 20 May "for the purpose of investigating the capabilities of landing craft including experimental loading of troops, vehicles and supplies with special emphasis on the landing on hostile shores of well-balanced combat teams."[37]

These experiments involved determining whether LSTs could carry a complete unit with all its equipment and how many LSTs would be required to transport and land a tank battalion, an antiaircraft battalion, and an armored infantry battalion, as well as other equipment. The tests would involve conducting amphibious operations under simulated combat conditions in the Solomon Islands, Maryland in the Chesapeake Bay, and at Little Creek, Virginia. Here, the LSTs would beach, discharge their tanks and other vehicles along with their crews, followed by a field exercise ashore before reembarking their vehicles, troops, and cargo on the same beach. Another task, added almost as an afterthought, was to determine whether a U.S. Army DUKW amphibious truck could be launched through the bow doors of an LST while underway at sea.

The investigative board, chaired by Navy captain Clifford G. Richardson, involved 10 LSTs earmarked for Operation Husky, the Allied amphibious operation designed to seize Sicily in July 1943. These brand-new ships were temporarily docked at the Norfolk naval base, where they were already being loaded with U.S. Army tanks, vehicles, and other cargo of the 45th Infantry Division earmarked to join the assault forces for Operation Husky. Richardson, who had helped write the Marine Corps' tentative landing manual in 1934 and who had taught Navy-Marine Corps coopera-

---

[36] Amphibious Force, U.S. Atlantic Fleet, Administrative Command, "Board to Investigate Loading of Landing Craft," 25 January 1943, Order FE25/A17-5, Serial 314, Naval Operating Base, Norfolk, VA.
[37] Amphibious Force, U.S. Atlantic Fleet, Administrative Command, "Capabilities of Landing Craft Type LST," 12 June 1943, Naval Operating Base, Norfolk, Reference Branch: Historical Amphibious File, Box 2, HAF 48, Marine Corps History Division, 1.

FIGURE 9
An LST launching a duplex-drive DUKW amphibious 2.5-ton truck at Guam.
Capt Clifford Richardson tested the ability of the LST to launch these vehicles while at sea,
giving rise to the proposal to use it to launch LVTs as well.
Source: official U.S. Marine Corps photo #87833

tion for two years at the Marine Corps' staff college in Quantico until 1939, drove the ships and their crews relentlessly from 12 February to 17 March 1943.[38]

Multiple landings and extractions by LSTs, the even newer Landing Craft, Tank (LCT), and other small craft were conducted at the Solomon Islands and Little Creek during those four weeks. Navy and Army crews gained an enormous amount of experience in operating these vessels, lessons that Richardson shared with the rest of the so-called "Gator Fleet," troops, and landing craft comprising the new amphibious force. The first recorded launch of a DUKW from an LST took place on 10 March, when *LST-400* launched nine of the Army's amphibious trucks near Little Creek in less than 20 minutes. When the tests were completed, Richardson provided the results of the evaluation board on 25 May 1943, to Admiral Kirk, who promptly forwarded the re-

---

[38] "Capabilities of Landing Craft Type LST," 12 June 1943, Historical Amphibious File, Box 2, HAF 48, Reference Branch, Marine Corps History Division, 1–2.

FIGURE 10
RAdm Charles M. Cooke, Adm King's principal planning officer. Cooke, considered a "vociferous advocate of expanded operations in the Pacific," was not a member of the Joint Chief's Planning Staff, but served unofficially as a sounding board for their ideas and concerns. According to Nimmer, Adm Cooke worked behind the scenes to help the Marine Corps acquire additional LVTs and other craft needed for Operation Galvanic, the amphibious assault at Tarawa in November 1943.
Source: *official U.S. Navy photo #NH 102845*

port to a number of senior headquarters staff, including Admiral King and his Navy staff, the Joint Chief's Joint Planning Staff, and the Commandant. A copy of this report would have landed on the desk of Colonel Nimmer as a matter of course.[39]

While much of the contents of the report might have only confirmed what Nimmer had already suspected, based on his previous experience and reading of the Operation Goldrush project report, one conclusion near the end that would have caught his eye, which stated in paragraph (i) that "the DUKW *can* be successfully loaded and unloaded from an LST while at sea" through its bow doors and that up to eighteen of the amphibious trucks could be carried aboard an LST at one time, with room left for additional equipment. To allow a DUKW to enter the water, the LST first had to open its bow doors and lower its ramp at 50 degrees below the horizontal; the DUKW would then slowly drive into the sea, having sufficient buoyancy to keep from sinking, and then engage its underwater propulsion system. For retrieval, the DUKW was at a disadvantage, in that it could not be turned around on the tank deck of the LST, which lacked a turntable platform and would have to be retrieved by its stern using ship's gear (i.e., its towing winch).[40]

The significance of this paragraph cannot be overemphasized. Richardson's tests proved beyond a doubt that an amphibious vehicle with roughly the same dimensions

---

[39] "Capabilities of Landing Craft Type LST," 12 June 1943.
[40] "Capabilities of Landing Craft Type LST," 12 June 1943, 5. Emphasis by author.

as an LVT *could* be launched from the bow of an LST while it was underway.⁴¹ To understand the implications of this discovery, a simple calculation revealed that the LVT-1 in use at that time could carry up to 20 fully equipped Marines and follow-on models of LVTs could carry even more than that. Thus, with the ability to carry up to 18 LVTs (although in practice 16 or 17 were usually carried), a single LST had the potential of launching up to 360 assault troops in one load, roughly one-half of an infantry battalion. Just as significantly, the LVT could continue moving inland, using its tracks like a tank to allow the assault troops to push even deeper into the enemy's defenses, thus expanding the beachhead even farther.

On 12 June 1943, the director, Division of Plans and Policies at Headquarters Marine Corps, Major General Keller E. Rockey, penned a memorandum on behalf of the Commandant to the commanding general, I Marine Amphibious Corps (I MAC), Major General Vogel (who was replaced in July 1943 by Lieutenant General Alexander A. Vandegrift) that simply stated that the results of Richardson's LST evaluation were furnished to him "for information." No other guidance was provided. Also on the distribution list for Rockey's memorandum were the commanders of the 1st, 2d, 3d, and 4th Marine Divisions, as well as the commander of Amphibious Forces, Pacific Fleet, Major General Holland Smith. But without an adequate understanding of the test's significance, and without any LSTs available in the Pacific for experimentation, as well as an overall shortage of LVTs (though the new more capable LVT-2s were coming off the assembly line), the report of Richardson's investigative board made hardly a ripple in the Fleet Marine Force being marshaled in the Pacific at that time. As for Captain Richardson, who had overseen the tests, after more than a year of service in the Atlantic Fleet's Amphibious Command, he finally was able to secure an assignment to the Pacific, where he commanded Transport Division 7, taking part in several amphibious operations, including the landings at Saipan, Tinian, and Leyte.⁴²

Strangely, no additional experimentation of the LST with the LVT by the Atlantic Fleet's Amphibious Force seems to have occurred after this test. Despite the distribution of the test's results to those commands that would have profited the most, no further tests would be conducted until just prior to their combat debut in November 1943, when the Pacific Fleet expressed interest. Perhaps this was due to the transfer of

---

⁴¹ "Joint Planning Staff (JPS) Document 205/1," 17 June 1943, Joint Staff Planners Operations against the Marshall Islands, Report by the Joint War Plans Committee, Enclosure A (Conclusion and Recommendations) and Appendix E, para. 2. Amphibious Tractors LVT(2), National Archives and Records Administration (NARA), College Park, MD, 11, 34–35. The dimensions of the DUKW were as follows (length x height x width, in feet): 31 x 8 x 7, versus that of an LVT-1, which was 21'6" x 8'1" x 9'10." The DUKW weighed 13,600 pounds empty, while the LVT-1 tipped the scales at 17,300 pounds. Thus, both vehicles were roughly similar in size and weight, inviting comparisons in their capabilities as amphibious landing craft. Norman Friedman, *U.S. Amphibious Ships and Craft: An Illustrated Design History* (Annapolis, MD: Naval Institute Press, 2002), 218.

⁴² Memorandum, Headquarters Marine Corps, Directorate of Plans and Policies, "Results of Evaluation of Suitability of Landing Ship, Tank (LST)," 12 June 1943, Historical Amphibious File, Box 2, HAF 48, "Capabilities of Landing Craft, Type LST, Report of," Reference Branch, Marine Corps History Division.

nearly all the leading personnel involved in the tests to fill combat leadership assignments during the next several months.

Additionally, the commencement of a series of amphibious operations in the Mediterranean (the landings at Sicily and Salerno) and in the Pacific (the landings at New Georgia and Bougainville) would have attracted everyone's immediate attention throughout the summer and early autumn of 1943. Nearly all available LSTs and LVTs also seem to have been put into service in support of these operations, leaving few available for noncombat use, such as testing or experimentation. Indeed, the record reveals that little, if any, additional testing and experimentation occurred with these two landing platforms after May 1943, with most of the attention thereafter being devoted to the introduction and initial testing of even newer platforms, such as the Landing Ship, Dock (LSD), the LVT(A)-1 "Amtank" (an LVT-2 fitted with a turret sporting a 37mm antitank gun), and the LVT-4.[43]

Back in the United States, if the potential of the LST-LVT combination could be proven, it would revolutionize how amphibious assaults were conducted. Colonel Nimmer considered the possibilities and thought about where this newly discovered capability might fit into the upcoming Central Pacific campaign, then in the initial planning stages. This, and other campaigns, would be subject of the Trident Conference that would be conducted in Washington, DC, during 12–25 May 1943. This event, attended by President Franklin D. Roosevelt, Prime Minister Winston S. Churchill, and the entire combined Allied chiefs of staff, would plot the war's future course, including strategy for the Pacific theater of operations. Nimmer and the rest of the Pacific theater planning team would attend and brief the senior leaders to lay out the fundamentals of what such a Pacific strategy would entail.[44]

One of the results of the Trident Conference was that the Joint Staff Planners were directed by the Joint War Planning Committee to estimate the forces required for an invasion of the Marshall Islands, the first major objective to be taken as part of the projected Central Pacific offensive, and to recommend possible dates. By 23 May 1943, the Joint Staff Planners had delivered a preliminary report suggesting that the invasion of the Marshall Islands should be carried out in three phases, starting with the Gilbert Islands, to eliminate the Japanese airfields there to protect the flank of the Marshall invasion force, which might be threatened by their opponent's still-considerable air and seapower.[45]

The Joint War Planning Committee further recommended that the Central Pacific operation needed to be initiated no later than the end of October 1943 to coincide with planned Burma operations to force the Japanese to disperse their troops. In its

---

[43] Joint Chiefs of Staff Memorandum JCS 311, "Mobility and Utilization of Amphibious Assault Craft, Report by Joint War Plans Committee," 15 May 1943, NARA, 1-2, 3-5.
[44] Combined Chiefs of Staff (CCS) Planning Memorandum CCS 239/1, "Operations in the Pacific and Far East in 1943-44," TRIDENT Conference, 23 May 1943, NARA.
[45] Combined Chiefs of Staff (CCS) Planning Memorandum CCS 239/1, "Operations in the Pacific and Far East in 1943-44," 130-31.

conclusion, the Trident Conference recommended that "operations again enemy positions in the Marshalls [should] consist of amphibious operations initially supported by carrier aircraft. The success of the operation will be greatly enhanced by the use of amphibious tractors which are capable of crossing coral reefs."[46] Undoubtedly, Nimmer had most likely drafted this statement or dictated it to one of his subordinates, as it very closely aligned with previous language he had used, his recent experience in the Southwest Pacific and his prior service as Major General Holland Smith's operations officer two years before.

Since the campaign against the Japanese-held Marshall Islands would be the first attempt in U.S. military history to assault fortified atolls, the Joint Staff Planners believed that "battle-tested shock troops with amphibious training," totaling one corps of two divisions would be needed for the campaign's first phase.[47] The Joint Planning Staff worked diligently for the next month on a detailed concept of operations, to include designating how many divisions, types of ships, and number of air wings would be needed to carry it out. Nimmer and the rest of his team on the Joint Planning Staff were encouraged in their endeavors by Rear Admiral Charles M. Cooke, Admiral King's principal planning officer at the Navy Department. Cooke, considered a "vociferous advocate of expanded operations in the Pacific," was not a member of the planning staff, but served unofficially as a sounding board for their ideas and concerns.[48] According to Nimmer, Admiral Cooke even helped the Marine

FIGURE 11
LtGen Victor H. Krulak.
While commander of the 2d Parachute Battalion, Krulak carried out a series of experiments during April 1943 in New Caledonia with the LVT-1 to determine its ability to cross barrier reefs with a full load of troops. The results were submitted to Commandant LtGen Thomas Holcomb, who ensured that they were quickly passed to Col Nimmer at the Joint Chiefs of Staff's Joint Planning Staff.
Source: official U.S. Marine Corps photo

---

[46] Combined Chiefs of Staff (CCS) Planning Memorandum CCS 239/1, "Operations in the Pacific and Far East in 1943–44," 139.
[47] "Operations against the Marshall Islands, 17 June 1943," Joint Planning Staff Report Nos. 205/1, NARA, 15.
[48] David Rigby, *Allied Master Strategists: The Combined Chiefs of Staff in World War II* (Annapolis, MD: Naval Institute Press, 2012), chap. 2.

Corps acquire additional LVTs and other craft needed for the upcoming offensive.[49]

Consequently, on 17 and 18 June 1943, Nimmer and the rest of his team issued their Joint Planning Staff Report numbers 205/1 and 205/2 titled "Operations against the Marshall Islands," that were then issued to Admiral Nimitz, commander in chief of the Pacific Fleet and Pacific Ocean areas, who would be responsible for the conduct of the campaign. One of the Joint Planning Staff's recommendations was that the new LSTs be used to transport assault troops and LVTs to the objective. The planning committee, echoing the Trident Conference report, stated in its own planning documents that the best assault craft for the invasion would be "amphibian" tractors that, when launched from tank landing ships outside the range of shore batteries, could "deploy and proceed shoreward without much danger of being stopped by the fringing reefs so abundant in that part of the world."[50]

Meanwhile, the combat debut of the LST in the Pacific took place on 30 June 1943, when several landed elements of the Army's 43d Infantry Division at Rendova in the northern Solomon Islands, one phase of the overarching New Georgia campaign. Since Rendova had narrow sandy beaches and no outlying coral reef, LSTs or LVTs were not involved in the initial stages of the landing, and the majority of the troops were landed via LCVPs launched by conventional assault transports. The few LVT-1s available were used primarily in their original logistical support role. Fortunately, the landings were unopposed and the tiny Japanese garrison was quickly overwhelmed, allowing the LSTs to land their cargo after successfully beaching. One of the challenges the planners for the assault on the Gilbert Islands (Operation Galvanic, the prelude to the Marshall Islands campaign), was that there were only 75 operational LVTs on hand in the 2d Marine Division, which had been chosen to seize Betio Island in the Tarawa atoll, the most important island in the Gilbert Islands. The Marine planners, backed up by Major General Julian C. Smith, commander of the 2d Marine Division, and the new commander of V Amphibious Corps, Major General Holland Smith, insisted that at least 125 LVTs would be needed to land the first three waves of assault troops, approximately 2,500 troops. The remaining waves would land using conventional landing craft.[51]

The great unknown about Betio and the coral reef encircling Tarawa were the tides and whether LCVPs would have enough freeboard (distance from the waterline to the upper deck) to cross when the assault waves attempted to land. The Marine planners feared that there would not be enough clearance (at least three feet were required) for the LCVPs, which would run aground and force the assault troops to dismount and wade the rest of the way to the shore, where they would then have to face the thoroughly alerted Japanese defenders. Therefore, the Marines' commanders

---

[49] *Nimmer Oral History*, 116.
[50] *Nimmer Oral History*, 11.
[51] Smith's Amphibious Corps, Pacific Fleet, was renamed on 25 August 1943 as V Amphibious Corps. Adm Richmond K. Turner was dual hatted at the time as the commander, V Amphibious Force, under which Smith's command was subordinated. Dyer, *The Amphibians Came to Conquer*, 547–48.

believed that having enough LVTs on hand was essential for the mission to succeed. There were 100 more of the new LVT-2s awaiting shipment on the West Coast, but it would take time to move them to the staging area in the South Pacific and issue 50 of them to the 2d Amphibious Tractor Battalion, while the Army's 27th Infantry Division would receive the rest for its Makin assault.[52]

Admiral Richmond K. Turner, who would command Task Force 54, the combined amphibious task force for Operation Galvanic, did not want LSTs to be included in his attack force for Tarawa, according to General Holland Smith's autobiography, because he believed their low 12-knot maximum speed would jeopardize his ship formations, which generally cruised at 18 knots or faster, and lose the element of surprise. He also stated that he would not wait for the arrival of more LVTs, since it would delay the operation even more. As Turner later related, "The capabilities of the LVT were not widely known at the time Galvanic was being planned" and it might be a mistake to depend too much on them for the attack's success.[53]

FIGURE 12
Adm Richmond K. Turner, commander of Task Force 54. Turner initially did not want LSTs to be included in his attack force for the Tarawa landing operation. The timely intervention of Gen Holland Smith averted the disaster that most certainly would have followed had the invasion gone forward with only Higgins Boats, most of which grounded on the island's barrier reef.
Source: official U.S. Navy photo #NH 80-G-309643

Doggedly, Holland Smith stood firm, telling Turner directly that "I've got to have those amtracks. We'll take a helluva licking without them. . . . No amtracks, no operation."[54] He got his way. Perhaps aided by the behind-the-scene machinations of Rear Admiral Cooke in Washington (as claimed by Nimmer in his 1970 interview), Smith arranged to have the 50 additional LVT-2s shipped from San Diego on 16 October 1943 via the USS *Carter Hall* (LSD 3), which dropped them off at the invasion staging area at Tutuila Island, American Samoa. After a 14-day journey from California, they were immediately delivered to the 2d Marine Division.[55]

---

[52] Croizat, *Across the Reef*, 86–87.
[53] Dyer, *The Amphibians Came to Conquer*, 656.
[54] Holland M. Smith and Percy Finch, *Coral and Brass* (New York, NY: Charles Scribner's Sons, 1949), 120.
[55] Deck Log, USS *Carter Hall* (LSD 3), 16 October 1943, World War II Diaries 1941–1945, Logbooks of U.S. Navy Ships and Stations, RG 24, NARA; and *Nimmer Oral History*, 115.

After the war, Turner disputed this account, stating that it was Admiral Nimitz who did not want LSTs carrying LVTs in the initial attack wave. Nimitz is also recorded as stating in August, three months before the assault on Tarawa, that he did not think that LVTs would be needed at all and that there would be sufficient freeboard for LCVPs to negotiate the reefs around Tarawa without having to prematurely disembark their assault troops.[56] Furthermore, Nimitz did not trust the results of the tests that Krulak had conducted the previous April, perhaps believing that the Marines had possibly rigged the test to demonstrate that the LVTs had the ability to negotiate a coral reef. In addition, Turner speculated after the war that the time spent having to wait on the arrival of the slow LSTs carrying the additional 50 LVTs might have been responsible for the torpedoing and loss of the escort carrier USS *Liscome Bay* (CVE 56), though he offers no evidence for this except for a statement to that effect in the U.S. Army's official history of the Gilbert Islands campaign, that on further examination reveals to be mere speculation on the Army's part.[57]

But more importantly, Nimitz demanded that another series of tests be conducted with LVTs to prove to his personal satisfaction that they could successfully cross a reef with a load of troops. According to the guidance laid down in the 30 August 1943 Joint Chiefs of Staff order for Operation Galvanic, the skeptical Nimitz's concerns were outlined almost as if he expected the LVT to fail. The order stated that "if [the LVT] trial shows that claims made for these craft are justified, it is planned to employ them in large numbers. Lacking these, it will be necessary to make the ship to shore movement in craft carried by attack transports (i.e., LCVPs), supplemented by additional landing craft."[58] That Nimitz could make such statements at that stage of the war was more of a reflection of the general lack of appreciation of what an amphibious assault against a fortified Pacific atoll would actually involve. No one at the time, including Nimitz, both Smiths, or Turner, knew what the true human and material costs would be or what capabilities the LTV would bring to the fight, if any.

Consequently, on 10 October 1943, Captain Fenlon A. Durand of the 2d Marine Division was ordered to take a detachment of LVTs from Company C, 2d Amphibious Tractor Battalion, then awaiting action in New Zealand, to Fiji aboard the attack transport USS *Harris* (APA 2) where he and his Marines would spend 13–17 October practicing crossing reefs using their LVTs. Proving once again that this could be done with minimum risk to the crew and cargo, the results of the test were provided to commander in chief, Pacific Fleet. Apparently satisfied, Nimitz immediately greenlighted their use for the invasion of the Gilbert Islands.[59] Interestingly, the results of Krulak's previous test had apparently not been shared with the 2d Marine Division, forcing Captain Durand to repeat the same tests and relearn the same lessons that

---

[56] Dyer, *The Amphibians Came to Conquer*, 655–56.
[57] Dyer, *The Amphibians Came to Conquer*, 679–80.
[58] Joint Chiefs of Staff Order 451-12, "The Seizure of the Marshall Islands: Report by the Joint Staff Planners," 30 August 1943, Enclosure B, NARA, 10.
[59] Croizat, *Across the Reef*, 87.

FIGURE 13
The final pairing: a U.S. Coastguard-crewed USS *LST-831* preparing to launch Marine Corps LVTs during the assault at Iwo Jima, February 1945. The LST normally carried up to 17 LVTs and 425 Marines and would slow to a speed of 5 knots prior to launching.
Source: official U.S. Marine Corps photo #4703

Krulak had six months earlier. The favorable tests results were also widely disseminated within the 2d Marine Division, which would be carrying out the attack.

Now that everyone involved in the invasion planning, including senior commanders, were convinced to their satisfaction that LVTs could cross a coral reef, such as the one encircling Tarawa with a load of troops, the last obstacle to combining with LSTs had to be overcome and that was the question as to whether LSTs could launch LVTs in the seaway. It already had been established the previous spring that LSTs could launch DUKWs, which were similar in weight and dimensions to an LVT-1, so the only task remaining was to carry it out using a real LVT. To be fair, the Navy's Bureau of Ships was also concerned that if launched improperly at sea, an LVT could cause irreparable damage to the LST's ramp, rendering the ship incapable of carrying out its primary function, as had been discovered in a previous test involving the DUKW.[60]

---

[60] Navy Department, Bureau of Ships message, "Record of Proceedings of Board to Investigate Loading of Landing Craft," 8 February 1943, Reference Library, Rare Book Room, Navy Historical and Heritage Command, Washington, DC, 1.

However, it was not unreasonable to assume that if a DUKW could successfully drive off the ramp of an LST with a trained crew, then an LVT could do it as well.

But could it? Before committing to using his scarce LSTs in this capacity, Admiral Richmond Turner wanted to know if launching and recovering LVTs from an LST was feasible and whether an LVT could be raised and lowered using the LST's elevator. Thus, the final stage of the evolution took place on 14 October 1943, when USS *LST-486* conducted a secret one-day test at Camp Pendleton, with the new, slightly heavier LVT-2 under the auspices of the Pacific Fleet's Amphibious Training Command. The loading and unloading tests were conducted in the Delmar boat harbor while anchored in nine fathoms of water.[61] No LVTs were lost in the test and *LST-486*, commanded by Lieutenant E. C. Shea, returned to its base at Port Hueneme, California, without incident.[62]

Though not quite as demonstrative as launching LVTs while underway, these limited tests proved that the LST could indeed launch the LVT while at sea. The results of the tests were transmitted to the Pacific Fleet almost immediately, followed by a spate of training exercises carried out by the Navy, Marine Corps, and Army in the weeks leading up to the invasion of the Gilbert Islands. The procedure quickly became a standard training subject within Training Command, Amphibious Force Pacific. In fact, on 2 November 1943, the Training Command, located in San Diego, ordered that ship-to-shore training for LSTs being prepared for service with V Amphibious Force in the Pacific include "training in loading and debarking LVTs over the ramp in open seaways."[63]

The first recorded instance of Marine Corps LVTs being launched from an LST while underway occurred on 5 November 1943 when USS *LST-243*, commanded by Lieutenant F. H. Blaske, was used by Marines from Company A, 2d Amtrac Battalion, to launch and recover LVTs in the harbor at Pago Pago, American Samoa. Captain Ray D. Horner, the company commander, oversaw his men's training on the loading and unloading of LVT-2s from the bow of the LST both in the harbor and while at sea until it was time to sail.[64] When they first arrived, they were met by a detach-

---

[61] Deck Log, USS *LST-486*, 14 October 1943, World War II Diaries 1941–1945, Logbooks of U.S. Navy Ships and Stations, RG 24, NARA; and Report of Commander, Task Force 13 (Commander Amphibious Training Command, Pacific Fleet), 14 October 1943, Research Library, Navy Historical and Heritage Command, 6.

[62] It is possible that others had already conducted this type of test using LVTs and LSTs, though as of this writing, no official or unofficial evidence has surfaced yet that could confirm this. Incidentally, MajGen Holland Smith by this time was commanding V Amphibious Corps, whose headquarters was colocated with that of Adm Turner at the Navy Yard in Pearl Harbor, HI; thus, Smith, a prominent LVT advocate, would have had ample opportunity to plead his case and convince Turner to order the tests be carried out at Camp Pendleton with *LST-486* before Operation Galvanic, but this remains speculation.

[63] "Department History of Training Command, Amphibious Forces, Pacific Fleet," in *Guide to United States Naval Administrative Histories of World War II*, comp. William C. Heimdahl and Edward J. Marolda (Washington, DC: Naval History Division, Department of the Navy, 1976), 25.

[64] Deck Log, USS *LST-243*, 5–7 November 1943, World War II Diaries 1941–1945, Logbooks of U.S. Navy Ships and Stations, RG 24, NARA.

ment from the 1st Amphibian Tractor Battalion, who had been combined with a few Marines from the 2d Battalion to form a new amtrac company for the 2d Battalion, giving it a total of three. It was here in American Samoa where these same 50 LVT-2s were fitted with improvised armor from boiler plate and machine guns to prepare them for the upcoming assault, since this model of the LVT-2 lacked any armor of its own.[65]

As the day of the invasion drew near, Marines, sailors, and soldiers drilled on the new procedures as much as possible until the time arrived when they would have to load aboard LSTs and cargo ships for the upcoming operation. Originally, the LSTs were supposed to arrive with sufficient time to transload the LVTs onto attack cargo ships, while the LSTs took on conventional loads, but their late arrival and last-minute training requirements precluded this. Finally, the slow LSTs (USS *LST-34*, *LST-242*, and *LST-243*) carrying the 2d Marine Division's new 50 LVT-2s sailed on 8 November for Ellice Island, the staging area for Operation Galvanic.[66] They finally joined the Tarawa assault force, Task Force 53, under Rear Admiral Harry W. Hill, at 0330 on 20 November 1943, only hours before the assault was to commence. This left no time to transfer the LVTs on board cargo ships to be launched per the usual procedure. So instead, the decision was made at the task force level to simply drive them straight out the bow of the LSTs as they had rehearsed in Samoa, a task that took 15 minutes, despite Japanese fire.[67] The 75 LVT-1s that traveled with the task force were launched conventionally from their host ships using the tried-and-true (though slow) boom and hoist method.

Having demonstrated their ability to negotiate the crossing of coral reefs, the 125 LVT-1s and LVT-2s of 2d Marine Division tipped the balance in the favor of Smith's assault troops at Tarawa. They proved to be the only craft that was able to get ashore after the LCVPs in the follow-on waves got hung up on the atoll's reef exposed during an exceptionally low neap tide. LVTs were also the only surface craft able to shuttle desperately needed supplies and ammunition to the beach and take wounded Marines back to the ships waiting offshore. The victory did not come cheap though, with nearly 66 percent of the LVTs used at Tarawa damaged or destroyed and their crews suffering proportionately.[68] The first Japanese-held island to be taken by an amphibious assault, Tarawa, though costly, was an unqualified success. According to Holland Smith, "This was our first frontal attack on a fortified enemy atoll and we

---

[65] Croizat, *Across the Reef*, 87; and "The Marines Amphibian," *Marine Corps Gazette* 37, no. 6 (June 1953): 45.
[66] Deck Log, USS *LST-34*, 8 November 1943, World War II Diaries 1941–1945, Logbooks of U.S. Navy Ships and Stations, RG 24, NARA.
[67] Report of Action, USS *LST-243*, Operation Galvanic, 29 November 1944, World War II Diaries 1941–1945, Logbooks of U.S. Navy Ships and Stations, RG 24, NARA, 5.
[68] Maj Henry G. Lawrence, "Report of Battalion Commander, 2d Amphibian Tractor Battalion, 2d Marine Division, 22 December 1943," in *Second Marine Division Report on Gilbert Islands Tarawa Operation*, FMFRP 12-90 (Washington, DC: Headquarters Marine Corps, 1991), 59–60.

were ignorant both of its capacity for resistance and of our own offensive limitations. The Marine doctrine of amphibious assault stood the test."[69]

At Tarawa, the LVT had also proven itself as an amphibious assault vehicle. Even though not intended to serve as such, the improvisation of armored plate and mounting of up to three machine guns by the Marines on the LVT-1s and LVT-2s tipped the balance in their favor. The Japanese had simply not anticipated their use and had taken no special defensive measures other than their usual ones (which were deadly enough) to prevent the LVTs from crossing the reef. Though losses had been heavy, without the LVT the Tarawa assault probably would have ended in failure. The majority of the troops embarked in LVTs survived to reach the shoreline; the follow-on waves of troops in LCVPs suffered far more heavily, especially when they were forced to disembark at the reef and wade nearly 400 yards to the beach, often in water up to their necks, as Japanese machine gun fire stitched the water around them.[70]

While few LVTs were able to cross the log barrier barring egress from the beach, in later amphibious assaults carried out in the Marshall and Mariana Islands during 1944, LVTs were able to penetrate far inland, often acting as personnel carriers. Additional modifications and newer LVT designs placed armored LVTs, LVTs with 37mm or 75mm gun turrets, LVTs with rear cargo ramps, or specialized LVTs such as command or recovery vehicles into the hands of troops. Reliability increased, making them more seaworthy, and as lessons learned were disseminated throughout the fleet, LVT operators became more skilled.[71] The LVT had truly met all expectations placed on it and in many cases exceeded them.

Nearly overlooked in the success of the operation was the fact that 50 LVT-2s had been launched directly into the sea from the 3 LSTs participating in the assault; an equal number were launched at the same time by the Army's 27th Infantry Division during its assault on neighboring Makin Island, where Japanese resistance was negligible. This was a significant tactical development and was duly remarked on in the official after action report, which stated that "this method of transportation proved highly satisfactory and simplified the execution of the initial ship to shore movement." Although the disadvantage of the LSTs low speed was noted, the report stated that it could be compensated for if the LST task group sailed earlier than the main body of the assault force, timed such that both forces arrived concurrently at the objective area.[72]

---

[69] Smith and Finch, *Coral and Brass*, 30.
[70] Capt James R. Stockman, USMC, *The Battle for Tarawa* (Washington, DC: Historical Section, Division of Public Information, Headquarters Marine Corps, 1947), 16.
[71] Col Joseph H. Alexander, USMC (Ret), *Across the Reef: The Marine Assault of Tarawa*, Marines in World War II Commemorative Series (Washington, DC: Historical Center, Headquarters Marine Corps, 1993), 12.
[72] Enclosure C, "Commander Fifth Amphibious Force Report of Galvanic Operations: General Notes on Atoll Attack," 4 December 1943, C5A/A16-3(3), Gray Research Center, Marine Corps University, 1–2.

While the other 75 LVT-1s on hand, all battered veterans of Guadalcanal, had been placed into the water alongside troopships and cargo ships using the slow and tedious hoist and boom method, running LVTs out of the lower hold or tank deck of an LST could be done quickly (on average, in about five minutes), as opposed to the four hours or more required to perform the same task from a troopship. Despite this advantage, the assault troops from the transports still had to be brought aboard their assigned LVTs while at sea, a risky task accomplished using LCVPs tying alongside the amphibious tractors. Except for this complication, the LST's only other disadvantage was its already remarked on low speed of 12 knots, which led their crews to nickname them "Large, Slow Targets." Despite the LST's disadvantages, after action comments were virtually unanimous in the opinion that launching from LSTs was the most preferable way to deploy LVTs during the conduct of amphibious assaults. This lesson was learned well; all subsequent assaults in the Central Pacific were to follow this procedure.[73]

With the LST having proven itself as a launch platform at Tarawa and Makin Island, the final evolution of the technique would be worked out during the next two campaigns. For example, during the invasion of the Marshall Islands that quickly followed on the heels of the Tarawa landing in 1–23 February 1944, all of the 340 LVTs used at Kwajalein and Eniwetok Atolls were launched from LSTs, with the assault troops transferring into them from LCVPs alongside.[74] This still proved a slow and dangerous process, especially when subjected to enemy fire. But when the invasion of the Mariana Islands took place five months later, not only were all 773 LVTs launched from 47 LSTs, but the assault troops were transferred on board the LSTs as early as six days prior to the invasion.[75] Once the command for "away all boats" was given, the assault troops, already crammed aboard their LSTs, would simply climb on board their LVTs lined up inside the LST's lower hold and await the command to launch.

With practice, the tempo of the assaults quickened. During the amphibious assault at Saipan in the Mariana Islands on 15 June 1944, for example, up to 17 fully loaded LTVs were launched from each LST within 10 minutes and less than 1,000 yards from the line of departure. This greatly sped up the pace of operations, since LTVs could simply drive off the ramp of the LST with its cargo of troops, supplies, or vehicles instead of being lowered by davits from assault transports.[76] Details were worked out between the Navy, Marine Corps, and Army regarding when and how as-

---

[73] Fifth Amphibious Corps, After Action Report, Enclosures G and K, 4 December 1944, 2d Amphibian Tractor Battalion, 22 December 1944; and USS *LST-243*, Report of Action, Operation Galvanic, 29 November 1943, HAF 48, Marine Corps Archives Branch.

[74] Jeter A. Isely and Philip A. Crowl, *The U.S. Marines and Amphibious War: Its Theory, and Its Practices in the Pacific* (Princeton, NJ: Princeton University Press, 1951), 274.

[75] Croizat, *Across the Reef*, 116.

[76] U.S. Pacific Fleet Amphibious Forces, Headquarters, Transport Doctrine, 18 September 1944, Reference Library, Rare Book Room, Navy Historical and Heritage Command, iv-5, para. 427.

FIGURE 14

An LVT launches from the bow of an LST during the invasion of Okinawa, ca. April 1945. By the Marianas campaign in June 1944, LSTs could launch 17 LVTs in 10 minutes. Within an hour, all 47 LSTs taking part in the operation had launched 773 LVTs carrying more than 20,000 Marines to the beach. Japanese defenders quickly learned to base their defenses farther inland, because any attempt to oppose this tactic would be rapidly overwhelmed.

Source: official U.S. Marine Corps photo #126-986

sault troops were loaded on board LSTs, since these ships were neither designed with sufficient berthing spaces nor life support for so many troops.

But the difference in the time between the old and new methods was striking. At Guadalcanal in August 1942, it took four to six hours for assault troops to climb into their LCVPs and reach the shore; by June 1944, this only took 10 minutes. Not only did the LST-LVT combination get troops to the beach faster and allow a mass of troops and materiel to build up and continue the push inland, but it also dramatically lessened the time that troops were exposed to enemy fire during the run-in to the beach. While a savings of four or six hours during an amphibious assault may not sound like much, in 1944 it spelled the difference between victory and defeat.

Just as important, this development facilitated the control of the ship-to-shore movement, always a daunting task even for veteran forces operating in ideal conditions. While the LSTs had to approach to within 6,000 yards of the beach before discharging their cargoes, the LVTs had to simply exit the LST in column formation and drive 1,000 yards to reach the line of departure. Once they had reached this imaginary line, the column of LVTs would be given the command by a nearby control ship to turn left or right. Here, they would then form into precise assault waves, orienting on the sea-lanes leading to the landing beach and drive the remaining 5,000 yards to the beach. Neighboring LSTs with their LVTs would do the same.

Instead of spending hours unloading and forming up for the assault, this new technique enabled the amphibious force commander to put thousands of troops ashore in a matter of minutes, even when assaulting a defended shoreline.[77] It also made achieving surprise far more likely, since the Japanese had little time to react once the assault began. Any attempt to stop the attack at the water's edge resulted in a quick defeat, with the defenders being quickly overwhelmed by the amphibious juggernaut, as they had been at Kwajalein and Eniwetok during January and February 1944. Though there were many errors made during the seizure of the Marshall Islands, the effectiveness of the LVT as an assault vehicle proved its worth. Additionally, once LSTs had discharged their loads, they could now recover and service LVTs, receive wounded, or shuttle additional troops from the transport area to the shore.

This technique, coupled with preinvasion naval and air bombardment, better communications, and more control ships brought even more success during the Mariana campaign. Having studied the lessons learned from the loss of the Marshall Islands, the Japanese commanders at Saipan, Tinian, and Guam had learned to avoid attempting a defense at the water's edge. Instead, they chose to defend farther inland, as they did at Peleliu, Iwo Jima, and Okinawa; and rather than face this amphibious juggernaut, they chose to fight a war of attrition, designed to make the Americans pay dearly for every inch of ground. In addition to the advantages LSTs provided in their ability to quickly launch waves of LVTs, LSDs also lent their weight. The medium tanks carried on board LSDs could be just as quickly landed from Landing Craft, Medium (LCMs) or LCTs once the first assault wave had secured a beachhead.[78] Paired with infantry carried by the LVTs, the work of reducing the enemy's inland defenses using their tank cannon or flamethrowers could begin in earnest once a beach foothold had been secured.

In summation, amphibious warfare had come a long way since 1934 with the drafting of the *Tentative Landing Operations Manual*. Advances in doctrine, ship construction, naval gunfire, close air support, communications, and landing craft had made the ship-to-shore movement the ultimate expression of the art of amphibious

---

[77] Edwin H. Simmons and J. Robert Moskin, eds., *The Marines* (Quantico, VA: Marine Corps Heritage Foundation, 1998), 196.
[78] LCMs were used during WWII and afterward. Affectionately known as "Mike" boats, they were one-third larger than the standard Higgins boat used to land infantry. The LSM was much larger and was a true seagoing vessel.

FIGURE 15
Amphibious juggernaut unleashed. Amphibious tractors, jammed with 4th Marine Division, churn toward Iwo Jima at H-hour. These troops served as the initial assault force, riding aboard LVT-4 Buffaloes.
Source: *official U.S. Marine Corps photo, History Division #110128*

warfare. It can be said with some degree of certainty that by 5 August 1945, amphibious warfare, as practiced by the United States armed forces in the Pacific theater of operations, had reached its highest state of development, far beyond anything that its earliest advocates could ever have envisioned. Central to the success of the Central Pacific drive from the invasion of the Gilbert Islands to its culmination at Okinawa a year and a half later, the LST-LVT combination was a one-two punch that paired a launch platform with a vehicle capable of negotiating a coral reef with a full load of assault troops.

That neither of these tools—the Landing Ship, Tank nor the Landing Vehicle, Tracked—was designed to perform these tasks, having been designed for completely different roles, speaks volumes for the ingenuity and improvisational genius of the Marine Corps–Navy team. With Marines such as Generals Smith and Nimmer, and sailors such as Admiral Richardson, as well as countless others, the Marine Corps and Navy's ability to improvise, adapt, and overcome the technological and doctrinal challenges they faced as well as its energy and drive to succeed, the generational aspiration of the Marine Corps and Navy to become the world's foremost amphibious assault force had become a reality. While both Nimmer and Smith retired from the

Marine Corps shortly after the war ended (Nimmer as a one-star general in 1947, Smith as a four-star general in 1946, and Richardson as a rear admiral in 1949), they left their mark on how the Marine Corps and the Navy would practice amphibious assault up to the present day.

POSTSCRIPT

After the war, the LST was replaced as an LVT launch platform by the Landing Ship, Dock (LSD), which was far superior in every specification, especially in speed. It could not only sail at 18 knots compared with the 12 knots of the LST, but the LSD could carry as many as 40 LTVs, as compared to the LST's 17. The relative scarcity of the LSD (fewer than 20 were commissioned before the war ended) and its utility as the primary launch platform for the Landing Craft, Tank (LCT) and Landing Craft, Medium (LCM), both of which carried tanks, meant that it would be dedicated to this purpose during World War II in the Pacific. Once the war ended, LTV operations shifted entirely to the role formerly carried out by LSTs, and the LST reverted to its original role as a logistics vessel. When the LTVs of the 1st Marine Division landed at Inchon in September 1950, they were all launched by LSDs.

CHAPTER SIX

The Union Defence Forces' Amphibious Invasion of German South West Africa, 1914

David Katz

INTRODUCTION

The Union of South Africa's amphibious invasion of German South West Africa (GSWA) in 1914 predates the Australian naval and military expeditionary force Battle of Bita Paka on the island of New Britain in September. However, the latter is known as the first amphibious operation of the First World War. Military historians have relegated the invasion of GSWA as African operations, far from the epicenter of the European conflict, to mere sidebars in the wider historiography of the First World War. Contemporary attempts to elevate their importance refer to the conflict outside of Europe as the "wider war." Historians must acknowledge that the African conflict was undoubtedly more than a minor curiosity for hundreds of thousands of its participants and victims.

Equally guilty of amnesia are contemporary South African historians—together with the various South African military academies and colleges—who have consigned South Africa's invasion of GSWA to the historiographical landfill. Readers will be hard-pressed to find details on the amphibious aspects of the operation in the secondary sources. Official historians deliberately protected reputations for political reasons and obfuscated the details of South Africa's amphibious operation.[1] Academic historians have fared little better, resorting to cross citations rather than engaging in the research process and consulting the primary evidence lying undisturbed in archives.[2]

This chapter breaks the trend by using primary documents from the National Archives of South Africa Pretoria (NASAP), the South African National Defence Force Archives (DODA) and the National Archives of the United Kingdom (TNA) and underutilized regimental histories to reconstruct the Union Defence Forces' (UDF) first amphibious operation. The narrative is pitched at the strategic and operational levels of war as the landings were unopposed. The strategic aspects of the campaign were rooted in a long, deep-seated desire for territorial expansion shared successively by the colonial government of the Cape Colony, the British Empire, and then by the newly formed dominion, the Union of South Africa in 1910.[3] This chapter aims to reveal the operational concepts underpinning the amphibious invasion, conceived by the British as early as 1902, and examine the final iteration of the operational plan developed by the UDF's defense minister, General Jan Smuts. Also under examination will be the performance of South Africa's fledgling UDF, formed a mere two years before the outbreak of war in 1912. The deeply politically divided UDF was an imperfect instrument of war in many ways, not least in possessing a contested doctrine represented by the former enemies who made up the UDF's numbers in 1914.

THE GENESIS OF THE PLAN TO INVADE GSWA 1902

As early as 1902, shortly after signing the Peace of Vereeniging treaty bringing the South African War (1899-1902) to an end, the British conceived a plan for the occupation of Swakopmund (Namibia) "in GSWA in the event of war with Germa-

---

[1] The official histories concerned include: BGen J. J. Collyer, *The Campaign in German South West Africa, 1914-1915* (London: Government Printing and Stationery Office, 1937; Nashville, TN: Battery Press, 1997 reprint); and *The Union of South Africa and the Great War, 1914-1918* (London: Government Printing and Stationery Office, 1924; Nashville, TN: Battery Press, 2004 reprint).

[2] Other contemporary historians who have tackled the subject of the invasion of German South West Africa but have preferred to concentrate on the operations after the amphibious landings include: I. van der Waag, *A Military History of Modern South Africa* (Johannesburg and Cape Town: Jonathan Ball, 2015); I. van der Waag, "The Battle of Sandfontein, 26 September 1914: South African Military Reform and the German South-West Africa Campaign, 1914-1915," *First World War Studies* 4, no. 2 (2013): https://doi.org/10.1080/19475020.2013.828633; R. Warwick, "Reconsideration of the Battle of Sandfontein:The First Phase of the German South West Africa Campaign, August to September 1914" (thesis, University of Cape Town, 2003); and Antonio Garcia, *The First Campaign Victory of the Great War: South Africa, Manoeuvre Warfare, the Afrikaner Rebellion and the German South West African Campaign, 1914-1915* (Warwick, UK: Helion, 2019).

[3] The Union of South Africa was created on 31 May 1910 and included four provinces: Cape Colony, Natal, the Orange Free State, and Transvaal. Bill Freund, "South Africa: The Union Years, 1910-1948-Political and Economic Foundations," in Robert Ross, Anne Kelk Mager, and Bill Nasson, eds., *The Cambridge History of South Africa* (Cambridge, UK: Cambridge University Press, 2011), 211-53, https://doi.org/10.1017/CHOL9780521869836.007.

ny."[4] Although it is beyond the scope of this chapter, it is interesting to speculate the underlying motivation for the United Kingdom's appetite for Germany's African territory given their recent costly and near-disastrous war with the Boer republics.[5] However, despite the remote possibility of a war with Germany in 1902, the British military in South Africa went ahead with a plan for invading GSWA in the event of war.

The GSWA terrain was particularly challenging for a landward or seaward invasion force (map 1). Any seaborne invasion along the GSWA coast would have to navigate through the harsh desert terrain of the Namib before reaching the more forgiving Central Region Plateau, where water and natural game were more plentiful (map 2).[6] The GSWA was bounded by the Kalahari Desert on the east, making any attempt to traverse it a tough challenge. The southern part of GSWA bordering the Union of South Africa was equally inhospitable, making a landward invasion a logistical nightmare. Map 1 shows the extensive railway system built by the Germans and still in existence in 1914. However, when the first plan was proposed in 1902, the only railway in existence ran from Swakopmund-Karibib-Windhoek and Lüderitzbucht was not fully developed as a port or linked by rail into the interior, making a British landing there pointless. The only viable option in 1902 for a seaborne operation was Walvis Bay/Swakopmund.[7]

The British identified Walvis Bay—a British colony on the GSWA coast—a mere 19 kilometers from the German coastal settlement of Swakopmund, as a preferable point of disembarkation. Besides the open nature of the anchorage at Swakopmund, compared to the well-sheltered anchorage at Walvis Bay, other considerations favored the prospects of the latter. The sheltered coastline of Walvis Bay extended for 22.5 kilometers, with suitable landing spots at any point. It would be impossible for the thinly spread Germans to successfully fortify or entrench across the entire length. The depth of the bay also allowed for the naval forces to anchor close enough to give the ground troops supporting fire. The bay was also eminently suitable to house the numerous logistic ships bringing in essential water supplies and other provisions needed to support the invasion force (map 2).[8]

The British could not count on the element of surprise since Walvis Bay and Swakopmund were the only two possible landing options on the GSWA coast. There-

---

[4] "Paper on the Occupation of Swakopmund German South West Africa," 17 October 1902, War Office (WO) 106-47, the National Archives of the United Kingdom (TNA).

[5] Boers, or Afrikaners, are settlers from as early as the 1600s of Dutch, German, or Huguenot descent who lived in Cape Colony, Transvaal, Natal, and Orange Free State.

[6] The Namib is waterless desert varying in width from 32 to 209 kilometers, covered with shifting sand dunes and without vegetation. See Evert Kleynhans, "A Critical Analysis of the Impact of Water on the South African Campaign in German South West Africa, 1914-1915," *Historia* 61, no. 2 (2016), http://dx.doi.org/10.17159/2309-8392/2016/v61n2a2. Kleynhans illuminates the fundamental role of water, its accessibility, and its protection in shaping the strategic and operational conduct of the campaign.

[7] "Paper on the Occupation of Swakopmund German South West Africa," 17 October 1902, 68-69.

[8] "Paper on the Occupation of Swakopmund German South West Africa," 17 October 1902, 62.

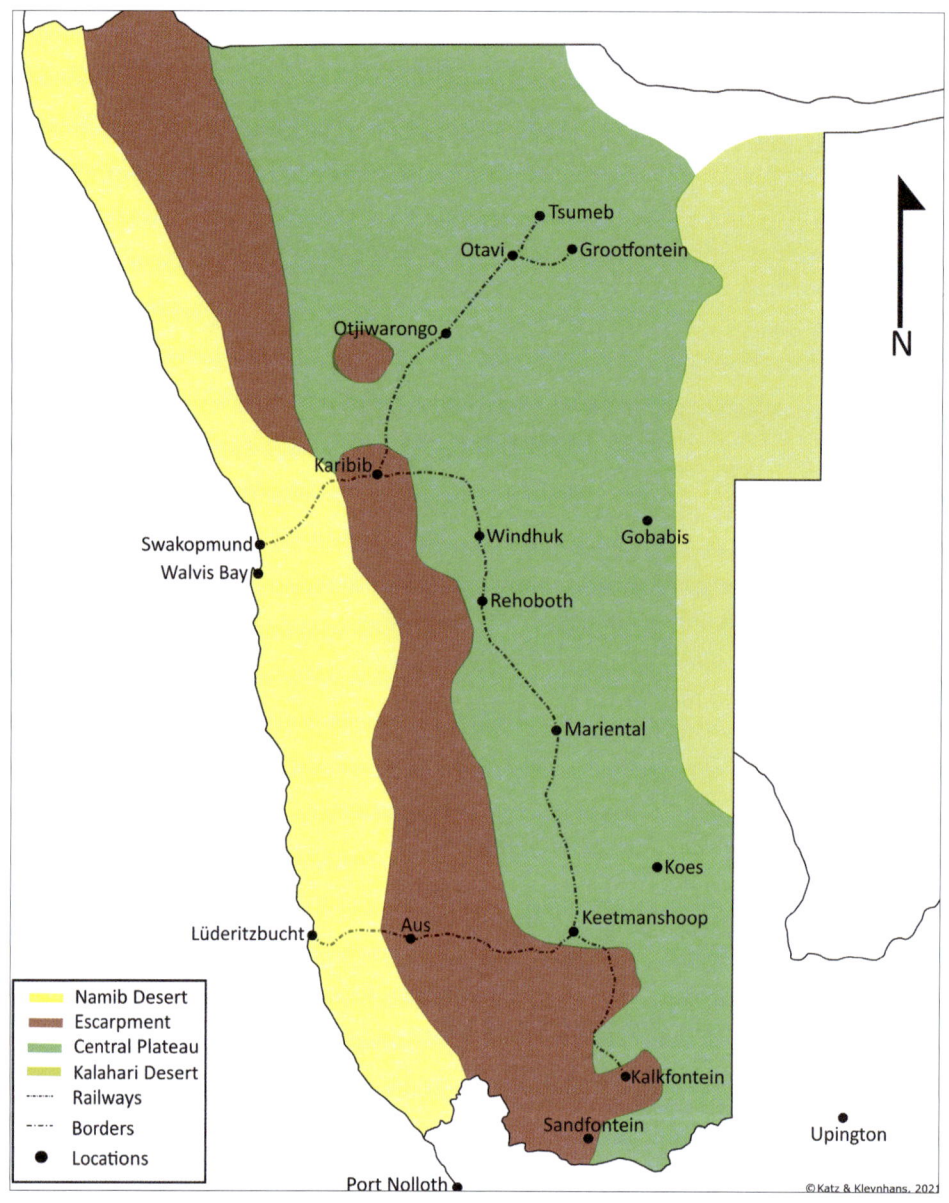

MAP 1
The harsh terrain of GSWA and the extensive railway line as it was on the eve of war in 1914.
Source: "Military Report on German South West Africa, 1906," WO 33-416, TNA

fore, the entire operation's feasibility depended on the effectiveness of naval artillery on the opposing German forces. The British expected that the Germans would oppose the amphibious invasion with between 1,000 and 2,000 mounted troops, two to three

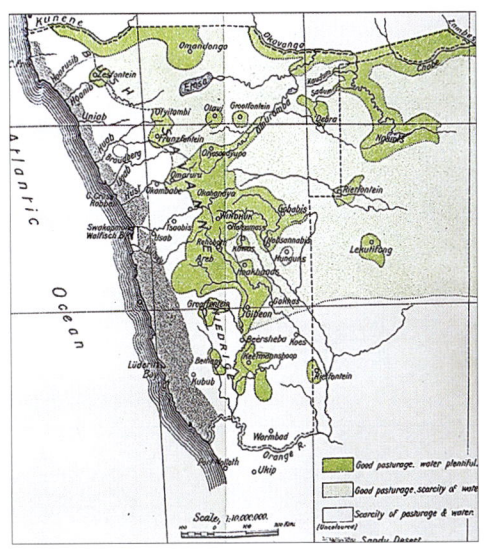

MAP 2
Availability of water and pasturage in GSWA.
Source: "Military Report on German South West Africa, 1906," WO 33-416, TNA

batteries of guns, and some friendly "native" militia.[9] There were no existing defense works or entrenchments at Walvis Bay/Swakopmund. Still, the British considered it prudent to plan for the area's maximum German concentration and defensive measures.[10]

The Germans could expect little assistance from the indigenous population due to their heavy-handed manner in dealing with the local population. Concentrating their forces against a British invasion would take a month due to the *Schutztruppe* (protection force) being scattered throughout GSWA. Consequently, the British did not expect much German resistance at Walvis Bay/Swakopmund, but rather that the Germans would retreat inland to a point of concentration. They aimed to confine the British at Walvis Bay/Swakopmund and then attempt to recapture one or both ports should the opportunity present itself. The British plan called for 1,500 mounted infantry accompanied by artillery to capture the ports of Walvis Bay/Swakopmund if the Germans had no time to concentrate their forces. Thereafter, the force would be reinforced with three battalions of regular infantry, three batteries of artillery, four machine guns, and a detachment of the Corps of Royal Engineers. Walvis Bay was preferred for the initial landing and thereafter Swakopmund would be occupied via an advance from the bay. The plan emphasized the difficulties of dealing with the scarcity and poor water quality. The invasion force would rely on receiving fresh water supplies and other logistics via the sea from the cape, as the area offered very little water or other supplies.[11]

THE REVAMPED PLAN TO INVADE GSWA, 1910

Eight years passed before the British revisited their GSWA invasion plan, just prior to

---
[9] The use of the term *native* is a colonial construct and in modern times is construed as being pejorative. The authors use of the word is limited to direct quotes from the primary sources.
[10] "Paper on the Occupation of Swakopmund German South West Africa," 17 October 1902, 64-65.
[11] "Paper on the Occupation of Swakopmund German South West Africa," 17 October 1902, 73-76.

the formation of the Union of South Africa on 31 May 1910. The German protection forces in GSWA were weak, and they could not expect any reinforcements from Germany in the event of a war. It remained unlikely that the *Schutztruppe* would be able to launch an invasion of South African territory.[12] Therefore, the strategic purpose of invading GSWA would be first to secure and deny the German Navy secure ports on the west coast of Africa, and second to acquire the entire territory for expansionist purposes.

General Paul Sanford Methuen, the British commander in South Africa in 1908, was determined that the British would not remain on the defensive regarding GSWA but would assume the offensive as soon as it was possible after the outbreak of hostilities. Although Methuen remained bullish on the prospects for a British offensive into GSWA, the fact remains that British forces in South Africa had experienced a steady reduction since the end of the South African War (a.k.a. Second Boer War) in 1902. Those British forces remaining in South Africa were earmarked for deployment to Egypt in the event of European hostilities. Any future offensive operation into GSWA would, of necessity, comprise troops belonging to the yet-to-be-formed Union of South Africa. The invasion of GSWA would need the cooperation of the British Royal Navy to establish sea superiority of the GSWA coast and provide the bulk of troop carriers, naval artillery support, and logistics, including the provision of fresh potable water.[13]

Methuen drew on the 1902 plan to invade GSWA and the fact that, because of the strain in British/German relations, plans to meet any threat emanating from GSWA were embedded in the Cape Colony western frontier defense scheme of 1907.[14] On 30 November 1908, Methuen addressed a letter to the naval commander in chief of the Cape Station, Admiral George Egerton. He focused on an offensive action by a joint naval and military expedition landing on the coast of GSWA. Methuen stressed the importance of Lüderitzbucht—developed by Germany as a significant port since 1902—and requested information on the naval policy in the event of war, particularly regarding the defense of Walvis Bay. Methuen expressed his strong disagreement with the defensive policy adopted in 1907, and on 27 December 1908, Egerton, agreed to taking offensive action against GSWA in the event of war. However, the Admiralty refused to guarantee assistance for any particular purpose, signaling discord between the views of the British General Staff and the Admiralty on naval policy. On 8 March 1909, Methuen addressed a letter to the secretary of the War Office that included a paper on preparing a plan of operations against the German forces in GSWA. He stressed the desirability of offensive action and cooperation of the imperial troops in

---

[12] "Memorandum on Project for the Despatch of an Expeditionary Force to GSWA," April 1910, WO 106-47, TNA.

[13] "Major General Ewart (Director of Military Operations) to Admiral Alexander Bethell (Director of Naval Intelligence)," April 1909, WO 106-47, TNA.

[14] "Memorandum General Methuen 'War with Germany: Operations in South Africa'," 5 March 1909, WO 106-47, TNA.

such action while admitting that the military strategy to be adopted must be subservient to the general policy of the British Empire.[15]

Methuen tried to force the hands of the imperial government and the General Staff, and by insisting on an offensive stance, he may have overstepped the mark. He should have constructed plans to meet all reasonable contingencies, offensive and defensive, by land or sea, with or without imperial troops. Instead, he meddled in the political aspects of the problem when protocol demanded that the Committee of Imperial Defence should lay down the overall strategic planning regarding German colonies.[16] Once the Command of Army Council reached a decision, they would inform Methuen of their policy and the forces at his disposal and then request an operational plan to meet the strategic objectives. There was a distinct lack of unity of command in 1910, even when the strategic and operational plans were a wholly British affair. The situation was certainly exacerbated when the South Africans took over the operational planning of the campaign but remained reliant on British naval support.[17]

A major development since the 1902 plan was the German construction of a serviceable port at Lüderitzbucht. The port contained reasonable landing facilities consisting of two piers with three five-ton cranes, good anchorage, and several tugs and lighters that could assist an invasion force with disembarkation. Water availability at the port consisted of three condensers yielding approximately 200 tons a day—wholly inadequate for sustaining an invasion force of any size. Water would have to be transported by sea from Cape Town. The port was linked to the interior via a railway line to Keetmanshoop (see map 1).[18] The Germans had also extended their railway line, thereby connecting their capital Windhoek with both ports and the far northern and southern interior of the colony. The extensive railway network would enable them to conduct an effective defense using interior lines of communication (see map 1).[19]

The British estimated that the German military strength then stood at a maximum of 7,379 personnel, including 935 indigenous troops, 170 artillery pieces of various caliber, and 27 machine guns. The report noted a steady improvement in German military efficiency as they became more accustomed to local conditions and colonial warfare. Further construction of the German railway line network had considerably enhanced their ability to concentrate their forces and meet an invasion at any point. Previous estimates of eight days to concentrate *Schutztruppen* at Swakopmund or Lüderitzbucht were now estimated at a fraction of that time.[20] The latest iteration

---

[15] "Precis of Correspondence in the Subject of Military Operations against GSWA," 28 April 1909, WO 106-47, TNA.
[16] For more on the records of the committee, see "Minister for the Co-ordination of Defense," Records of the Cabinet Office, CAB 64, TNA.
[17] "Precis of Correspondence in the Subject of Military Operations against GSWA," 28 April 1909.
[18] "Memorandum on Project for the Despatch of an Expeditionary Force to GSWA," April 1902, WO 106-47, TNA, 12–15.
[19] "Memorandum on Project for the Despatch of an Expeditionary Force to GSWA," April 1902, 16.
[20] "Memorandum on Project for the Despatch of an Expeditionary Force to GSWA," April 1902, 37–44.

The Union Defence Forces' Amphibious Invasion

of the plan contained a fundamental change in *Schwerpunkt*, as the main body of the invasion force would emanate from the south converging on Kalkfontein from two different directions (see map 1). A seaborne landing at Port Nolloth would advance to Sandfontein and then Kalkfontein, while a second southern prong would proceed from Upington and follow a northwesterly route toward Ukamas-Kalkfontein.

The British considered that the invasion of Lüderitzbucht would have little prospects of success owing to the lack of water at that port. However, the forces advancing from the south on Keetmanshoop could transfer its line of communication to Lüderitzbucht. The seaborne landings at Walvis Bay threatening Swakopmund would serve as a diversion and hopefully compel the Germans to detach some troops to its defenses. Furthermore, the occupation and blockade of the GSWA ports would cut the Germans off from all communication with Europe. Therefore, using the landings at Walvis Bay/Swakopmund merely as a diversionary tactic, the first operational objective would be the surprise seizure of Kalkfontein to prevent a rapid concentration of German troops in the south. The plan called for 2,500–3,000 mounted troops for the first portion of the invasion, accompanied by two batteries of field artillery, 12 machine gun detachments of two guns each, two mountain batteries, one company of bridging and railway engineers, and a company of signalers. Interestingly, the order of battle called for an additional two camel corps of 1,000 troops each.[21]

THE PENULTIMATE PLAN, 1914

Britain declared war on Germany on 4 August 1914. The next day, the offensive subcommittee of the Committee of Imperial Defence agreed that there were significant strategic and political advantages to be gained by capturing GSWA and destroying the German wireless stations situated at Lüderitzbucht, Swakopmund, and Windhoek. Great importance was attached to securing the cooperation and participation of the Union of South Africa. On 6 August, His Majesty's Government approached the South Africans to render a "great and urgent Imperial service" and "seize such part of GSWA as would give them the command of Lüderitzbucht, Swakopmund, and the wireless stations there and in the interior."[22]

The latest British conceptualization of the plan relied on a limited invasion of GSWA to seize the German wireless stations at the coast and in the interior as the initial objective. They were more interested in depriving the German Navy of its communications and ports in the southern oceans rather than a land grab of German territory. Their plan, reliant on limited naval resources, called for the speedy deployment of a relatively small task force. The South Africans, however, envisaged an operation on a much grander scale that would encompass nothing less than the complete conquest of the entire GSWA. Therefore, the manpower and resources conceived for the

---

[21] "Memorandum on Project for the Despatch of an Expeditionary Force to GSWA," April 1902, 44–49.
[22] "Operations in the Union of South Africa and GSWA August 1914–August 1915—Narrative of Events," 5 August 1914, CAB 44-2, TNA.

South African plan would significantly exceed that required to complete the limited British objectives or seizing the ports and rendering the wireless stations inoperable. The different South African and British objectives would lead to confusion and frustration between the parties as soon as the initial operations got underway.[23]

The urgent seizure of the wireless stations at Lüderitzbucht and Swakopmund could only be achieved in a reasonable amount of time by undertaking a joint naval and military expedition. The British considered that capturing the wireless station at Windhoek would be a more serious military undertaking requiring much time in preparation. They suggested that such an operation should only be undertaken after the wireless stations on the coast had been destroyed or seized and that an operation into the interior should form a separate expedition altogether. The British were adamant that the operational details must be left to the Union government and the naval aspects to the senior naval officer at the Cape Station. The South Africans were encouraged to work together with the senior naval officer in formulating a joint plan of operations, as they possessed no naval assets of their own, and would be completely reliant on the British Navy.[24]

The first indication that the South Africans intended to launch an operation beyond that which the British expected was a request by the Union on 11 August that they retain in South Africa either the whole or part of the imperial artillery, which was then under orders to move to the United Kingdom. Intelligence sources revealed on the same day that the Germans had evacuated Swakopmund, blown up its jetty, and scuttled its tugs. A similar situation was said to exist at Lüderitzbucht.[25] On 12 August, the secretary of state for the colonies informed the union government that the imperial artillery serving in South Africa was urgently required in the United Kingdom. However, under the false impression that the South Africans were limiting their operation to the seizure of the points of main importance, the British were "confident that the Union forces would, as they stood, proved equal to the task."[26] The British believed that German resistance and either Swakopmund or Lüderitzbucht was improbable, and they were anxious that the expedition to seize these ports should be expedited.[27]

The South Africans informed the British on 17 August of their arrangements for a force of 1,600 troops with artillery to land at Swakopmund and Lüderitzbucht. An additional force of 1,600 personnel was to land at Port Nolloth and thereafter proceed

---

[23] "Operations in the Union of South Africa and GSWA August 1914–August 1915–Narrative of Events," 6 August 1914, CAB 44-2, TNA.
[24] "Operations in the Union of South Africa and GSWA August 1914–August 1915–Narrative of Events," 8 August 1914, CAB 44-2, TNA.
[25] "Operations in the Union of South Africa and GSWA August 1914–August 1915–Narrative of Events," 11 August 1914, CAB 44-2, TNA.
[26] "Operations in the Union of South Africa and GSWA August 1914–August 1915–Narrative of Events," 12 August 1914, CAB 44-2, TNA.
[27] "Operations in the Union of South Africa and GSWA August 1914–August 1915–Narrative of Events," 13 August 1914, CAB 44-2, TNA.

to Steinkopf and then to the Orange River on the GSWA border. Another force of 1,000 troops was to march from Upington to the Orange River. The South Africans would construct a railway from Prieska to the GSWA border.[28] On 20 August, the South Africans once again informed the British of their intention to increase the force earmarked for Swakopmund and Lüderitzbucht to 5,000 dismounted fighters and a force of 3,000 soldiers to Steinkopf.[29]

Smuts's ambitious invasion plan, designed to deliver the entire GSWA territory into South Africa's hands, drove the feverish quest for military assets and manpower. The original seaborne invasion allowed for four separate columns to converge on Windhoek (map 3). The plan called for the C Force under Colonel P. S. Beves with approximately 2,000 troops to land at Lüderitzbucht, and with the help of the British Royal Navy, its primary task was to destroy critical infrastructure such as the wireless station.[30] The next objective for this group would be to advance inland toward Aus along the railway line with the objective of capturing Seeheim/Keetmanshoop. Farther south, Brigadier General Henry T. Lukin commanded A Force with 2,500 troops, and he would land at Port Nolloth and threaten the southern border of the colony. The capture of Sandfontein thereafter would provide Lukin with a gateway into southern GSWA, since this first staging post had excellent water resources. A farther advance northward to Kalkfontein would take the A Force to the southern terminus of the German railway system (map 3.) Lukin's next objectives were Warmbad and then farther along the railway line, to join forces with Beves's column at Seeheim/Keetmanshoop.[31] Joining Lukin and protecting his right flank would fall to the B Force under General Salomon Gerhardus Maritz with 1,000 mounted troops. He would invade GSWA from the southeast, with Upington as his base of operations, and he would protect Lukin's exposed right flank.[32]

The most significant and crucial formation in terms of size and its ultimate role was D Force commanded by Colonel Duncan McKenzie with 4,000 troops. He was to land at Walvis Bay, capture Swakopmund, and then advance toward the final

---

[28] "Operations in the Union of South Africa and GSWA August 1914–August 1915—Narrative of Events," 17 August 1914, CAB 44-2, TNA.

[29] "Operations in the Union of South Africa and GSWA August 1914–August 1915—Narrative of Events," 20 August 1914, CAB 44-2, TNA.

[30] Ivan S. Uys, *South African Military Who's Who, 1452-1992* (Germiston, South Africa: Fortress, 1992), 18. Beves served in the Rand Pioneer Regiment and started his military life in the UDF as the commandant of cadets.

[31] "Lukin's Report on A Force, 19 August 1915," DC Group 2, Box 252, Folio 17138, South African National Defence Force Archives (DODA). Lukin describes his strength on 25 August 1914 as 135 officers, 2,463 other ranks, 522 Black troops, 12 field guns, and 12 machine guns.

[32] *The Union of South Africa and the Great War, 1914-1918*, 13. See also Collyer, *The Campaign in German South West Africa, 1914-1915*, 28-29.

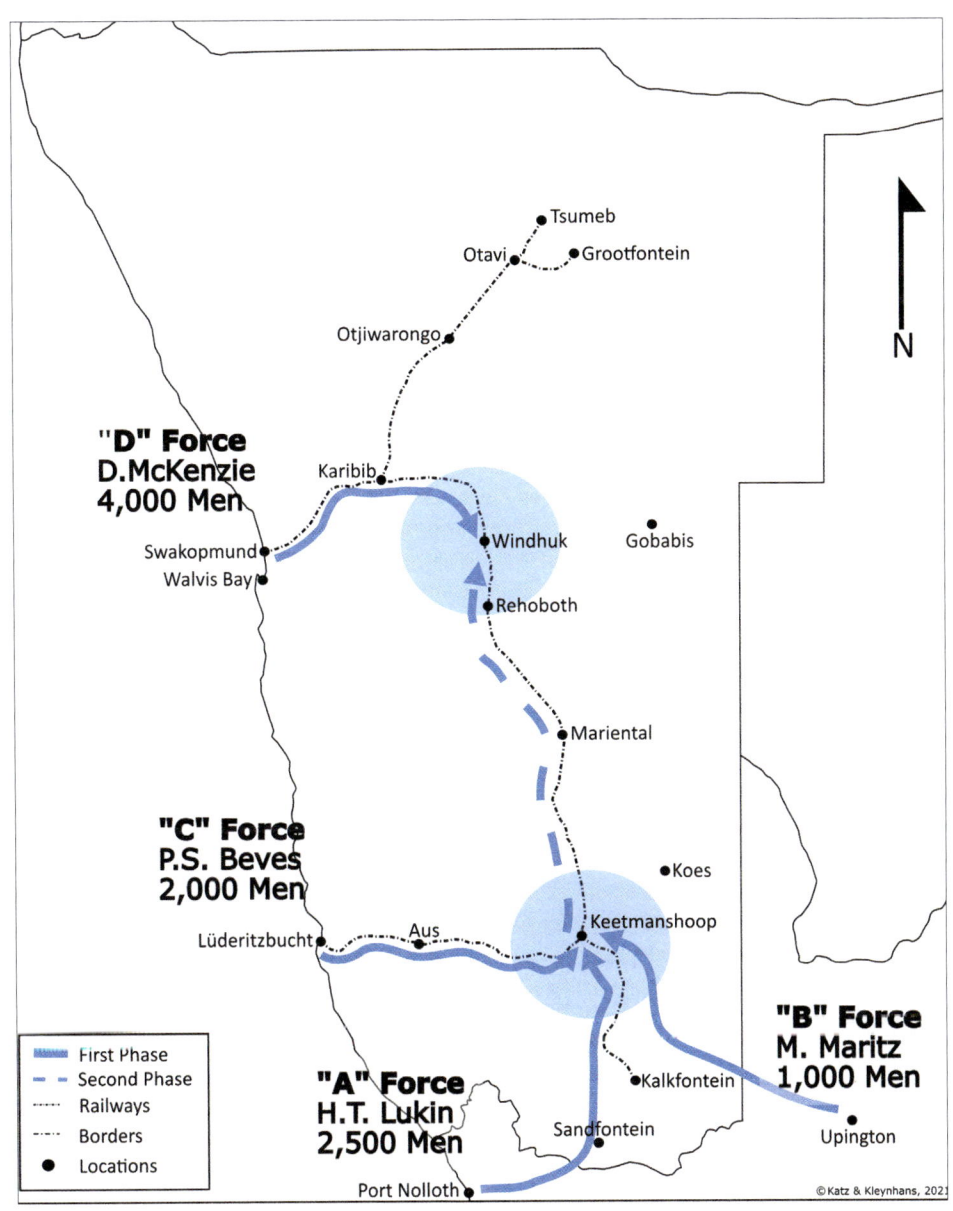

MAP 3
Smuts's original bold plan for the seaborne and landward invasion of GSWA.
Source: courtesy of the author, adapted by MCUP

objective—Windhoek.³³ The capture of Windhoek would sever the rail links to Keetmanshoop in the south and render the German defense south of Windhoek untenable (see map 3). Opposing the 9,500 converging Union forces would be the 5,000 *Schutztruppe* defending GSWA territory. D Force was fundamental to the success of the entire delicately balanced operation. Failure to land the D Force and seize Swakopmund would allow the Germans unfettered opportunities to concentrate their forces either against Lukin in the south or Beves at Lüderitzbucht. For as long as McKenzie held Swakopmund in force, thereby threatening Windhoek, the Germans would have to second guess any intentions of attacking either Beves or Lukin.

It is D Force and its intended deployment that official historians have wittingly —and later contemporary historians unwittingly—written out of the history books. Its omission renders Smuts's original plan nonsensical and obfuscates the mechanics of the operational plan he intended. Brigadier General J. J. Collyer, in the official history, identified the need for close cooperation and for a "simultaneous" advance of A, B, and C Forces to overcome a concentrated enemy over any one of the advancing forces. Collyer also identified the need before any forward movement in the south of GSWA be undertaken until that force was either considerably reinforced or for "arrangements made and put into effect for a diversion elsewhere which would compel the enemy to detach heavily." The diversion Collyer refers to is of course the missing D Force to be landed at Walvis Bay/Swakopmund (see map 3).³⁴

The British grew impatient with the steady increase in expedition manpower and chose to remind the South Africans of the urgent nature of the expedition. They demanded to know whether the proposed increase in troop numbers on the account of the additional transports required would cause any delay in the sailing of the force.³⁵ The South Africans replied that they expected no delay, and that the expedition would be ready to sail on 5 September. However, a lack of escorts meant that the senior naval officer could only sail on 12 September.³⁶ The South Africans estimated that, given the size of the expedition, the naval escorts should not be less than one warship and one armed merchant cruiser. Sailing without the required escorts

---

³³ "Buxton to Harcourt," 8 October 1914, ADM 137-13, Folio 50, TNA. See W. S. Rayner and W. W. O'Shaughnessy, *How Botha and Smuts Conquered German South West: A Full Record of the Campaign* (London: Simpson, Marshall, Hamilton, Kent, 1916), 9. The official histories make no mention of D Force in their initial lineups and orders of battle. Mention of D Force can be found in the work of the embedded journalists and the primary sources. This is perhaps the reason for historians overlooking its existence. S. Monick, *A Bugle Calls: The Story of the Witwatersrand Rifles and Its Predecessors, 1899-1987* (Johannesburg: Witwatersrand Rifles Regimental Council, 1989), 87. The regimental history identifies that the reinforcements received at Lüderitzbucht were originally designated D Force and destined for Walvis Bay; however, the outbreak of the rebellion led to a revision of Smuts's plans.
³⁴ Collyer, *The Campaign in German South West Africa*, 30-31.
³⁵ "Operations in the Union of South Africa and GSWA August 1914-August 1915—Narrative of Events," 23 August 1914, CAB 44-2, TNA.
³⁶ "Operations in the Union of South Africa and GSWA August 1914-August 1915—Narrative of Events," 24 August 1914, CAB 44-2, TNA.

would pose a considerable risk to the expedition as the enlarged task force would provide the German Navy with a tempting target.[37] In an attempt to maximize the use of scarce naval assets, the British suggested that the expedition sail together with the transport ship HMHS *Dover Castle* (1904) transporting the Essex and East Lancashire regiments to the United Kingdom. Both forces would be escorted by the cruiser HMS *Astraea* (1893), which, after seeing the expeditions safely landed at Swakopmund and Lüderitzbucht, would then proceed with the *Dover Castle* to the United Kingdom.[38]

Meanwhile, Lukin's expedition to Port Nolloth landed on 31 August and immediately experienced delays in disembarkation partly because of the state of the port and partly to disorganized staff work.[39] The UDF could expect the same or longer delays at Lüderitzbucht and Swakopmund.[40] Lack of planning, organization, and experienced staff officers took an early toll on efforts.[41] Chaotic disembarkation procedures at Port Nolloth—10 days to land the stores—delayed the rest of the GSWA expedition a few days beyond 12 September. The expectation was that disembarkation at the other ports would be a lengthy process too. The regimental history alludes to the chaos of disembarkation at Port Nolloth. It seems that it took the Transvaal Horse Artillery from 31 August to 9 September to fully assemble at the port before making their way to Steinkopf. The regimental author attests to the disorganization that accompanied the embarking at Cape Town and disembarking at Port Nolloth. The move to Steinkopf began on 4 September, but Lukin only established his headquarters there on 10 September. Lukin reports that all troopships had arrived by 3 September, and their disembarkation was not completed until 17 September.

Roland M. Bourne, the secretary of defense, belatedly formed a Joint Operational Command in Pretoria on 9 September to alleviate the logjam, which a senior naval

---

[37] "Operations in the Union of South Africa and GSWA August 1914–August 1915—Narrative of Events," 31 August 1914, CAB 44-2, TNA.
[38] "Royal Navy Log Books of the World War I Era: HMS Astraea," Naval History Homepage, updated 29 September 2017. The HMS *Astraea* was a light cruiser of 4,360 tons with an armament of: two 6-inch, eight 4.7-inch, eight 6-pounders, and three 18-inch guns. Its armor consisted of a 2-inch deck and 4.5-inches for the guns. "Operations in the Union of South Africa and GSWA August 1914–August 1915—Narrative of Events," 1 September 1914, CAB 44-2, TNA.
[39] "Lukin's Report on A Force," 19 August 1915, DC Group 2, Box 252, Folio 17138, DODA. Disembarkation, according to Lukin, took more than two weeks and was not completed before 16–17 September 1914. The major delay occurred with the disembarkation of the animals that were slung twice from ship to lighter and then lighter to shore. See "Letter from Secretary for Defence to Unknown," 19 September 1914, DC Group 2, Box 252, DODA, which refers to the great difficulties of disembarkation. See also "Methods and Points to be Observed in Embarking and Disembarking," GSWA Group, Box 14, DODA.
[40] "Telegram Officer Advising Gov of SA to H. B. Jackson," 2 September 1914, ADM 137-9, TNA.
[41] "Letter from Secretary for Defence to Unknown," 19 September 1914. The letter refers to great loss of equipment for lack of care and the unsuitability of donkeys compared with mules. There was the problem of inferior quality equipment such as artillery harnesses.

and army officer and a senior representative of the South African railways staffed.[42] Delayed timetables clashed with the British efforts to repatriate their garrison forces using the same scarce shipping resources. The British informed the South Africans on 7 September that they would not delay the departure of the ships conveying troops back to the United Kingdom beyond 14 September. The South Africans were encouraged to make suitable arrangements to meet the deadline.[43]

The British determination to keep to a strict timetable exasperated the South Africans. When Sydney C. Buxton assumed the role of governor general on 8 September, he sent an impassioned plea to the British that the nonavailability of naval escorts would scupper the whole expedition with disastrous effects on public opinion. Smuts asked personally and informally whether the repatriation of the imperial garrison could be delayed by a few days to facilitate the GSWA expedition.[44] Political pressure forced the British to weigh the cost of delaying the repatriation of the imperial garrison for a couple of weeks, against dampening enthusiasm for the expedition within the Union. Admiral Henry B. Jackson, the advisor on overseas expeditions and planning attacks on Germany's colonial possessions, concluded that the importance of the expedition outweighed any benefits of early repatriation of the imperial garrison.[45] The considerable benefits of destroying three German radio stations compared to repatriating one-and-a-half battalions to the United Kingdom won the day.[46] The British unequivocally decided on 9 September that HMS *Astraea* would be available for escort duties for the expeditions to Lüderitzbucht and Swakopmund and the repatriation of the cape garrison to the United Kingdom would not be allowed to interfere.[47]

With the GSWA expedition back on track, and British patience restored, the next problem on the horizon was of the considerable delays at Port Nolloth. The South Africans claimed that bad weather caused the delays in disembarkation.[48] However, large-scale disorganization meant that the Walvis Bay part of the expedition would take place one week after the landings at Lüderitzbucht on 14 September.[49] A combination of bad weather and worse planning intervened, delaying the departure for the

---

[42] "Joint Naval and Military Operations, Secretary of Defence," 9 September 1914, DC Group 2, Box 252, DODA.
[43] "Telegram S. S. for Colonies to Gov of SA," 7 September 1914, ADM 137-9, TNA.
[44] "Telegram Governor of Union of SA to Secretary of State for the Colonies," 8 September 1914, ADM 137-9, TNA.
[45] "Telegram C-in-C Cape to H. B. Jackson," 9 September 1914, ADM 137-9, TNA.
[46] "Telegram S. S. for Colonies to Gov of SA," 9 September 1914, ADM 137-9, TNA.
[47] "Operations in the Union of South Africa and GSWA August 1914-August 1915—Narrative of Events," 9 September 1914, CAB 44-2, TNA.
[48] "Operations in the Union of South Africa and GSWA August 1914-August 1915—Narrative of Events," 12 September 1914, CAB 44-2, TNA.
[49] "Telegram C-in-C Cape to H. B. Jackson," 11 September 1914, ADM 137-9, Folio 426, TNA.

Walvis Bay/Swakopmund to 26 September.[50] The landing at Walvis Bay/Swakopmund would only be complete by 11 October, causing considerable delay to the repatriation of the imperial garrison.[51]

Further delays at Port Nolloth meant the naval transports could only get back to Cape Town by 17 September, which delayed Beves's occupation of Lüderitzbucht to 18 September.[52] There was also no luxury of a wharf in Lüderitzbucht as late as March 1915, and horses disembarking there had to swim to the shore. Colonel James Irvine-Smith of the British Army Veterinary Division reports that, by 17 March 1915, disembarkation of horses had improved by avoiding slinging and using a special gangway, allowing 900 animals to be offloaded in 10 hours.[53]

Adding significantly to the rapidly thickening fog of war—before Beves set out for Lüderitzbucht—the South African political horizon became increasingly clouded following the resignation of General Christiaan Frederik Beyers, a senior member of the Union Defence Force and chief of its Active Citizen Force (conscripts) with another senior UDF officer, Jan Kemp, on 13 and 15 September, respectively. Their resignations and the worsening political situation in the Union, which included the looming prospect of rebellion, cast a shadow on the GSWA campaign.[54]

FIASCO AT SANDFONTEIN, 26 SEPTEMBER 1914

The failure to secure Walvis Bay/Swakopmund placed Beves at Lüderitzbucht in a precarious position. The occupation of Swakopmund would have placed the Germans in

---

[50] "Telegram Botha to Buxton," 11 September 1914, PM 1/1/32, File 4/95/14-4/97/14, Minute no. 868, Correspondence file, National Archives of South Africa Pretoria (NASAP). Gen Louis Botha cautioned that it was unlikely the landing at Walvis Bay would be completed before 30 September 1914.

[51] "Telegram Botha to Buxton," 12 September 1914, PM 1/1/32, File 4/95/14-4/97/14, Minute no. 875, Correspondence file, NASAP; and "Telegram Governor of Union of SA to Secretary of State for the Colonies," 12 September 1914, ADM 137-9, Folio 434, TNA.

[52] Collyer, *The Campaign in German South West Africa*, 28–29; and "Letter from Rear Admiral H. K. Hall to the Secretary of the Admiralty," 15 October 1914, ADM 137-8, TNA. The harbor at Lüderitzbucht was reported as excellent and the piers, lighters, and cranes were all intact. The Germans failed to destroy the facilities. The navy provided three 4.7-inch guns to protect the port from sea and land attack. It was regarded as a protected port and a secure land base.

[53] "Telegram C-in-C Cape to H. B. Jackson," 5 September 1914, ADM 137-9, TNA; Neil Orpen, *The History of the Transvaal Horse Artillery, 1904–1974* (Johannesburg: THA Regimental Council, 1975), 14; F. B. Adler, *The History of the Transvaal Horse Artillery* (Johannesburg, South Africa: Specialty Press, 1927); "Lukin's Report on A Force," 19 August 1915, DC Group 2, Box 252, Folio 17138, DODA; Mark Coghlan, *History of the Umvoti Mounted Rifles, 1864–2004* (Durban, South Africa: Just Done Productions, 2012), 1162; and "Veterinary Services GSWA Campaign and Rebellion August 1914 to July 1915, Report by Colonel James Irvine-Smith," AG 14, Box 13, File 2, DODA, 15.

[54] "Telegram, Buxton to Harcourt," 15 September 1914, ADM 137-9, Folio 472, TNA. Beyers published his manifesto on his resignation in Gen James B. M. Hertzog's newspaper, *Otago Daily Times*. See Piet van der Byl, *From Playgrounds to Battlefields* (Cape Town, South Africa: Howard Timmins, 1971), 92, for a physical description of Maritz.

a dilemma. They could now concentrate their forces on Lüderitzbucht. Smuts's predicated his plan on the ability to advance his forces simultaneously on exterior lines thereby preventing the concentration of German forces using interior lines.[55] Beves, facing the might of the *Schutztruppe* alone, would have to rely on Lukin ensconced at the southern border to create a diversion to distract the Germans. The threat of a flank attack by Lukin prevented German concentration against the port. However, Lukin faced problems of his own beside the prospect of moving his troops over many kilometers of inhospitable, arid terrain. Maritz with B Force guarded Lukin's vulnerable right flank, but he grew increasingly hostile to the idea of invading GSWA.

A perfect storm was brewing that placed Beves in considerable jeopardy. The government took note of Maritz's recalcitrant behavior and, coupled with delays in the seaborne operations, an uneasiness descended on the entire operation.[56] Smuts cajoled Lukin to proceed with his advance to discourage the Germans and keep Maritz onside. Cooperation between these two forces would be crucial as Maritz would protect Lukin's right flank. The advance along exterior lines called for Lukin to strike through Raman's Drift on the Orange River and successively capture the towns of Warmbad and Kalkfontein. The latter was the southern terminus of the German railway system (map 4). Such a thrust by Lukin would further thwart any German intentions of invading the Union.[57]

It soon became apparent that Maritz would not cooperate in covering Lukin's flank. Furthermore, strong indicators emerged that he was about to declare open rebellion. Instead of his force bolstering Lukin's right flank, it began instead to menace him. Maritz posed a real danger if he could add his force to the enemy, thereby destroying the delicate balance of fighting power. Instead of his usual decisiveness, Smuts took no action to remove Maritz immediately despite all the evidence of his wavering attitude.[58] Instead, he ordered Maritz to advance to Schuit Drift from Kakamas and then head to Ukamas to assist and cooperate with the force under Lukin on 23 September. Smuts's decision to test Maritz's loyalty rather than replace him is a testament to the challenging political climate, where his usual decisiveness and indeed, ruthlessness occasionally gave way to expediency.

---

[55] "Letter Smuts to McKenzie," 6 January 1915, DC Group 2, DODA. Smuts stressed to Duncan McKenzie, commanding the Central Force at Lüderitzbucht, of the need to advance his forces "simultaneously" with those under Botha at Walvis Bay/Swakopmund of the Northern Force. This is strong evidence of Smuts's intention of the simultaneity of advances.

[56] *Judicial Commission of Inquiry into the Causes of and Circumstances Relating to the Recent Rebellion in South Africa: Minutes of Evidence, December 1916* (Cape Town, South Africa: Cape Times, 1916), 11-16.

[57] "Slaag van Sandfontein," 26 September 1914, AG 14, Box 13, File 7, DODA, 1. The after action report clearly states that the operational objective of A Force was the capture of Warmbad and then Kalkfontein. It was "anticipated" that this would lessen the chances of an invasion from GSWA and "materially assist" the forces landing at Lüderitzbucht. See "Lukin's Report on A Force," 19 August 1915, DC Group 2, Box 252, Folio 17138, DODA. Lukin states his immediate objective was Warmbad.

[58] Earl Buxton, *General Botha* (London: John Murray, 1924), 45. Buxton asserts that the rebellion came as a complete surprise to the South African government and that no preparations were made to meet it.

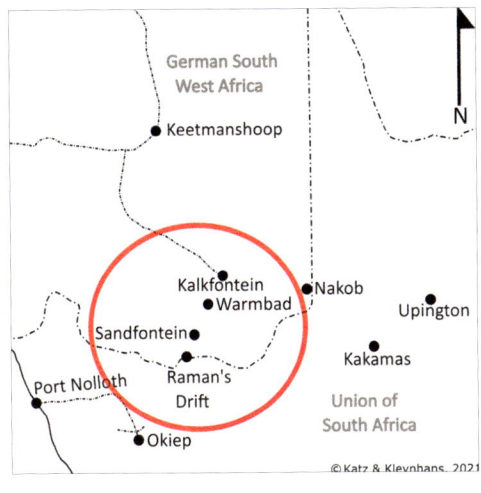

MAP 4
Sandfontein: southern gateway into GSWA.
Source: adapted from Gerald L'ange, Urgent
Imperial Service: South African Forces in German
South West Africa, 1914-1915 (Johannesburg:
Ashanti, 1991), 85

Smuts pressed Lukin to advance to Sandfontein expeditiously to create a diversion to relieve Beves of pressure at Lüderitzbucht and allow reinforcements to arrive at the scene. The official history describes Smuts's communication with Lukin as one verging on a request for "self-sacrifice."[59] Lukin was under no illusion as to the precariousness of the situation. He possessed reliable intelligence showing the enemy was determined to oppose his advance to Kalkfontein and that they would use the railway to concentrate considerable forces against him.[60] However, Lukin paints a different picture in his report, blaming his predicament on intelligence failure. He felt that scouts should have detected the large force of 1,800 Germans and 10 guns. If he knew of the impending attack by such a force, he would have withdrawn the Sandfontein force within three or four hours. Lukin pointed out that the disaster would have been greater had the Germans delayed their attack and allowed him to advance on Warmbad.[61] When Lukin did advance, contrary to what Smuts expected, he did so with only a fraction of the force available to him.

Inevitably and not unexpectedly, Maritz disobeyed Smuts's order to advance, leaving Lukin alone deep inside German territory. Maritz was aware that 2,000 Ger-

---

[59] *The Union of South Africa and the Great War*, 14; and Collyer, *The Campaign in German South West Africa*, 32, 48. Collyer goes to great lengths to explain that Lukin must have expressed his reservation to division headquarters on being ordered to Sandfontein. The extent of his reservation is contained in this line: "Headquarters had to request high pressure to the verge of self-sacrifice on the part of General Lukin to which he most loyally responded." Collyer cites the fact that Lukin did not receive vital intelligence that the Germans were gathering a force in proximity to him because of a bungle at headquarters. Lukin is quoted as saying that if he received this intelligence in time, he would have been apt to withdraw from Sandfontein promptly. The fact is that Lukin should have expected a strong German response to his advance in any event, and he did not provide a sufficient force forward.

[60] "Buxton to Harcourt," 25 September 1914, ADM 137-9, Folio 580, TNA; and Collyer, *The Campaign in German South West Africa*, 48. Collyer has a different take on the events pertaining to the intelligence of a German threat to Sandfontein. Collyer, who was Lukin's brother-in-law, blames a nameless staff officer at headquarters who posted instead of telegraphed the intelligence summary to Lukin, so it only reached him on 7 October 1914. See Ian van der Waag, "The Battle of Sandfontein, 26 September 1914: South African Military Reform and the German South-West Africa Campaign, 1914–1915," *First World War Studies* 4, no. 2 (2013): 22n84, https://doi.org/10.1080/19475020.2013.828633.

[61] "Lukin's Report on A Force," 19 August 1915, DC Group 2, Box 252, Folio 17138, DODA.

man troops were advancing on Lukin's forces, and he rebuffed Lukin's request for reinforcements. Smuts immediately summoned Maritz to Pretoria on 30 September and instructed him to step down from command. Maritz flatly refused to cooperate, and Smuts eventually transferred the command of the A and B Forces, including the troops under Maritz, to Colonel Coen Jacobus Brits on 2 October. His action effectively deducted 1,000 soldiers from the UDF strength and added them to that of the Germans.[62]

On 26 September, Lukins A Force, unsupported by Maritz and understrength for the task allotted, suffered a severe defeat at the hands of the Germans at Sandfontein.[63] The advance in such small numbers to Sandfontein was an operational error considering the uncertainty of Maritz's allegiance, knowing that the Germans were in force in the vicinity and contrary to the rules of concentration. Furthermore, Lukin committed grave tactical errors such as the lack of adequate reconnaissance. Smuts revealed the overall strategic concern of the operation when he pressed Lukin to hold the Orange River and not retire farther south in the wake of the Sandfontein fiasco. Smuts was concerned that Lukin would no longer pose a threat to the German flank, thereby leaving them free to deal with the forces at Lüderitzbucht. Further communication instructed Lukin to move most of his forces from Steinkopf to the Orange River and adopt an aggressive posture to keep the enemy away from an increasingly vulnerable Beves. In the wake of these developments, Smuts finally abandoned the Walvis Bay/Swakopmund expedition on 29 September and despatched McKenzie's D Force, originally earmarked for Walvis Bay, to bolster Lüderitzbucht on 30 September.[64]

The delayed landing at Walvis Bay was a combination of an initial lack of British naval escorts, the UDF's disorganization at the staff level, and finally the outbreak of the Afrikaner Rebellion a few weeks after the Sandfontein debacle.[65] The German naval fleet roaming rampant in the South Atlantic in mid-December added to the heightened alarm later in the campaign but was not the principal reason for delays in August/September. Smuts together with Buxton concurred with the suggestion of the Vice Admiral Herbert King-Hall, the naval commander in chief of the Royal Navy's Cape Station, to abandon the idea of the Walvis Bay/Swakopmund expedition on 29

---

[62] *Judicial Commission of Inquiry*, 19-21.
[63] Collyer, *The Campaign in German South West Africa*, 36-49.
[64] "Buxton to Harcourt," 29 September 1914, ADM 137-9, Folio 624, TNA; and "Telegram Buxton to Naval C in C Cape Station," 28 September 1914, PM 1/1/32, File 4/95/14-4/97/14, Correspondence file, NASAP. A further indicator that the operation to Walvis Bay was abandoned was a suggestion by Smuts and Buxton that HMS *Kinfauns Castle* (1899) remain at Walvis for a few days longer to fool the Germans that it was proposed to land a force there. This would alleviate some of the risk Beves at Lüderitzbucht faced in light of the Sandfontein fiasco and the cancellation of the Walvis Bay landing.
[65] "Telegram Buxton to Secretary of State," 8 September 1914, PM 1/1/32, File 4/95/14-4/97/14, Correspondence file, NASAP. Smuts informally through Buxton called for another warship, HMS *Cumberland* (1902), to be dispatched to the area and cover the landings at Walvis Bay.

September.⁶⁶ King-Hall believed that a landing would be untenable considering the chaos experienced at Lüderitzbucht and the difficulty of protecting Walvis Bay from the sea because of its vast defensive perimeter. Furthermore, he could not account for all German shipping in the area.⁶⁷

The British were growing increasingly concerned with poor organization and bungled logistics at Lüderitzbucht. The rapidly deteriorating political situation within the Union, soon to experience open rebellion, coupled with the disastrous reversal at Sandfontein, did not inspire confidence. Finally, on 28 September, the British suggested altering the plan. Smuts concurred that these factors, together with a rapidly developing Afrikaner Rebellion in the Union, which included Maritz's treachery, placed Lüderitzbucht in a precarious position.⁶⁸ There was thus little option but to bolster the defenses of Lüderitzbucht with D Force formerly earmarked for Walvis Bay/Swakopmund. Smuts would only reinstate the expedition to Walvis Bay/Swakopmund on 25 November after he and Botha registered decisive successes against the rebels.

POSTSCRIPT

The invasion of GSWA was an ambitious undertaking, and more so, as the amphibious aspects added a layer of complexity. The Smuts plan called for a simultaneous landing of South African forces at three ports. South Africa lacked naval resources and would have to rely on the British Navy to transport and protect the amphibious landings. The plan called for a joint operation in its true sense, and furthermore, it involved the military assets of two nations, South Africa and the United Kingdom. An amphibious operation of this nature requires the highest communication and cooperation between the participants. At the outset of the invasion, The South Africans and United Kingdom possessed differing intentions, with the United Kingdom having limited objectives while the South Africans sought to conquer GSWA in its entirety. Sound communication between the participants, a prerequisite in amphibious operations, remained poor during the planning and operational phases of the initial invasion.

The South African objective required manpower and resources that overburdened the limited British naval assets earmarked for the amphibious operation. The South African requirement for a simultaneous amphibious landing at Port Nolloth, Lüderitzbucht, and Walvis Bay to overwhelm the German defenders was impossible

---

⁶⁶ "Telegram Buxton to Naval C in C Cape Station," 28 September 1914, PM 1/1/32, File 4/95/14-4/97/14, Correspondence file, NASAP.

⁶⁷ "Telegram Naval C in C Cape Station to Buxton," 27-28 September 1914, PM 1/1/32, File 4/95/14-4/97/14, Correspondence file, NASAP.

⁶⁸ "Buxton to Harcourt," 5 October 1914, ADM 137-13, Folio 32, TNA; and "Buxton to Harcourt," 8 October 1914, ADM 137-13, Folio 50, TNA. Buxton cites the reversal at Sandfontein, Lukin's challenges regarding water and transport, and Maritz's "unreliability" and the delays on disembarkation at the landings as "destroying all possibility of simultaneous action."

given the limited British naval resources. Exacerbating the problem was the UDF's poor planning and preparation for the operation. Amphibious operations require a high degree of staff work. The UDF did not possess sufficient staff officers, and those they had were poorly qualified and inexperienced for the job. Poor discipline and planning played havoc at the landings, and the disembarkation at the ports took much longer than anticipated, which in turn, tied down British naval assets.

The GSWA campaign's postponement allowed the original Smuts plan to be revised between 5 and 8 October. The plan retained most of Smuts's original objectives, but this iteration contained a massive fourfold increase in numbers deployed.[69] The whole operation depended on the availability of British Royal Navy ships to support the extended operation.[70]

The new incarnation of the Smuts plan contained fundamental differences from the original. Besides Maritz's former B Force, Smuts initially relied mainly on the UDF's Active Citizen Force units. These possessed a distinct colonial/British structure and doctrine. They had a formal rank structure, trained in British methods, and were led by English officers with a distinctly British command style. The Boer commandos were more informal, led by Afrikaners, with a less rigid structure and a directive command style that encouraged initiative and the devolution of decision-making down to the lower levels of command. Smuts boosted the invasion's second iteration by adding the Boer Republican-style commandos of the second line ACF Reserve (Class B) Rifle Association members. These units played a significant role in extinguishing the Afrikaner Rebellion a mere few weeks before their deployment to GSWA. Predominantly Afrikaner and veterans of the South African War, these mounted infantry forces were earmarked for deployment to Walvis Bay/Swakopmund (Northern Force) to be commanded by Botha, leading from the front.[71] Botha and Smuts decided that the commandos, who proved loyal in extinguishing the Afrikaner Rebellion, could now be used to good effect in GSWA. Once nimble and supported by 10,000–12,000 colonial/British-orientated units, Smuts's plan became bloated with a cumbersome compliment of 40,000 troops.

Unlike the original plan, Smuts now consulted the British on matters connected with the expedition.[72] Smuts was impatient about reinvigorating the stalled proceedings and proposed that the Walvis Bay expedition launch date be 12 December. The British issued a cautionary note that the expedition should not start until the

---

[69] "Letter Smuts to Crewe," 18 December 1914, JSP, Box 196, Folio 156, NASAP. In this letter, Smuts confirms the appointment of J. L. van Deventer to command the whole Orange River and the raising of six further mounted brigades for GSWA. The hand of Smuts in directing and recruiting for the campaign was everywhere.

[70] "Buxton to Harcourt," 8 October 1914, ADM 137-13, Folio 51-53, TNA; "Memorandum Admiral H. B. Jackson," 8 October 1914, ADM 137-13, Folio 88, TNA; and "Telegram Botha to Buxton," 7 October 1914, PM 1/1/32, File 4/95/14-4/97/14, Minute no. 994, Correspondence file, NASAP.

[71] "Appointment of Botha," AG 1914–1921, Box 8, Folio G5/305/9199, DODA.

[72] "C in C Cape to Admiralty," 9 October 1914, ADM 137-13, Folio 70, TNA.

German naval squadron in the Southern Ocean (Antarctic Ocean) was located and neutralized.[73] Smuts insisted that further delays would have severe repercussions for the campaign and morale on the home front.[74] Despite Smuts's obvious irritation, the British delayed the expedition by an additional two weeks. In the interim, Duncan McKenzie, commanding the forces at Lüderitzbucht, received a further 2,000 reinforcements.[75] Intelligence sources confirmed that the German fleet set sail from South America and made its way to South Africa on 7 December.[76] However, the German naval threat was finally eliminated in the Battle of the Falklands on 8 December, when the SMS *Gneisenau* (1906), SMS *Scharnhorst* (1906), SMS *Leipzig* (1905), and SMS *Nurnberg* (1906) were sunk by the British Royal Navy.[77]

Colonel P. C. B. Skinner, on loan from the British Army, with two infantry brigades under his command disembarked at the undefended harbor of Walvis Bay on Christmas day 1914.[78] Skinner oversaw the invasion until Botha assumed overall command of the Northern Force.[79] The invaders immediately set about building a defensive line around Walvis Bay.[80] The landing surprised the Germans and went unopposed. The Germans, who had long since abandoned Walvis Bay/Swakopmund in favor of making their defense farther into the interior, allowed for a bloodless occupation.

---

[73] "C in C Cape to Admiralty," 25 November 1914, ADM 137-13, Folio 573, TNA; and "C in C Cape to Admiralty," 27 November 1914, ADM 137-13, Folio 621, TNA. The British Royal Navy had four duties regarding the expedition to GSWA that involved the conveyance of troops to Walvis Bay, to protect Walvis Bay, to cover and protect Lüderitzbucht, and to guard the lines of communication from the cape to Lüderitzbucht and Walvis Bay. The British were reluctant to split their forces or undertake the expedition until such time as the enemy force were dealt with.
[74] "C in C Cape to Admiralty," 30 November 1914, ADM 137-13, Folio 649, TNA.
[75] "Buxton to Harcourt," 30 November 1914, ADM 137-13, Folio 651, TNA.
[76] "Sir R. Tower, Buenos Ayres to Admiralty," 7 December 1914, ADM 137-13, Folio 710, TNA. The tip-off was received from a correspondent of the *New York Times*.
[77] "India Office to Admiralty," 9 December 1914, ADM 137-13, Folio 728, TNA.
[78] The expeditionary force consisted of the Imperial Light Horse, Grobbelaar's scouts, and an artillery brigade. Col P. C. B. Skinner, formerly of the Northumberland Regiment, was loaned from the British government to support Botha, and during the GSWA campaign, Botha asked him to set up a general staff. He was previously the commandant of the South African Military Academy.
[79] Rayner and O'Shaughnessy, *How Botha and Smuts Conquered German South West*, 164.
[80] "Letter of Proceedings from Captain of HMS *Astraea* to C in C Cape Station," 29 January 1915, ADM 123/144, general letters and proceedings Walvis Bay, TNA. British seapower would form an integral part of the early defense of Walvis Bay and Swakopmund, with the ship guns and the infantry cooperating in a firing scheme should the Germans approach the beachhead.

The Union Defence Forces' Amphibious Invasion

CHAPTER SEVEN

Operation Albion

The German Amphibious Landing on the Baltic Islands, 12–17 October 1917

Eric Sibul

INTRODUCTION

Operation Albion, the German amphibious landing on the Baltic (Estonian) Islands during 12–17 October 1917, was an important and unique operation in the First World War warranting greater historical examination.[1] Despite the fact that the Russian Provisional Government granted Estonia autonomy on 12 April 1917, substantial Russian forces remained on Estonian soil to defend the maritime approaches to Petrograd (now St. Petersburg). Estonia was also important for Triple Entente offensive naval operations from the port of Tallinn (Reval).[2] A goal of Operation Albion was to end the Anglo-Russian submarine threat to German iron ore traffic from Sweden to Germany.

In autumn 1917, the German General Staff had the greater strategic problem of quickly ending the war on the eastern front to shift resources westward. In the Russian maritime defense scheme, positions on the Estonian islands and Estonian shore

---

[1] These islands included Saaremaa (Ösel), Muhu (Moon), and Hiiumaa (Dagö). Referred to by Germans as the Baltic Islands, referred to by the Estonians as the Estonian Islands. On 12 April 1917, the Russian Provisional government endorsed the law drafted by Estonian leaders for the autonomy of Estonia. This law joined the islands that had been administered as part Livonia to Estonian administration as most of the population were Estonians.

[2] The *Triple Entente* refers to the formal association between Russia, France, and Great Britain during World War I. Michael Wilson, *Baltic Assignment: British Submariners in Russia, 1914-1919* (London: Leo Cooper, 1985), 38-39.

MAP 1
Defense positions, coastal artillery, and minefields.
Source: official Estonian Navy map, adapted by MCUP

of the Gulf of Finland were the "hinge to the door" to the Russian capital of Petrograd. Operation Albion was successful as it caused panic and government collapse in Petrograd.[3]

Operation Albion illustrates the importance of an armed forces' ability to adapt to new situations quickly. The German armed forces planned and executed Operation Albion in a few weeks despite having no amphibious doctrine or experience. It was

---

[3] E. Laaman, "Langemine 20 aasta eest" [The Fall of Saaremaa 20 Years Ago], *Sõdur*, 40–41 (1937): 978; and William S. Lind, "Operation Albion," *On War* #318, Defense and National Interest, 19 October 2009.

also perhaps the first true joint operation including selection of a joint air commander. The success of the operation had the ultimate strategic result of collapsing the Russian Provisional Government and ending the war on the eastern front on German terms in 1917.

TRIPLE ENTENTE SUBMARINE OPERATIONS

During the first months of the war in 1914, Russian submariners, although professionally competent, were handicapped by old and dated vessels. By the end of 1914, Russian submarines had made 14 patrols but failed to have any success against German shipping. Therefore, the British Admiralty decided that best way they could immediately assist their Russian ally was by reinforcing the Russian submarine fleet in the Baltic. In addition to carrying out maritime reconnaissance and attacks on German warships, an important aim for the British submarines was to disrupt the vital traffic of high-grade hematite iron ore from the Swedish port of Luleå on the Gulf of Bothnia to north German ports. Luleå was the site of the Svartön ore docks and the terminus of the railway line to the Malmberget ore fields in northern Sweden.[4]

In October 1914, two British *E*-class submarines ran the Danish Belts and Sounds and operated out of Tallinn.[5] A third *E*-class successfully made the run through the Belts and Sounds, one was lost as German antisubmarine warfare techniques improved. Autumn 1915 was perhaps the most successful period for British submarine flotilla operating in the Baltic. The HMS *E8* (1913) sank the armored cruiser SMS *Prinz Adalbert* (1901) and HMS *E19* (1915) sank four German flagged ore carriers. To follow up on the success of 1915, the British Royal Navy opted for a risk-free route from Archangel, Russia, via inland waterways to the Gulf of Finland to reinforce their Baltic flotilla in July 1916 with four small *C*-class submarines.[6]

RIGA FRONT

At the end of August 1914, the Russians suffered a huge defeat with the invasion of East Prussia. Half of the Russian 2d Army Corps was annihilated, 92,000 troops were captured, and large stocks of artillery and transport equipment were lost.[7] A series of

---

[4] "Sweden Aiding Germany with Iron Ore, Claim," *Chicago Daily Tribune*, 19 July 1917, 5; "Electric Railway in Sweden," *Railway Age Gazette* 59, no. 21 (November 1915): 942; and Capt Donald Macintyre, "A Forgotten Campaign—IV: Forlorn Hope," *RUSI Journal* 106, no. 624 (1961): 65, https://doi.org/10.1080/03071846109420730.

[5] What have historically been known as the Belts and Sounds are also known as the Danish Straits, which are narrow, shallow, island-dotted sea areas that lie between the Baltic and the North Sea. They are a classic maritime chokepoint. The Belts and Sounds comprise three general areas, there is the 5 km-wide Öresund between the island of Zealand on which Copenhagen is situated and the western coast of Sweden, the Great Belt, which has a width of 18 km, and the Little Belt. Malcolm W. Cagle, "The Strategic Danish Straits," U.S. Naval Institute *Proceedings* 86, no. 10 (October 1960): 36.

[6] Macintyre, "A Forgotten Campaign—IV," 66; and Wilson, *Baltic Assignment*, 38–39.

[7] Edgar Anderson, "The Military Situation in the Baltic States," *Baltic Defence Review* 6, no. 2 (2001): 117.

follow-up battles kept the Russians off balance until the spring of 1915. Lithuania was largely occupied and Kurzeme (Courland) fell to the Germans, the broad Daugava (Düna, Western Dvina) River would hold the German advance to the northeast for an extended period. Latvian territory was cut in two by the front lines. Latvian territory falling fully under German control did not bode well for the future of Latvia and its national leaders pushed tsarist authorities for the formation of Latvian units under the command of Latvian officers. They were able to achieve the creation of national Latvian light-infantry units. Starting in August 1915 with two battalions, the Latvian units grew to eight combat regiments and one reserve regiment that were combined in two brigades for a short period in one division. These units, commanded by Latvian officers, broke through the German front lines several times by using new infiltration style tactics. Due to their bravery and success, they drew the attention not only of Triple Entente military observers, but also of the international press. These Latvian units holding the Daugava line paid a heavy price in lives during the battles of July 1916, Christmas 1916, and New Year 1917.[8]

OPERATION ALBION

In spring 1917, the coordinated offensive of Triple Entente Powers had broken down. On the western front, the great spring offensive of the British at Arras, and that of the French along the Aisne River, failed. On the eastern front, continuous military setbacks since August 1914, and the stress of the war on the economy and society forced Tsar Nicolas II to abdicate in March 1917 and a republican provisional government was formed under liberal Petrograd lawyer Aleksandr Kerensky. Kerensky's provisional government kept Russia in the war against the Central Powers, starting an offensive in July on the banks of the Dniester River.[9] It was so successful that it caused a crisis for the Central Powers on the Austrian front. Therefore, German troops were rushed to the Dniester region in the support of the Austrians. These German troops were able to go on the counteroffensive advancing some 144 kilometers within 13 days. With the Russian Empire increasingly in internal chaos since March 1917, the German General Staff planned to strike decisive blows against the Russians to bring about complete disruption of their war effort. The capture of Riga was a key step in striking these blows. The northern portion of the eastern front was along the Daugava (Düna, Western Dvina) River with German forces holding the southern bank and Russian forces holding the northern bank. The Russian 12th Army, consisting of eight divisions, was still holding a bridgehead on the south bank of the Daugava west of Riga. As it presented a threat, the German General Staff decided to try to eliminate the bridgehead, but it could not be done by merely launching a frontal attack. For this

---

[8] Anderson, "The Military Situation in the Baltic States," 118.
[9] Aleksandr Kerensky (1881–1970) was a moderate socialist revolutionary who served as head of the Russian Provisional Government from July to October 1917. Before becoming the leader of the provisional government, he had been a member of the Duma and a prominent lawyer, frequently defending revolutionaries accused of political offenses.

reason, the Germans decided to cross the river east of Riga, at a point in the river that was 410–500 meters in width, with the idea of capturing the city and cutting off the bridgehead garrison from a northeasterly direction. The river crossing was successfully carried out on 1 September and Riga fell to the Germans soon after. However, the Russian 12th Army was able to withdraw from the bridgehead and establish new front lines along the Gauja (Livländische Aa) River north of Riga. The northern portion of the front was secured by the Germans and the capture of Riga was a considerable blow to the Russian side due to the long and stout defense put up by the Latvian rifle regiments of the 12th Army. Anger in Latvian ranks grew as they blamed the Russian High Command and the Provisional Government for the disaster. While the flank on the land was secured for Germans, they did not control the Gulf of Riga and their advance northward could still be menaced from the Baltic Sea. There were two channels affording access to the Gulf of Riga: Irbe Strait, approximately 27 kilometers in width and situated between the southernmost point of the Island of Saaremaa and the Kurzeme (Courland) coast.[10]

As the situation unfolded in September 1917, the Russians decided they would concentrate all available naval forces into Muhu (Moon) Sound and the Gulf of Riga as the means to disrupt German offensive land operations northward. The Russian right flank of the land front was protected from the sea, while the situation of the Germans was comparatively difficult because their left flank, ending on the coast, was constantly under the danger of being attacked from the sea. To eliminate this danger, the Germans had to obtain the control of the Gulf of Riga. For this, it was necessary to be the master of the two entrances: the Irbe (Irben) Strait and the Suur (Great) Sound. Capture of Saaremaa (Ösel) and Muhumaa (Moon) would enable German control of the two straits. Thus, on 19 September 1917, German emperor Wilhelm II issued the following order:

> *In order to control the Gulf of Riga, and for the purpose of affording protection to the flank of the field forces in the east, the islands of Ösel [Saaremaa] and Moon [Muhu] will be captured in a joint attack by the land and naval forces; moreover, the Great [Suur] Sound will be blocked so that hostile naval forces cannot pass through it.*[11]

Lieutenant General Oskar von Hutier, commander of the German *8th Army*, was charged with the overall direction of the operation, for which was given the codeword "Albion." While the orders mentioned only the protection of the flank of field forces, the German leadership were looking for Operation Albion to have a larger opera-

---

[10] Erich von Tschischwitz, *The Army and Navy in the Conquest of the Baltic Islands in October 1917*, trans. Henry Hossfield (Fort Leavenworth, KS: Command and General Staff School Press, 1933), 2–3; and Anderson, "The Military Situation in the Baltic States," 118.
[11] von Tschischwitz, *The Army and Navy in the Conquest of the Baltic Islands in October 1917*, 5.

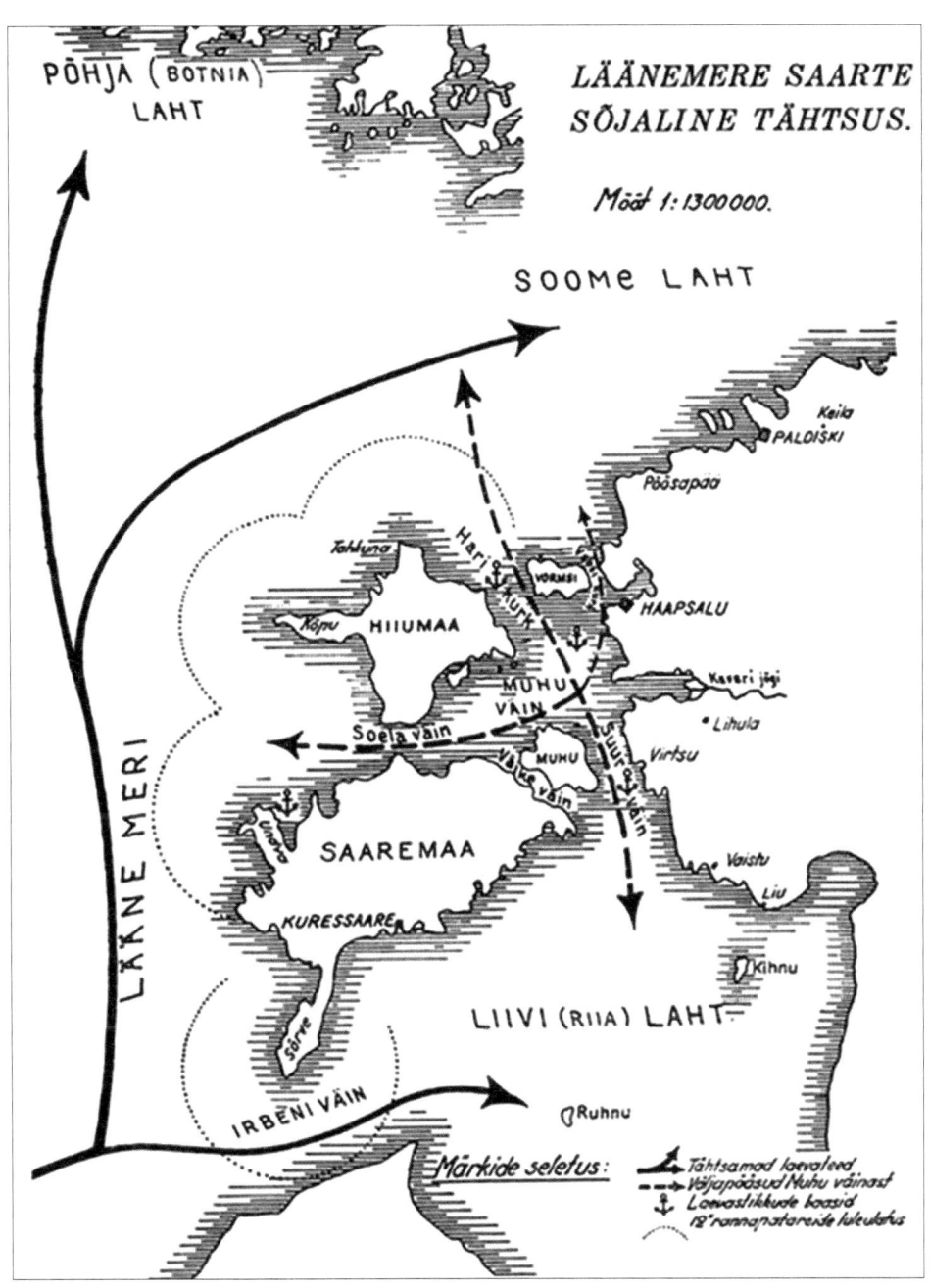

MAP 2
Sealanes, channels, and defensive positions.
*Source: Nikolai Reek, Saaremaa Kaitsmine Ja Vallutamine A. 1917
(Tallinn: Tallinna Eesti Kirjastusühisus, 1937)*

tional and strategic effect, ending the submarine threat to ore supplies coming from Sweden and ultimately ending the war on Germany's eastern front.[12]

THE MUHU SOUND FORTIFIED POSITION

The area of the main Estonian islands—Saaremaa (Ösel), Muhu (Moon), and Hiiumaa (Dagö)—equaled 3,972 square kilometers, with the largest of the islands Saaremaa at 2,714 square kilometers. Saaremaa (Ösel) had a series of peninsulas jutting out to sea, allowing for construction for coastal artillery positions that project fire far into the Baltic. For the defense of the islands, the Russian armed forces formed an extensive defensive organization that consisted of coastal batteries, land forces, and naval minefields. The land forces, known as the Muhu Sound Fortified Position (*Moonzundskaya ukreplennaya positziya*), were a joint force in the command structure of the Russian Baltic Fleet. The position's commander was Rear Admiral Dimitry Aleksandrovich Sveshnikov, a former cruiser captain and his chief of staff was the army staff captain Nikolai Reek, a 27-year-old native of Tallinn. Reek would go on later to have an influential career with the Estonian Army. The Estonian islands had approximately 60,000 inhabitants, with the Estonians forming the greater part of the population and the Baltic Germans forming a smaller minority. Both the Baltic Germans and the Estonians had been loyal to the Russian Empire's war effort, but by autumn 1917, Estonians were looking toward national independence and the Baltic Germans increasingly saw their future with the German Empire.[13] The Estonians formed their first national military unit on 25 April 1917, the 2d Naval Fortress Regiment that in May 1917, became the 1st Estonian Infantry Regiment, two battalions of which would take part in defense of the islands.[14] The islands had no great elevations, so Saaremaa (Ösel) as the largest island that afforded the ability to deploy sizable ground forces with space to maneuver. Total land forces on Saaremaa consisted of two infantry divisions and 20 heavy coastal guns.[15] The heavy coastal guns on Sõrve Peninsula represented a critical defensive capability of fortified position as it controlled the Irbe Strait, the gateway to the Gulf of Riga; thus, the gun positions were organized into the special

---

[12] Nikolai Reek, *Saaremaa Kaitsmine Ja Vallutamine A. 1917* [The Defense and Conquest of Saaremaa in 1917] (Tallinn: Kindralstaab IV Osakond, 1937), 5-6; and von Tschischwitz, *The Army and Navy in the Conquest of the Baltic Islands in October 1917*, 5.

[13] The Baltic Germans were decedents of Teutonic Order and formed the ruling aristocracy and landowners. By 1914, the Estonians owned their own farms vice being tenant farmers and entered a growing professional and mercantile middle class. The inhabitants of the islands made their livings from raising livestock and crops, fishing, boat building, shipping, and commerce. Prior to Estonian autonomy, the islands were administratively under the province of Livonia. Zigmantas Kiaupa, *The History of the Baltic Countries* (Tallinn, Estonia: Avita, 1999), 130; and von Tschischwitz, *The Army and Navy in the Conquest of the Baltic Islands in October 1917*, 12.

[14] Anderson, "The Military Situation in the Baltic States," 118.

[15] Land forces on Saaremaa consisted of 9 infantry battalions, 4 cavalry squadrons, equipped 108mm machine guns, 24 trench mortars, 6 heavy mortars, 42 light field guns, 4 heavy field guns, 44 antiaircraft guns, and 3 companies of marine guards with 4 machine guns. Reek, *Saaremaa Kaitsmine Ja Vallutamine A. 1917*, 19-20.

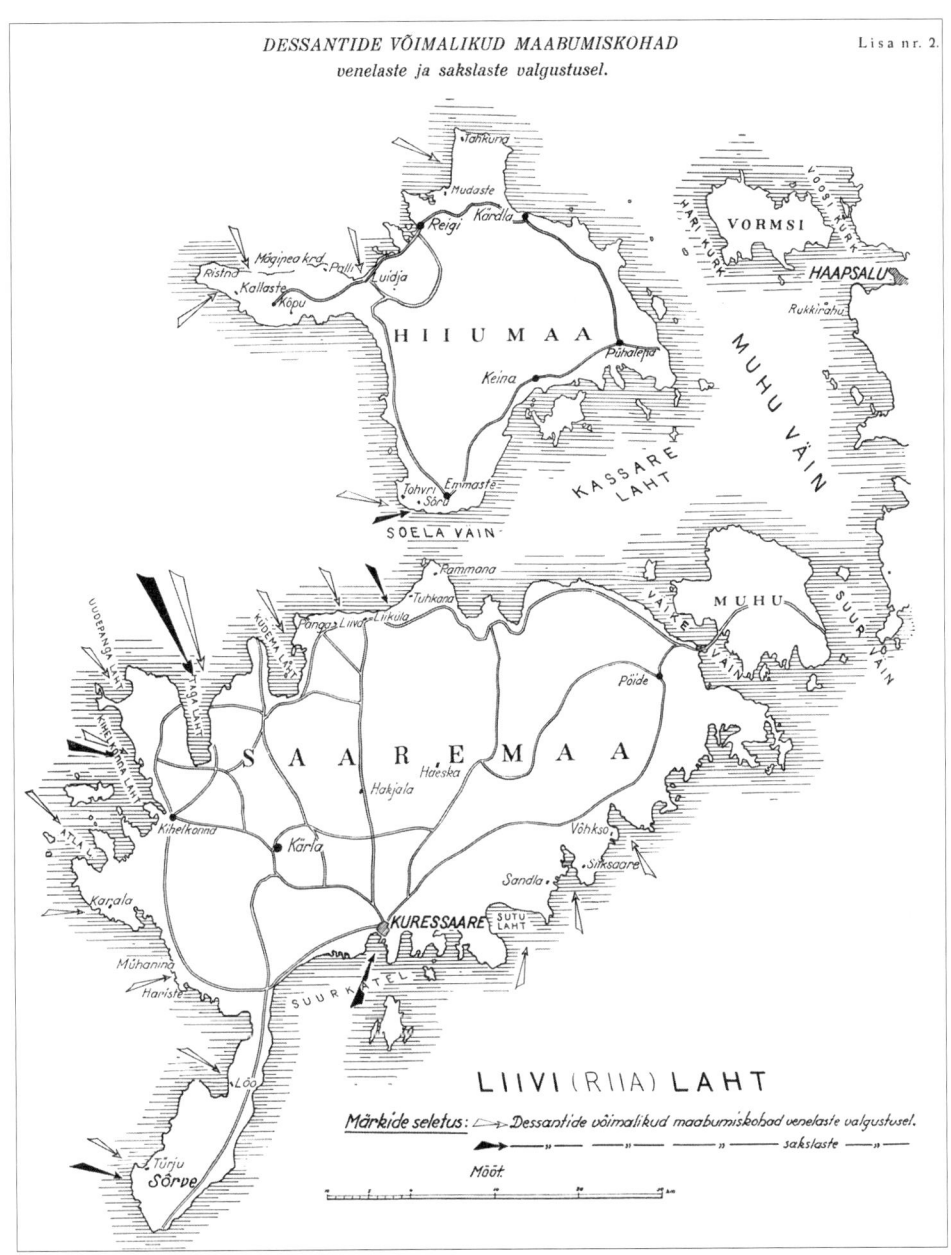

MAP 3
Road network with white arrows marking potential German landing zones as wargamed by Russian staff and black arrows marking German planned landing zones.
Source: Reek, Saaremaa Kaitsmine Ja Vallutamine A. 1917

autonomous sector. The special autonomous sector was under a separate commander and logistics were organized separately so that it would be able to operate independently if it was cut off from the rest of Saaremaa by the Germans. The units in the special autonomous sector consisted of 4th Coastal Defense Artillery Battalion and

Operation Albion

the marine guards.[16] The commander of the special autonomous sector, Russian Navy captain M. G. Knüpfer, had the task of defending the Irbe Strait and harmonizing his action with the activities of the Russian fleet. The guns were exposed, with the magazines and shelters protected only against smaller aerial bombs.[17] While the special autonomous sector had its self-contained logistics, supporting the rest of Saaremaa was a difficult task due to the comparatively large size and long coastline of the island making movement and communications difficult.

The hub of transportation and communication on Saaremaa was in Kuressaare (Arensburg), the provincial capital. It was the only sizable town on Saaremaa with approximately 5,000 inhabitants. Kuressaare had an electric power plant and submarine telegraphic cables connecting it with Pärnu on the mainland. Kuressaare was directly served by two harbors: the old harbor and the port at Roomassaare. The old harbor was shallow and filled with silt, thus suitable only for launches and fishing boats. Roomassaare was about 2.2 kilometers south of the Kuressaare with a new quay accommodating vessels with a draught up to three meters. Five highways led from Kuressaare, two in the direction of Muhu; one north toward the Pammana Peninsula; one to Kihelkonna, situated on the northwest coast; and one on the southernmost point of the Sõrve Peninsula. The most direct route from the road hub in Kuressaare to the mainland was the post road to Orissaare, from where the 3.5-kilometer stone causeway took it to Muhu. The post road continued from the causeway to the east coast of Muhu, where a steam ferry ran 7.2 kilometers across Muhu Sound to Virtsu on the mainland.[18] This route was the main route of supply and reinforcement for Russian forces and defensive position on Saaremaa. No railways were constructed on Saaremaa or Muhu, and the main supply route was dependent on eight often-inoperable motor trucks and three widely dispersed horse transport units that lacked healthy horses. Consequently, it was not possible to use the transport units for quickly moving reserves. Telephone and telegraphic communications were also a problem because of shortages of materials and skilled technical personnel. The very long wires connecting outlying units were difficult to repair quickly and messages overloading the submarine cable to the mainland made communications very difficult.[19]

Not all communications and movement problems were due to a lack of material or infrastructure. The war weighed heavily on the leadership of the Russian Army. Junior leadership suffered particularly because of high casualties and replacement officers had to be quickly trained. The company commanders in the two Russian divisions on the islands were mainly ensigns who had finished officer training during

---

[16] 4th Coastal Defense Artillery Battalion with four 12-inch coastal guns. Reek, *Saaremaa Kaitsmine Ja Vallutamine A. 1917*, 21.
[17] Reek, *Saaremaa Kaitsmine Ja Vallutamine A. 1917*, 14-16.
[18] Reek, *Saaremaa Kaitsmine Ja Vallutamine A. 1917*, 9; Karl Schlossmann, *Estonian Curative Sea-Muds and Seaside Health Resorts* (London: Boreas, 1939), 36-37; and von Tschischwitz, *The Army and Navy in the Conquest of the Baltic Islands in October 1917*, 13.
[19] Reek, *Saaremaa Kaitsmine Ja Vallutamine A. 1917*, 102-8.

the war. The number of experienced senior officers was very limited. The swampy and thickly forested terrain on Saaremaa presented a high requirement for unit maneuver for which the leadership was unprepared.[20]

Support of air operations were also affected by movement problems between outlying air stations and depot/workshop facilities and shortages of materiel and skilled mechanics. Aircraft operating in the Muhu Sound Fortified Position's area belonged to the Russian naval air service. The two main air stations in the area were at Kihelkonna on west coast of Saaremaa and the other at Haapsalu (Hapsal) on the Estonian mainland. Haapsalu served as the headquarters and depot for seaplane stations at Tahkuna and Kõrgessaare on Hiiumaa (Dagö). The Kihelkonna Air Station was a well-prepared facility, it served as the headquarters and depot for naval aviation on Saaremaa.[21] Aircraft engines requiring major repairs were brought to the workshops at Kihelkonna from other air units on the island. The station was defended by antiaircraft guns, which were placed so that they could also fire out to sea on surface targets. However, the station was situated outside of the positions of land forces on Saaremaa, so it was not actually defended in the event of a German landing elsewhere. Aircraft maintenance was difficult as there were deficiencies in technical training of personnel and a shortage of skilled mechanics, supplies, spare parts, and tools. While there were a good number of skilled pilots among naval aviation personnel, they often could not put their abilities to use due to the constant aircraft maintenance problems. The morale of aviation officers was low, according to a British Royal Flying Corps report in August 1917, due to "the entire absence of authority on the part of any senior officer at any station, and . . . slackness and indifference shown by other officers."[22] The Russian naval vessels committed to support the Muhu Sound Fortified Position suffered from similar maintenance problems to the air arm.[23]

The Russian Baltic Fleet committed to operations in the Gulf of Riga and Muhu Sound with 121 vessels of different types. The largest ships were the older battleships *Slava* (1905) and *Grazhdanin* (1903, originally *Tsesarevich*). They were older and smaller predreadnought battleships that could go through the dredged channel in Muhu Sound due to their small draught. Attached to the Russian Navy were also three British C-class submarines. With the limited range of the C-class boats, the British had established an advanced base at Rohuküla (Rogokul) on the Estonian mainland six kilometers south of Haapsalu. Rohuküla was nearer than Tallinn to the Muhu chan-

---

[20] Reek, *Saaremaa Kaitsmine Ja Vallutamine A. 1917*, 35–36, 105.
[21] Facilities included large hangars, workshops, storehouses, a radio station, an independent water works, an electric power plant, and petroleum stores. Reek, *Saaremaa Kaitsmine Ja Vallutamine A. 1917*, 106.
[22] "Osel Island Naval Air Station, Reports on, with General Remarks on Russian Air Services," 5 November 1917, AIR 1/36/15/1/241, United Kingdom National Archives, hereafter Osel Island Naval Air Station reports.
[23] Reek, *Saaremaa Kaitsmine Ja Vallutamine A. 1917*, 105; and Osel Island Naval Air Station reports.

nel and Gulf of Riga.[24] While the submarines could sortie the central Baltic to operate against German shipping, the primary task of the rest the naval force was to maintain control of the Gulf of Riga. Minefields closing the Irbe Strait were key in this task. The Russian fleet laid mines at the beginning of the war and continuously renewed and improved the minefields. However, by autumn 1917, maintaining the minefields and other Russian naval operations were nearly paralyzed due to disorder within the ranks and a lack of morale. Crews did not put the necessary emphasis on maintenance equipment and materiel. Discipline was entirely lacking, and crews did not trust their officers. Daily shipboard political meetings and negotiations by semaphore and signal lamps with other ships took away from critical tasks and kept things in a constant state of tension. Incapable of establishing and maintaining the discipline, many naval officers had become apathetic or abandoned the ships, leaving others to take on an overwhelming number of additional tasks.[25] The Imperial German Navy had its own morale problems, and this became a consideration in the German planning for Operation Albion.

GERMAN PLANNING

One of the considerations for launching Operation Albion as a major amphibious effort was to engage the German fleet, as much of it was inactive in port and the morale among the ranks was plummeting. Germans had little experience with amphibious operations to draw on for planning. Crossing the Daugava in September 1917 and the Danube and other rivers did give them some experience moving a large force across a body of water relevant to the disembarkation of forces, such as using horse boats that had a ramp in the bow. The horse boats provided a comfortable platform for landing horses or vehicles.

Each boat could carry 70 soldiers with full equipment, or 10 horses, or 2 field guns with ammunition, or a 6-inch artillery piece. Difficulties in landing a large number of horses or motor vehicles led to the use of bicycle-equipped light infantry, which could be loaded easily in conventional landing boats pulled by lighters. Once on Saaremaa, they could move quickly over relatively large distances. The deployment of bicycle troops was relatively new and had never been used by an amphibious landing force.[26]

As the German joint staff considered the places for landing on Saaremaa, Taga Bay (Tagalaht) was quickly determined to have the most advantages. Capturing Roomassaare quay was initially considered as ships could be readily off loaded at the port facility. However, Sõrve coastal batteries would first have to be silenced to penetrate the Irbe Strait and thus surprise would have been lost. Therefore, the idea

---

[24] Described as "an unlovely place," meager support facilities at Rohuküla consisted of a pier and fuel and ammunition storehouses in Wilson, *Baltic Assignment*, 106; and Reek, *Saaremaa Kaitsmine Ja Vallutamine A. 1917*, 102–3.
[25] Reek, *Saaremaa Kaitsmine Ja Vallutamine A. 1917*, 102–8; and Wilson, *Baltic Assignment*, 160.
[26] Bruce I. Gudmundsson, *On Armor* (Westport, CT: Greenwood Publishing, 2004), 42–43.

FIGURE 1
German troops and their horses disembarking from a horse boat at Taga Bay.
Source: Imperial War Museum photo IWM Q 87079

of landing at Roomassaare was abandoned. On the western coast of Saaremaa Taga Bay was determined as the best location as it allowed troop transports to approach very near to shore and it allowed ready fire support from the fleet. Taga Bay was sheltered against the dominating autumn westerly winds. Furthermore, the beach in the bay and the terrain features immediately inland were favorable for the landing as good roads led to Kuressaare. Taga Bay was 300 kilometers away from Liepāja (Libau) in Latvia (the main embarkation port), 60 kilometers overland from Kuressaare, and 120 kilometers from the southern-most tip of the Sõrve Peninsula. A drawback of landing at Taga Bay was that it presented a long open sea crossing for the transport fleet. Additionally, the entrance was guarded by two Russians coastal batteries, which had to be silenced before landing operations could commence. The second alternative considered for the landing was the Pammana region. Pammana was quite favorable in terrain for a landing, but it was more open to the winds than Taga Bay. From Pammana, the roads went toward Kuressaare and the causeway to Muhu. Given these considerations, the German concept of operations was as follows: (1) main landing in Taga Bay with forces moving inland in the direction of Kuressaare; (2) secondary landing at Pammana with forces moving inland in the direction of Orissaare and light forces to cut the causeway to Muhu; and (3) naval bombardment of Kihelkonna and on the western coast of the Sõrve Peninsula as a deception to keep Russian forces off balance. The landings required surprise, speed, and decisive action once on shore and very strong cooperation between the army and navy.[27] Therefore, the Germans formed a special army and navy command that included a joint air command. The army component consisted of an expeditionary corps and had as its nucleus the *42d Infantry Division*, which had participated in the crossing of the Daugava in September 1917. The light infantry bicycle companies of this force would push quickly along roads to interdict Russian reinforcement or withdrawal and serve as a mobile reserve. Portable radio transmitters would allow units to coordinate actions once landed. The 24,600 personnel, 8,500 horses, 2,500 vehicles, and 55 guns of the expeditionary corps would be transported in two echelons.[28] In addition, the expeditionary corps needed great quantities of ammunition and engineering material as well as subsistence for 30 days, which represented 2,300 tons alone.[29] To gain sea control and land the expeditionary corps, the German Navy organized a force of 181 ships, 124 small motor vessels, 94 aircraft, and 5 airships. The backbone of the naval force was 10 of the most modern battleships of the *König* and *Kaiser* classes. Because of their heavy caliber guns, they were effective against coastal batteries. The additional value of the battleships was that they had well engineered watertight compartments, thus mine explosions presented minimal damage to them. The main tasks for the air component included reconnaissance, close air support, bombing, and screening. The air command made well-organized arrangements for dropping messages from airplanes to ground troops so that air reconnaissance information could be quickly relayed to ground forces. With their air strength, the Germans would have continuous surveillance over the area of operations. Airships gave strategic reach to the reconnaissance and provided the capability for long-range bombing raids. The large Friedrichshafen FF41A seaplanes would carry out long range reconnaissance and bombing as well as aerial mining and even the carrying of troops and supplies.[30]

OPERATION ALBION EXECUTED

The troops of the expeditionary corps embarked at Liepāja, which had harbor facilities adequate for the purpose of accommodating not only the fleet of transports but also the numerous mine-hunting and mine-sweeping flotillas, together with other

---

[27] Reek, *Saaremaa Kaitsmine Ja Vallutamine A. 1917*, 112–13; and Cdr William C. I. Stiles, "The German Operation against the Baltic Islands" (thesis, Army War College, 1930), 6; and von Tschischwitz, *The Army and Navy in the Conquest of the Baltic Islands in October 1917*, 30–31.

[28] Weaponry of the expeditionary corps included 220 machine guns and 84 mortars. Stiles, "The German Operation against the Baltic Islands," 6.

[29] Capt G. von Kobinski, German Navy (Ret), "The Conquest of the Baltic Islands," U.S. Naval Institute *Proceedings* 58, no. 7 (July 1932): 976.

[30] "The conquest of the Baltic Islands: Translation of Vice-Admiral Schmidt's dispatch November 1919," ADM 186/594, United Kingdom National Archives; and Reek, *Saaremaa Kaitsmine Ja Vallutamine A. 1917*, 109.

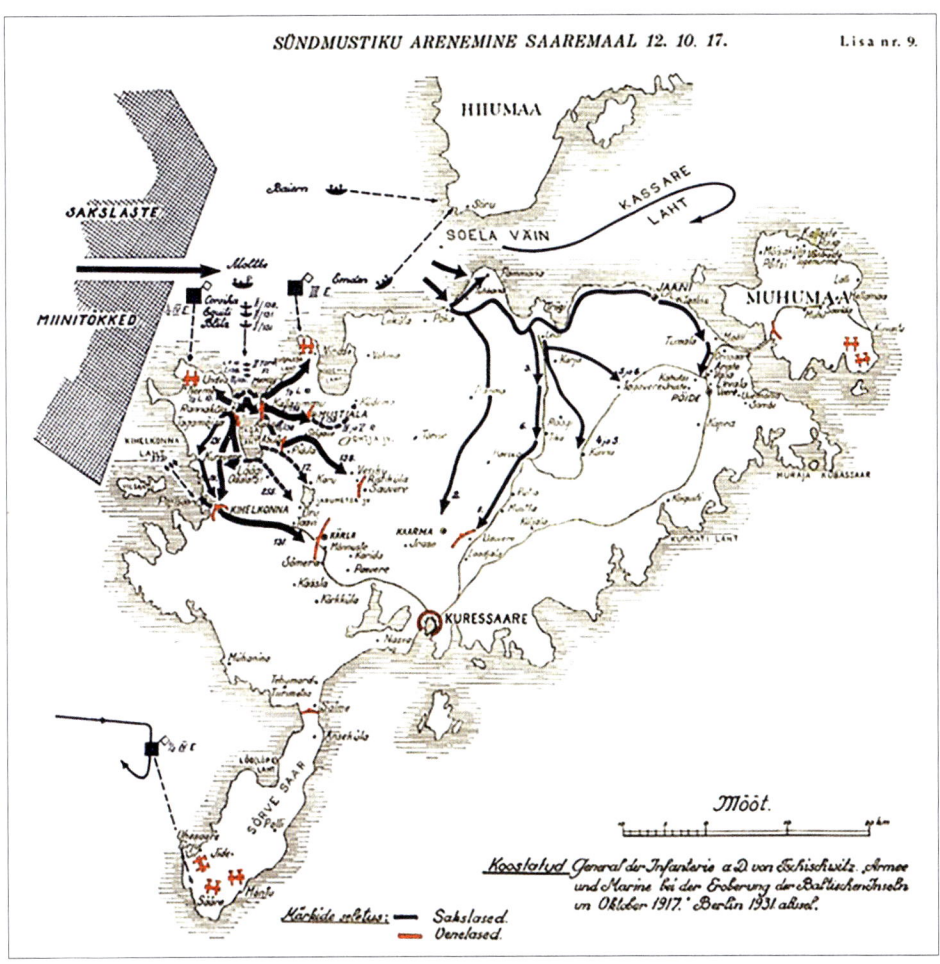

MAP 4
German landings, 12 October 1917, with German forces in blue, Russian forces in red.
Source: Reek, Saaremaa Kaitsmine Ja Vallutamine A. 1917

units.³¹ The preparatory period and concentration of forces took place between 21 September and 8 October, while the actual embarkation of forces took place 8-10 October. German intelligence spread information that the actual destination of the expeditionary force was Kronstadt and, as preparations were taking place in Liepāja, German aircraft bombed gun positions on the Sõrve Peninsula. Airships kept observation over the Gulf of Riga and the entrance to the Gulf of Finland as well as carrying out air raids on Pärnu and Viljandi on the Estonian mainland. On 11 October,

---

³¹ Liepaja had served as principal base of the Russian Baltic Fleet before its capture by the Germans. At Liepaja, facilities offered protection against attack, and the available wharves ensured rapid loading during the process of embarkation.

FIGURE 2
German battleship SMS *Bayern*, accompanied by a Zeppelin on scouting duties, en route, October 1917.
Source: *Imperial War Museum photo IWM Q 87082*

the naval forces with transports and supply ships departed Liepāja for the crossing to the islands. The advance through a channel cleared of minefields went without incident, and the transport and escorts arrived at Taga Bay at 0300 on the morning of 12 October. At 0530, the landing began when German battleships opened fire on the Russian coastal batteries defending the entrance of Taga Bay. The batteries were quickly silenced and in the hands of German troops.[32]

At 0845, the transports were ordered into Taga Bay, and by 1000 the disembarkation was in full swing. As the transports entered Taga Bay, German destroyers opened fire on the Kihelkonna Air Station. About the same time, two battleships opened fire on the western coast of the Sõrve Peninsula. German aircraft also joined the operations against Kihelkonna. Despite naval bombardment and presence of German aircraft, some Russian seaplanes succeeded in taking off to attack the German ships, but they were quickly driven off. Nevertheless, the Russians were able to carry out air reconnaissance and, based on the location of German forces, the Russian headquarters at Kuressaare was able to discern the German concept of operations.[33] The reaction of the Russian command was to try to reinforce from the mainland to undertake a counteroffensive to throw the German expeditionary corps back into the sea. As the landings started, Admiral Sveshnikov left Kuressaare for Haapsalu to organize reinforcements, leaving Captain Reek in Kuressaare to direct command up to the last possible moment. Reek knew the German course of action as the Russian staff had wargamed German landings on Saaremaa and actions generally followed a predictable course. The Germans came to their decision to use bicycle troops through their own wargaming of the problem. The only great surprise was the use of bicycle troops, which gave unexpected speed to the German advance to the causeway.[34] Two battalions of bicycle troops went ashore along with the *18th Shock Company* and a naval landing party with secondary landing force landed in the Pammana region between Liiküla and Tuhkana.[35] The landing started at 0830 and was not met with any opposition. The cyclists moved quickly in the general direction of the town of Orissaare and the bridgehead to the Muhu causeway. To advance more quickly, the *18th Shock Company* rode carts taken from the local residents.[36]

By the evening of 12 October, Germans landed four infantry regiments, three bicycle battalions, and one artillery battery ashore. These forces advanced 10-12 kilometers from the beachhead and captured the Kihelkonna Air Station. Meanwhile, German forces continuously came ashore. The bicycle battalions held Orissaare while the Russians still held the causeway. In the next two days, German forces advanced southward from Taga Bay toward the Sõrve Peninsula. Moving quickly, the Germans succeeded in cutting off the peninsula, which caused the main Russian forces on the island and the two infantry divisions in Kuressaare to withdraw in disorder toward

---

[32] Reek, *Saaremaa Kaitsmine Ja Vallutamine A. 1917*, 119-24; and von Tschischwitz, *The Army and Navy in the Conquest of the Baltic Islands in October 1917*, 30-31.
[33] Reek, *Saaremaa Kaitsmine Ja Vallutamine A. 1917*, 125.
[34] Reek, *Saaremaa Kaitsmine Ja Vallutamine A. 1917*, 135-36.
[35] A total of 1,900 troops landed.
[36] Reek, *Saaremaa Kaitsmine Ja Vallutamine A. 1917*, 126-31; and von Tschischwitz, *The Army and Navy in the Conquest of the Baltic Islands in October 1917*, 61-62.

FIGURE 3
German troops going ashore at Saaremaa.
Source: German Federal Archives, Sammlung von Repro-Negativen (Bild 146)

Muhu. The German cyclists and shock troops held their position in Orissaare with the Russians holding the Muhu end of the causeway. Withdrawing Russian forces from Kuressaare concentrated in the Pöide region just south of Orissaare, which initiated desperate actions by the Germans to hold out at Orissaare and for the Russians to breakthrough and hold the causeway. However, Russian forces around Pöide gave up hope as the Germans closed in from two directions. On the afternoon of 15 October, the commander of the Russian 107th Infantry Division gave permission to all officers for their units to surrender. Those who did not want to surrender could attempt to penetrate the German lines and escape.[37]

By 16 October, Saaremaa fell entirely under the control of the Germans. The continuous naval shelling and air attacks had demoralized the Russian forces trapped on the Sõrve Peninsula. That morning, the Russian 425th Infantry Regiment surrendered, allowing the Germans to gain control of the entire Sõrve Peninsula, its coastal batteries, and the Gulf of Riga. The next morning, the Germans completed sweeping the Irbe Strait so they could send a force of 28 ships, including the battleships SMS *König* (1913) and *Kronprinz* (1914), into the Gulf of Riga. To boost the morale of defenders on Muhu, the Russian vessels on the gulf engaged the German Navy, despite their material inferiority, before withdrawing through Muhu Sound. The battleships

---

[37] Reek, *Saaremaa Kaitsmine Ja Vallutamine A. 1917*, 191–92; and von Tschischwitz, *The Army and Navy in the Conquest of the Baltic Islands in October 1917*, 99–102.

Slava and *Grazhdanin*, the armored cruisers *Admiral Makarov* (1908) and *Bayan* (1902), and 13 destroyers engaged in a running battle. The battleship *Slava* received a series of effective hits, dropped out of the line, and ran aground in Suur Sound. The crew abandoned the ship and blew up the remaining ammunition. The Russian force withdrew, blocking the Muhu Sound by sinking a number of blocking ships in the channel. Despite the fact that the bulk of the Russian naval force had successfully escaped, the naval battle had a paralyzing effect on the defenders of Muhu. Seeing the ships retreating, some of which were burning, the defenders lost the last bit of hope. On the morning of 18 October, the five battalions defending Muhu surrendered to the Germans.[38] The Russians planned to evacuate their forces from Hiiumaa. However, as Muhu forces surrendered, Hiiumaa's defenders left their positions to await transports arriving on the eastern shore of the island to take them to the mainland. With the delay of the evacuation transports, panic set in and the Russian forces surrendered to German forces that had landed at Pammana. The German capture of the two other small islands warrants mention. Ruhnu, in the center of the Gulf of Riga, and Abruka, south of Kuressaare, were occupied on 13 and 15 October, respectively. Friedrichshafen FF41A naval aircraft accomplished this by landing troops in perhaps the first air assault in history.[39]

Operation Albion ended with German losses of about 400 troops, including naval personnel. The Russian casualties were relatively light as well, despite losing a strategically key location. The Germans captured 20,000 Russian prisoners and 140 artillery pieces. Russian naval losses were light as only the battleship *Slava* and the destroyer *Grom* (1916) were lost.[40]

AFTERMATH

Despite its low cost in lives and material, Operation Albion achieved great strategic effects. Capturing the islands opened the route to the Russian capital of Petrograd, which was the ultimate German strategic goal associated with Operation Albion. In danger of attack from the rear, Tallinn's fortifications protecting the entrance to the Gulf of Riga were evacuated. On 19 October, the Provisional Government made the announcement that the Russian capital was to be moved from Petrograd to Moscow illustrating the gravity of the situation for the Russians. German possession of the Estonian islands put Petrograd within range of German air attacks.[41] According to historian Eduard Laaman, who the witnessed events,

The Russians saw this assault on the Estonian islands as the opening of the gates

---

[38] The defenders of Muhu included two battalions each of the 470th Infantry and the 471st Infantry, and two battalions of the 1st Estonian Regiment and the Death Battalion, which consisted of volunteers only.
[39] von Tschischwitz, *The Army and Navy in the Conquest of the Baltic Islands in October 1917*, 193.
[40] Reek, *Saaremaa Kaitsmine Ja Vallutamine A. 1917*, 195-206; and von Tschischwitz, *The Army and Navy in the Conquest of the Baltic Islands in October 1917*, 184-93.
[41] "Peace with Russia May Be German Goal: Operations in Baltic Possibly Have This End in View as Well as the Influencing of Sweden by Seizing Aland Islands," *New York Times*, 21 October 1917.

to Petrograd. A mindless panic seized the centers of Russian state power, the Bolsheviks took advantage of this and carried out their coup d'état a few weeks later and then immediately asked for a truce.[42]

With control of the Gulf of Riga, the German ore shipments from Sweden vital for war industries were protected from Allied interference. It also opened the Äland (Ahvenanmaa) Islands to Swedish occupation, which culminated with landing operations from 18 February to 2 March 1918. Despite the 1856 dictates of the Treaty of Paris, the Russians had established a submarine forward operating base in the Älands to project power to the mouth of Gulf of Bothnia, which was now lost.[43]

Control of the Gulf of Riga also secured the left flank of the German land forces on the Riga front, while endangering the right flank of the Russian lines manned by the Latvian riflemen along Gauja (Livländische Aa) River. In late October 1917, the Latvian riflemen were partially pulled off the front lines and were now in Petrograd, Moscow, and on various important points on the Russian railway network. Despite widespread embitterment in the ranks at the Russian Provisional Government for heavy casualties and the disastrous fall of Riga, the Latvians, for the most part remained a disciplined force as the Russian Army largely disintegrated. Bolshevik leader Vladimir I. Lenin was able to convince Latvian riflemen commander Jukums Vācietis to support the Bolshevik power play. As the Bolsheviks seized key buildings in Petrograd, the Latvian rifle regiments took control of key railway junctions to prevent the movement of troops to Petrograd to thwart the coup d'état. Due to the demoralized condition of the Russian Army, the Latvians accomplished their task with ease.[44] On 25 November 1918, Bolshevik leader Leon Trotsky negotiated the Trea-

---

[42] Laaman, "Langemine 20 aasta eest," 978.
[43] Part of Finland, ethnic Swedes inhabited the Åland Islands, which the 1856 Treaty of Paris had demilitarized. However, in 1914, the Russian Empire disregarded the treaty and fortified the islands with 10 coastal artillery positions, two piers for submarines, two airfields, barracks, and a telephone system connecting the installations. The Russians established a forward operating base for submarines supported by submarine tender *Svjatitel Nikolai* for the use of British and Russian navies in the archipelago. This treaty violation was greatly resented in Sweden, where there was growing pro-German sentiment. On 15 February 1918, the Swedes landed 700–800 troops on the strategic islands using the icebreaker *Isbrytaren I*, the gunboat *Thor*, and the transport steamer *Runeberg*. They were later reinforced by the gunboats *Sverige*, *Svenkund*, and *Oscar II*. Some 1,200 Russian soldiers were disarmed in the islands. The British submarine campaign in the Baltic came to an end, as on 1 April 1918 Germans landed a force of 13,000 troops under Prussian general Rüdiger von der Goltz in western Finland. The British submarines had been harbored in Helsinki and, with the Germans on the way, LtCdr Francis Newton Allen Cromie oversaw the towing of the seven submarines out of Helsinki harbor into the Gulf of Finland, where they were scuttled between 3 and 8 April 1918. The final 30 British bluejackets in Finland departed by rail to Murmansk for evacuation back to Britain. "Peace with Russia May Be German Goal"; Lauri Sauramo, "Ahvenanmaan sotilaallinen ja sotilaspoliittinen merkitys" [The military and defense significance of the Ahvenanmaa Islands], *Tiede ja Ase* 5 (1937): 198–99; and Macintyre, "A Forgotten Campaign—IV," 559.
[44] Edgar Anderson, "The Role of the Latvian Riflemen during the Russian Civil War," *Strenlnieks*, nos. 34-35 (1974): 7-10; and Uldis Ģērmanis, "Zemgallian Commander: Colonel Vācietis and the Latvian Riflemen in World War I and the October Revolution," *Jaunā Gaita*, no. 92 (1973).

ties of Brest-Litovsk with the Germans, which allowed Germany to move the bulk of their forces from the eastern front to the western front in December 1917.[45] Although Russia was now knocked out of the war, it came too late for Germany as the entrance of American manpower and resources tipped the balance to the Triple Entente side.

THE LEGACY OF OPERATION ALBION

Although peace with Russia did not result in a German victory in World War I, Operation Albion was widely studied in 1920s and 1930s as an amphibious operation. It stood in sharp contrast to the failed British amphibious operations at Gallipoli from 25 April 1915 to 9 January 1916. As it was the most successful example of amphibious landings in the war, the Americans—Army and Marine Corps—the British, Germans, Argentinians, Swedes, Danes, Estonians, Soviets, and Japanese all studied Operation Albion during the interwar period. What makes it somewhat unique as subject of study in professional military education is that chiefs of staff for both sides, Erich von Tschischwitz and Nikolai Reek produced detailed, operationally oriented accounts available in the English language.[46] These records later served as the basis of case studies and battlefield staff ride prereading. During World War II, various powers heeded the example of Operation Albion to differing extents. American planners were perhaps more under the influence of French *bataille conduite* (methodical battle) concepts adopted into American doctrine.[47] With the post–Vietnam War military reform movement in the U.S. armed forces, interest in Operation Albion was renewed, notably from reform movement luminaries, military theorist William S. Lind and

---

[45] von Kobinski, "The Conquest of the Baltic Islands," 984.

[46] The Army War College offers a translation of von Tschischwitz, *The Army and Navy in the Conquest of the Baltic Islands in October 1917*, from the German, as does the Army Command and General Staff School version, and the original in German was published in 1931. The original version of Reek, *Saaremaa Kaitsmine Ja Vallutamine A. 1917*, is published in Estonian; the English translation of Reek, an unpublished typescript, most likely for the benefit of the British and perhaps the Japanese was recently published as an English translation of Reek's account of events in Art Johanson, *General Nikolai Reek Writings Including Operation Albion and Battle of Cēsis* (Tartu: Baltic Defence College, 2021). Reek became an important figure in the Estonian Army and was heavily involved in professional military education. In the immediate aftermath of Operation Albion, the Russian High Command gave him the task of writing the after action report for which he had copious notes and collected material. Secondary works on Operation Albion include Michael B. Barrett, *Operation Albion: The German Conquest of the Baltic Islands* (Bloomington: Indiana University Press, 2008); and Gary Staff, *Battle for the Baltic Islands 1917: Triumph of the Imperial German Navy* (Barnsley, UK: Pen & Sword Maritime, 2008).

[47] *Bataille conduite* (methodical battle) emphasized infantry advances in slow stages covered by massive artillery support. The United States replicated the concept during the interwar years and its accompanying process-focused education. Donald E. Vandergriff, "The US Army Culture Is French!," *Small Wars Journal*, 16 June 2018. As according to Mark E. Grotelueschen, "the results of the Army's Field Manual Project, begun in 1927 by then chief of staff Charles Summerall, led to the creation in 1930 of the *Manual for Commanders of Large Units* a document that relied heavily on French doctrine and advocated the French Army's firepower-based concept of 'methodical battle'." Mark E. Grotelueschen, "The AEF Way of War: The American Army and Combat in World War I" (PhD diss., Texas A&M University, 2003), 286.

Marine Corps colonel Michael D. Wyly. According to Lind, American amphibious operations in World War II were characterized by landings that came in waves to take a beachhead, followed by stopping and building up combat power for an advance resembling World War I land tactics such as those used at the Battle of the Somme (1916) applied to landing operations.[48] According to Wyly, such American operations usually focused on terrain and attrition, while Operation Albion focused on maneuver and the destruction of the enemy's will to resist. As a result, American casualties were often high despite heavy advantages in sea and airpower, manpower, firepower, and logistics.[49] In the Marine Corps Amphibious Warfare School (later Expeditionary Warfare School), for which Lind and Wyly developed curriculum, Operation Albion provided a ready example of German operational art, *Sturmtruppen* (storm trooper) tactics that led to so-called "blitzkrieg" tactical concepts and the practice of the German command philosophy of *auftragstaktik* or mission command in an amphibious environment. This shift came at a time when the Marines were adopting these German concepts and moving away from detailed command and *bataille conduite*. Operation Albion remains a relevant example of how armed forces can adapt to new and unfamiliar situations quickly. The operational improvisation of the Germans also remains a relevant example, as with the decline of specialized amphibious fleets, the improvised use of commercial shipping will be likely in future landing operations. As the development of modern antiaccess/area-denial (A2/AD) weapon systems have made World War II-style contested amphibious assaults and mass parachute drops largely obsolete, the "indirect approach" of Operation Albion, which focused on entering permissive landing zones and isolating enemy strong points, has taken on new value as an operational planning example.

---

[48] William S. Lind, "Operation Albion," *On War* #318, Defense and National Interest, 21 October 2009.
[49] Michael Duncan Wyly, "Landing Force Tactics: The History of the German Army's Experience in the Baltic Compared to the American Marines in the Pacific" (thesis, George Washington University, 1983), 717.

CHAPTER EIGHT
Beyond Cold Shores
Inland Maneuver in Historical Polar Amphibious Operations

Lance R. Blyth

The Arctic and Antarctic polar regions, along with their near-polar contiguous areas, are at risk of becoming sites of conflict, potentially requiring military forces to conduct polar campaigns.[1] Polar geography—the Arctic surrounds an ocean, Antarctica is surrounded by ocean—combined with limited infrastructure and the distance from power projection points means any polar campaign will include amphibious operations. Landings on such cold shores will require forces able to survive and be mobile in the extreme environment to maneuver inland. This chapter examines inland maneuver during three historical polar or near-polar amphibious operations: the Germans at Narvik, Norway, in 1940; the Americans and Canadians in the Aleutians in 1943; and the British in the Falklands in 1982. In each case, the author analyzes how the forces survived, maintained mobility, and maneuvered inland. Each case study reveals that the better a landing force was trained for mountain warfare under winter conditions, the better it performed polar and near-polar inland maneuver.

NARVIK, 1940

The Narvik landing force, consisting of the German *3d Mountain Division* Staff com-

---

[1] Ryan Patrick Burke, *The Polar Pivot: Great Power Competition in the Arctic and Antarctic* (Boulder, CO: Lynne Rienner, 2022). Also see Ryan Burke and LtCol Jahara Matisek, "The Polar Trap: China, Russia, and American Power in the Arctic and Antarctica," *Journal of Indo-Pacific Affairs* (October 2021): 36–64.

manded by General Eduard Dietl and the *139th Mountain Infantry Regiment* (a.k.a. *Battle Group Dietl*), embarked on 10 destroyers in the port of Hamburg on 6 April 1940, headed for Narvik, located more than 160 kilometers (km) north of the Arctic Circle. After a rough sea voyage, due to using a storm front for concealment from the British Royal Navy, during which waves swept the regiment's infantry guns off the decks, the naval task force entered the fjord leading to Narvik on the night of 8 April.[2] Finding no coastal defenses blocking their way, the destroyers steamed onward, laying off Narvik in the early morning of 9 April, sinking two Norwegian coastal defense ships and delivering the landing force directly onto the Narvik quay. Within two hours, the *2d Battalion, 139th Mountain Infantry*, secured Narvik, the initial objective of the amphibious landing, without incurring a single casualty.[3] The remaining two battalions and regimental staff landed north of Narvik and seized a Norwegian supply depot. However, British naval counterattacks on 10 and 14 April sunk all the German destroyers, leaving the landing force isolated.[4]

Battle Group Dietl continued to advance inland, seizing control of the iron ore railway from Narvik to the border crossing to neutral Sweden at Bjørnfjell by 16 April, a strategic goal of the campaign, and pushed farther north.[5] As they did, the mountain troopers found themselves in "a pure alpine landscape in an artic environment."[6] The mountains ran from sea level to peaks of more than 4,600 feet, with tree line at 2,000 feet, and all covered with 3–6 feet of snow. There were cliffs and glaciers, canyons with mountain streams and lakes, and only a thin layer of soil over granite rocks. North of the Arctic Circle, Narvik's nights were bright, and the sun would stay above the horizon until the end of May. The deep snow, storms, and cold would all turn into rain and damp by that same time.[7]

Cut off hundreds of kilometers from reinforcements, facing stiffening Norwegian resistance, and concerned with the possibility of an allied landing, *Battle Group Dietl* went on the defensive from mid-April.[8] The *139th Mountain Infantry*, with 2,000 troops, deployed two battalions to the north and one to the south by Narvik. The 2,600 beached sailors, organized into naval battalions armed with seized Norwegian

---

[2] Alex Buchner, *Narvik: The Struggle of Battle Group Dietl in the Spring of 1940*, trans. Janice W. Ancker (Philadelphia, PA: Casemate, 2020), 1–22.
[3] Henrik O. Lunde, *Hitler's Pre-Emptive War: The Battle for Norway, 1940* (Philadelphia, PA: Casemate, 2009), 151–87, 194–217, 263–70; and Earl F. Ziemke, *The German Northern Theater of Operations, 1940–1945*, Army Pamphlet 20-271 (Washington, DC: Department of the Army, 1959), 44–48.
[4] For the entire amphibious invasion of Norway in 1940, see James K. Greer, "Operation Weserubung: Early Amphibious Multidomain Operations," in Timothy Heck and B. A. Friedman, eds., *On Contested Shores: The Evolving Role of Amphibious Operations in the History of Warfare* (Quantico, VA: Marine Corps University Press, 2020), 186–99, https://doi.org/10.56686/9781732003149.
[5] Buchner, *Narvik*, 35–38.
[6] Buchner, *Narvik*, 24.
[7] Buchner, *Narvik*, 23–35.
[8] Lunde, *Hitler's Pre-Emptive War*, 274–310.

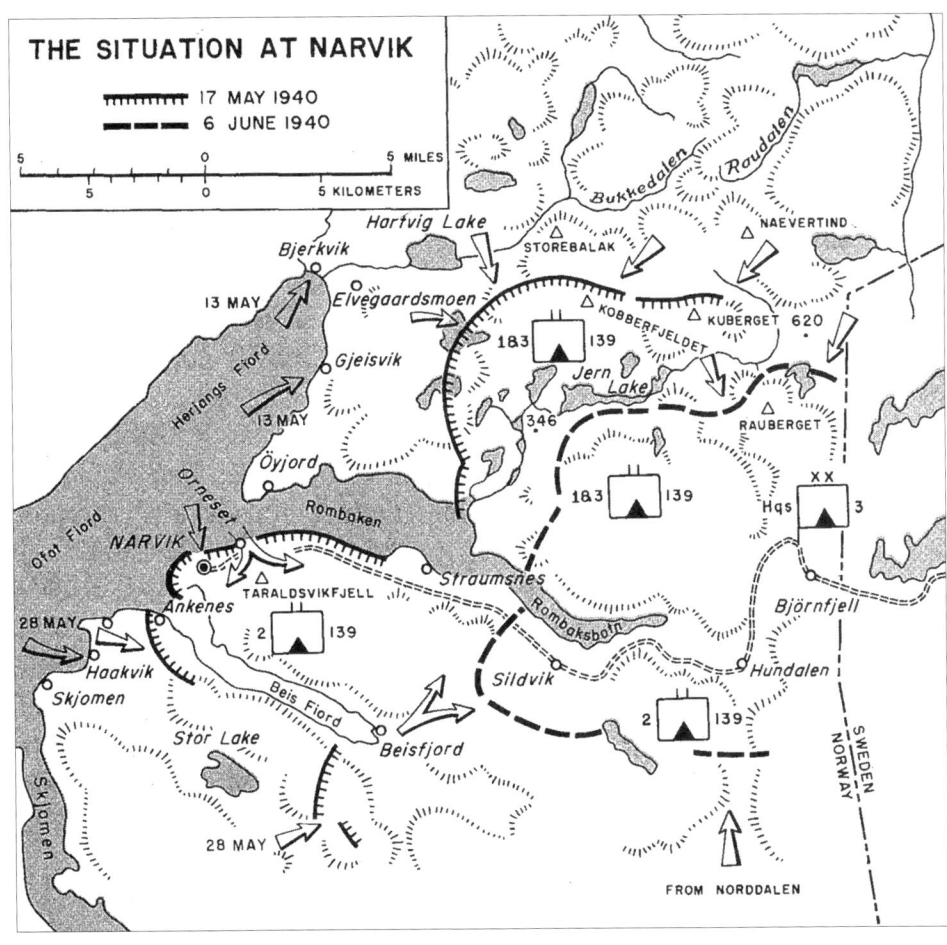

MAP 1
The Situation at Narvik.
Source: Earl F. Ziemke, The German Northern Theater of Operations, 1940–1945 (Washington DC: Department of the Army, 1959), 100

weapons and uniforms, held positions scattered along the coasts and the railway.[9] From late-April, the battle group held off British, and then Polish and French, attacks from the south, and Norwegian, and then French, attacks from the north. As the Germans did not have enough troops to cover all potential landing sites, the Allied amphibious landings in mid- and late-May flanked them, forcing the battle group to pull back in the north and give up Narvik in the south. A trickle of reinforcements, namely hastily trained mountain troops and airborne infantry, parachuted into the pocket.[10] By the start of June, *Battle Group Dietl* was pressed back along the Swedish border. But on 8 June, the Allies, reacting to German successes in the Battle of France

[9] Lunde, *Hitler's Pre-Emptive War*, 346–73.
[10] Lunde, *Hitler's Pre-Emptive War*, 404–36.

Beyond Cold Shores
165

(10 May–25 June 1940), withdrew their forces and the Norwegians were forced to surrender. The battle group held, if just barely, for two months.[11]

Why was this ad hoc amphibious force successful in their inland maneuver? In large part, it was because they were familiar with the mountain environment. The mountain troops at Narvik, in the words of their chronicler, "looked upon the massive arctic mountains and immediately felt at home."[12] Most of the German mountain troops of the *139th Mountain Infantry* were in fact Austrians, recruited from the mountainous provinces of Styria and Carinthia. The *3d Division* had its headquarters in Graz, Austria, which sits in a basin on the edge of the Eastern Alps.[13] The *139th Regiment* was based in and around Klagenfurt, Austria, between the Karawanken Mountains to the south and the Gurktal Alps to the north. The entire *3d Division* had been built after the Anschluss (annexation of Austria) in 1938 on the foundations of Austrian mountain troops based in the region.[14]

The Austrian mountain troopers also possessed a deep doctrinal knowledge of mountain warfare from the experiences on the Alpine front during World War I.[15] From the first winter of the war, the Austro-Hungarians issued a series of directives to units serving in the Alps, which were consolidated into mountain warfare manuals in 1917 and 1918.[16] The German military, drawing on their mountain warfare experiences in the Alps and Carpathians, also possessed mountain warfare doctrine, including the *Provisional Training Instruction for Mountain Troops of 1935*.[17] *Battle Group Dietl* certainly followed the precepts of these mountain warfare doctrines. The mountain troops organized their defenses into a string of machinegun positions emplaced for mutual support on any potential avenue of approach: paths, passes, and ridge junctions. Counterattacks by reserves proved decisive, but they had to be kept close to the front.[18] Delaying actions proved crucial at Narvik, with withdrawing troops establishing numerous consecutive firing points, as the battle group fell back on three separate occasions to avoid being outflanked and to shorten the lines to ensure reserves were available.

The basing of the mountain troops and their doctrine allowed for training in

---

[11] Lunde, *Hitler's Pre-Emptive War*, 461–513.
[12] Buchner, *Narvik*, 26.
[13] James Lucas, *Hitler's Mountain Troops: Fighting at the Extremes* (London: Cassell, 1999), 16, 199–200.
[14] Roland Kaltenegger, *Die Geschichte der deutschen Gebirgstruppe 1915 bis heute: vom Deutschen Alpenkorps des Ersten Weltkrieges zur 1. Gebirgsdivision der Bundeswehr* (Stuttgart, Germany: Motorbuch Verlag, 1980), 89, 99–100.
[15] Alexander Jordan, *Krieg um die Alpen: Der Erste Weltkrieg im Alpenraum und der bayerische Grenzschutz in Tirol* (Berlin: Duncker & Humblot GmbH, 2008); and Mark Thompson, *The White War: Life and Death on the Italian Front, 1915–1919* (New York: Basic Books, 2010), 193–206, 294–327.
[16] Adams Carter, trans., *Manual for Service in the Mountains* (Vienna: War Ministry, 1917); and Adams Carter, trans., *Mountain Warfare* (Vienna: War Ministry, 1918), 26–33.
[17] Kaltenegger, *Die Geschichte der deutschen Gebirgstruppe 1915 bis heute*, 16–78.
[18] War Department, *German Mountain Troops* (Washington, DC: Military Intelligence Division, 1944), 17–20.

the high mountains, where conditions replicated those in the Arctic surroundings of Narvik.[19] Mountain troops trained to cross and climb ice and snow, using crampons, ice axes, and ropes, and operations in the face of potential avalanches, skills useful in the Norwegian wilderness. About 25 percent of each unit received ski training, but this proved adequate as, once in Norway, each battalion only mounted one or two platoons on skis for reconnaissance, counterattack, and rear-guard actions.[20] The remainder, holding defensive positions, had to dig their own paths or posthole (sink through the snowpack) through the snow. The mountain troops learned to make improvised shelters in the rocks and spend winter nights in snow shelters at high altitudes, allowing them to survive in the Arctic mountains of Norway. Mountain units trained to deliver supplies via vehicle, then cart, then pack animal, then porter, supplemented by air-delivery.[21] Sailors functioning as porters and air-dropped supplies ultimately sustained the frontline mountain troops in Narvik.[22]

The mountain troops also had trained with the individual combat equipment needed for the high mountains and, in Norway, for operations in the Arctic.[23] They knew to dress in layers, wearing little on the move and then adding clothing once bivouacked. In addition to the standard army issue, the mountain troops received specialized caps, shirts, sweaters, wind jackets, anoraks, overmittens, trousers, and boots. Mountaineering equipment also included sun goggles, snowshoes, sleeping bags, and air mattresses, all of which were carried in a rucksack.[24] However, the mountain troops who landed at Narvik only had their mountain caps, boots, and rucksacks.[25] The Swedes did allow three rail cars of clothing to cross the border on 26 April and captured Norwegian Army stocks provided many other articles, resulting in a motley appearance.[26] The battle group also purchased or stole from Norwegian civilians many items, particularly skis and anything white that could be used for snow camouflage.[27] While the mountain troopers may not have had all their specialized equipment, they knew what they needed, why they needed it, and how to use it.

The ability of mountain troopers to adapt reflected not just their training but also their leadership. General Dietl, the battlegroup commander, was an experienced civilian mountaineer and skier, captaining the 1936 German Olympic ski team.[28] He had combat experience as a company commander on the Western Front during World War I. Dietl also had more than a decade of experience as a mountain troop

---

[19] Wilhelm Hess, *Arctic Front: The Advance of Mountain Corps Norway on Murmansk, 1941*, trans. Linden Lyons (Havertown, PA: Casemate Publishers, 2021), 51.
[20] War Department, *German Mountain Troops*, 54–62, 125–31, 156–67.
[21] War Department, *German Mountain Troops*, 10–12, 63–78.
[22] Buchner, *Narvik*, 51–53, 60, 66, 158.
[23] War Department, *German Mountain Troops*, 11–15.
[24] War Department, *German Mountain Troops*, 84–90.
[25] Buchner, *Narvik*, 29.
[26] Ziemke, *The German Northern Theater of Operations*, 88; and Lunde, *Hitler's Pre-Emptive War*, 291.
[27] Buchner, *Narvik*, 32, 36.
[28] "Olympic Winter Games Garmisch-Partenkirchen 1936," Olympics.com, accessed 3 August 2023.

FIGURE 1
Battle of Narvik: German mountain troops.
Source: German Federal Archive, Bundesarchiv, Bild 183-2005-1202-500

commander between the wars. Dietl's experiences made him a calm, controlled, inspiring commander.[29] Similarly, the *3d Division's* junior leaders had combat experience from the 1939 invasion of Poland in the High Tatras Mountains and on the plains around Lemberg (Lviv, Ukraine).[30] The battlegroup also benefited from another form of leadership. Drawing on Austrian experiences and practices, German mountain units gave a portion of their personnel, ideally one in four, more mountaineering and ski training, designating them military mountain guides (*Heeresbergführer*).[31] The primary purpose of the military mountain guides was to serve as specialists in moving units through mountain terrain, while managing mountain risks. These guides led patrols, emplaced mountaineering routes or ski tracks, took communication teams to high points, or served as assault unit commanders.[32]

Furthermore, the Narvik battlegroup was able to operate with the air and naval services. The mountain troops worked well with the navy during their initial landing, but the sinking of the destroyers ended any more cooperation. Airpower ultimately proved crucial for the mountain troops. A battery of field artillery air-landed on a

---

[29] Lunde, *Hitler's Pre-Emptive War*, 152–53. A Bavarian, Dietl was an early supporter of the Nazi Party.
[30] Lucas, *Hitler's Mountain Troops*, 18–26.
[31] War Department, *German Mountain Warfare*, 79–83.
[32] Kurt Pflügl, "Soldaten im Hochgebirge (III)," *Truppendienst, Folge 293* (Ausgabe 5/2006).

frozen lake in mid-April, which soon melted, eliminating it as an airfield for further resupply. By early May, the German Air Force (*Luftwaffe*) occupied air bases within range of Narvik, bombing Allied warships and supply depots, disrupting their build-up.[33] And, as noted earlier, reinforcements arrived by air from the end of May in the form of a parachute battalion and two mountain troop companies, quickly trained as parachutists, totaling nearly 1,000 men. *Luftwaffe* operations allowed the German mountain troops at Narvik to hang on, just long enough.

ALEUTIANS, 1943

On 11 May 1943, four battalions of the U.S. 7th Infantry Division came ashore on Attu in the Aleutians, an island chain that, while south of the Arctic Circle, is generally considered part of the Arctic.[34] Two battalions landed in the northeastern part of the island, while two landed in the southwest, aiming to link-up and attack the Japanese garrison at the eastern end. Ultimately reinforced by another four battalions during the following week, the American infantry struggled up basins covered by muskeg—an impassible, spongy soil of moss over water-logged peat and mud—beneath ridges as high as 3,000 feet, many covered with snow.[35] For a week, the two landing forces painfully advanced, hindered by the terrain, the weather, a lack of supplies and support, and dogged Japanese resistance from dug-in positions below the military crest on ridges, which were regularly obscured by fog. Finally linking up on 18 May, the now-unified force turned east and spent another week fighting its way up, onto, and down snow-covered ridges and across rain-soaked valleys. The Japanese fell back in good order but, with no relief forthcoming and refusing to surrender, they launched a counterattack on the night of 29 May. When that failed, most of the survivors committed suicide with hand grenades, U.S. forces only took 28 Japanese captives, and Attu fell on the next day.[36]

While the American landing force captured its objective, the inland maneuver in this near-Arctic environment was less than successful. The landing force ultimately totaled 15,300 troops, sustaining 3,829 casualties. Cold injuries—mainly frostbite and trench foot—made up the single largest category of losses at 1,200, exceeding the 1,148 wounded in action.[37] The force encountered a cold and wet environment on Attu. The air was continuously cold, with constant wind and regular strong gusts. Light rain and snow fell regularly during the attack, and fog for eight hours a day was not

---

[33] Ziemke, *The German Northern Theater of Operations*, 88, 92, 94.
[34] Niels Einarsson et al., *Arctic Human Development Report* (Akureyri, Iceland: Arctic Council, 2004), 17–18.
[35] Stetson Conn, Rose C. Engelman, and Byron Fairchild, *The Western Hemisphere: Guarding the United States and Its Outposts*, U.S. Army in World War II, CMH Pub 4-2 (Washington, DC: Center of Military History, 2000), 279–95.
[36] Brian Garfield, *The Thousand-Mile War: World War II in Alaska and the Aleutians* (Fairbanks: University of Alaska Press, 1995), 273–340.
[37] *Cold Injury, Ground Type* (Washington, DC: Medical Department, Office of the Surgeon General, Department of the Army, 1958), 84–85. In addition to the 1,148 wounded in action and 1,200 cold injuries, the landing force lost 549 troops killed in action, 614 to diseases, and 318 to other nonbattle injuries.

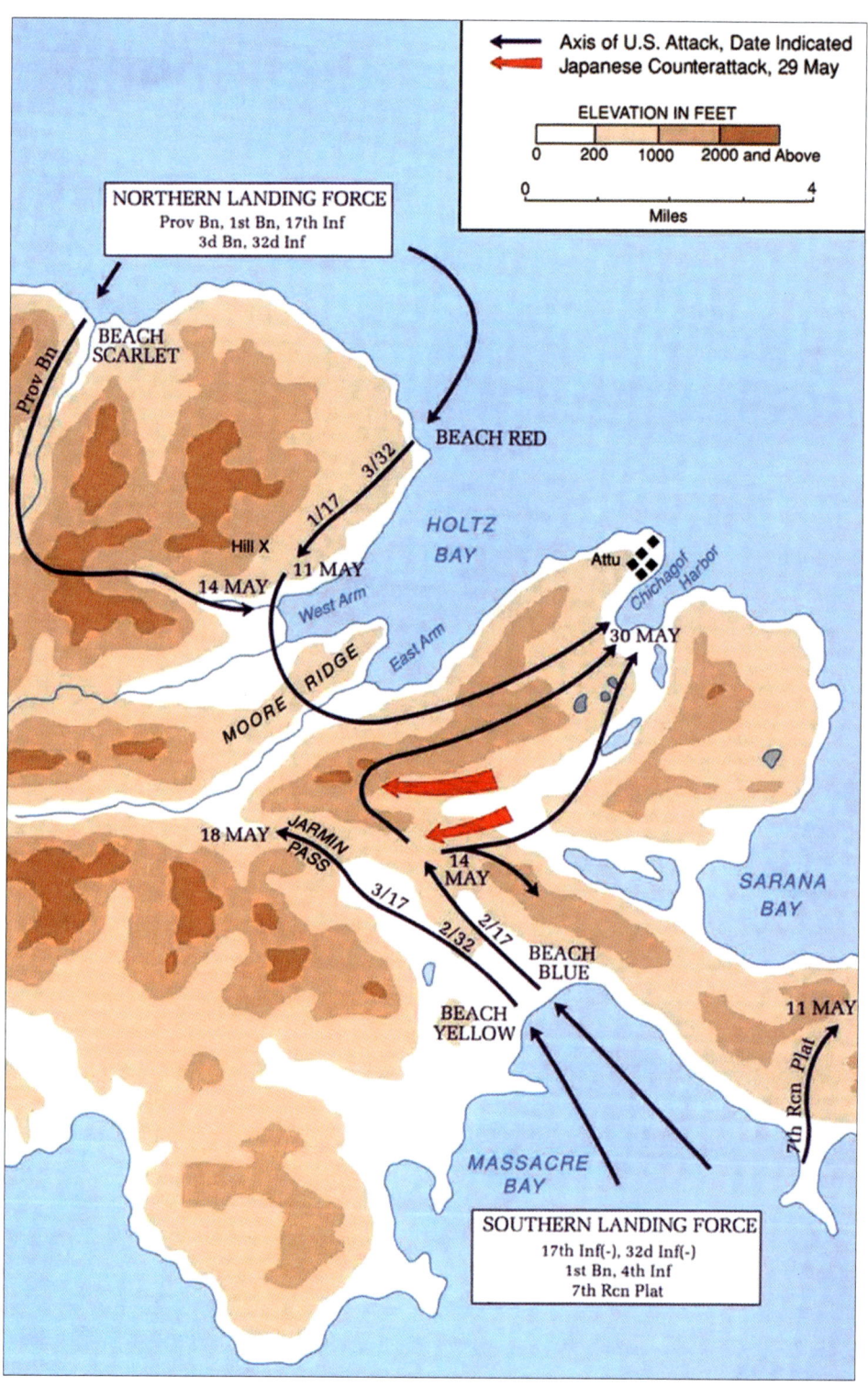

MAP 2
Capture of Attu, 1943.
Source: George L. MacGarrigle, *Aleutian Islands, 3 June 1942–24 August 1943*, U.S. Army Campaigns of World War II (Washington, DC: U.S. Army Center of Military History, 2019)

uncommon. Temperatures in the valleys ranged from 25° to 30° F, but between 10° and 24° F on the ridges, where much of the combat took place.[38] The landing force was constantly exposed to cold and dampness for days at a time for which they were unprepared.

The U.S. Army did have doctrine by 1943 that could have prepared them. *Operations*, Field Manual 100-5, published in 1941 had sections on "Mountain Operations" and "Combat in Snow and Extreme Cold." The former insisted that "mountainous terrain offers no insuperable obstacles to the conduct of military operations, even in cold weather, if the troops are properly equipped, clothed, supplied, and trained."[39] The section on snow and extreme cold opened with the admonition that "severe weather conditions handicap movement and require special tactical and logistical measures for successful operations."[40] While there was no mountain operations field manual at the time—it was under preparation by the staff of the Mountain Training Center at Camp Hale, Colorado, and would not be issued until 1944—there were other amplifying manuals.[41]

The U.S. Army's first-ever *Operations in Snow and Extreme Cold* field manual, a slim volume of 85 pages, noted three major problems for operating in snow and extreme cold, pertinent for the Attu invasion:

(a) Keeping men and animals warm.
(b) Moving troops across snow and ice.
(c) Transporting and preserving supplies and equipment.[42]

Additionally, the U.S. Army Air Corps prepared a two-volume *Arctic Manual*, likely to prepare flyers for potential survival situations in the far north. Drawing on the work of Arctic explorers, it included chapters on geography, food and drink, clothing and personal equipment, health, accident, and disease, travel, and transportation.[43] However, there is no evidence that the 7th Infantry Division made any use of any of these sections or manuals.[44]

Part of the reason the 7th Infantry Division did not look at the mountain operations or the cold weather sections in *Operations* or the *Arctic Manual* was that it

---

[38] *Cold Injury, Ground Type*, 86-88.
[39] *Operations*, Field Manual (FM) 100-5 (Washington, DC: War Department, 1941), 213, emphasis in original.
[40] *Operations*, 225.
[41] John C. Jay, *History of the Mountain Training Center*, Study no. 24 (Fort Monroe, VA: Historical Section, Army Ground Forces, 1948), 91-93. *Mountain Operations*, FM 70-10, was not published until December 1944.
[42] *Operations in Snow and Extreme Cold*, FM 31-15 (Washington, DC: Government Printing Office, 1941), ii.
[43] *Arctic Manual*, 2 vols. (Washington, DC: Army Air Corps, U.S. Army, 1940).
[44] Maj Joshua D. Walters, USA, "The Impact of Training and Equipment at the Battle of Attu, Aleutian Campaign-Historical Study and Current Perspective" (master's thesis, U.S. Army Command and General Staff College, 2015), 26-31, 33.

had been busy conducting desert, motorized training.[45] The Army assigned the 7th Infantry Division to the Attu landing not because of its training, but because it was near full strength and was near amphibious training sites along the California coast.[46] With only three months to prepare, the division staff understandably focused on training for the unfamiliar amphibious landing. On the beaches and off the coast of Monterrey, under Marine Corps trainers, the division practiced embarkation, wet and dry net debarkations, and boat landings.[47] Training to get on the shore took up most of the time available, so little effort was devoted to moving beyond the shore, never mind how cold and wet that shore would be.

The clothing issued to the division reveals that lack of preparation and understanding of the near-Arctic environment. The olive-drab woolen trousers were chemically treated to be water-repellent, but it wore off and, in a few days, they were soaked. Men received an Arctic M41 field jacket that was hip-length and made of wind-proof, water-repellent cotton with a wool lining, but it did not provide full protection as it was too short and lacked a hood. Despite the jackets being designed to be worn over sweaters, as described in the supply catalog, the division did not issue any sweaters or additional layers.[48] Footwear was a 12-inch-high Blucher boot, which did not keep the feet warm, was not waterproof, and whose soles wore out quickly. Once wet, the boots could not be dried under field conditions and then shrank, constricting blood flow to the feet, exacerbated by the fact that the boots were issued at the proper size, making it impossible for soldiers to wear the recommended two pairs of socks.[49] The Arctic sleeping bags issued were too bulky, consisting of two down-filled cases, and so were left in rucksacks, along with rain gear, to be brought up to the landing troops later. Few men ever received their rucksacks. Eventually, sleeping bags were pushed forward, but only after four or five days.[50]

The landing force had not received any training on how to use this equipment or how to deal with the climate they would face beyond the beach. In particular, the soldiers received no training on how to care for their feet. They had not been instructed to remove their boots as often as possible, to change socks, and dry the insoles. Many men on Attu did not remove their boots for five days after the landing. Some threw away their wet socks without trying to dry them.[51] They had not been taught to dry their sleeping bags after use, and many discarded their cold-weather clothing to

---

[45] Bruce Gardner and Barbara Stahura, *Seventh Infantry Division, 1917–1992: World War I, World War II, Korean and Panamanian Invasion—Serving America for 75 Years*, rev. ed. (Nashville, TN: Turner Publishing, 1997), 10.
[46] Conn, Engelman, and Fairchild, *The Western Hemisphere*, 277-78.
[47] Walters, "The Impact of Training and Equipment," 23-36.
[48] *Cold Injury, Ground Type*, 90; and *Quartermaster Supply Catalog*, Sec. 1, *Enlisted Men's Clothing and Equipment*, OQMG Circular no. 4 (Washington, DC: Army Service Forces, 1943), 8.
[49] *Cold Injury, Ground Type*, 90; and *Quartermaster Supply Catalog*, 3.
[50] *Cold Injury, Ground Type*, 90; *Quartermaster Supply Catalog*, 31; and Walters, "The Impact of Training and Equipment," 30-31.
[51] *Cold Injury, Ground Type*, 92-93.

FIGURE 2
Hauling supplies on Attu.
Source: official Department of Defense photo

lighten their combat loads.[52] A key part of why none of this training was provided is that the division executed a deception plan for the landing, giving lectures on tropical diseases and issuing summer clothing, while the specialized cold-weather equipment was loaded onto ships in sealed crates, only to be opened at sea.[53]

Further, the division did not take full advantage of Aleutian bases. The United States began the campaign with a base at Cold Bay on the tip of the Alaska Peninsula, a naval facility at Dutch Harbor on Unalaska Island, and an airfield farther east on Umnak.[54] Support facilities were pushed forward to facilitate landings, with island bases established on Adak in August 1942 and then Amchitka in January 1943.[55] But when the Attu landing force steamed into Cold Bay on 24 April aboard five cramped transports, they stayed on ship. Only the division's Provisional Scout Battalion, organized to protect the flank of the Northern Landing Force, debarked, as it was to board submarines for the landing. The battalion spent a week training in the snow

---

[52] *Cold Injury, Ground Type*, 93–94.
[53] Walters, "The Impact of Training and Equipment," 32.
[54] Conn, Engelman, and Fairchild, *The Western Hemisphere*, 223–76.
[55] Department of the Navy, *Building the Navy's Bases in World War II: History of the Bureau of Yards and Docks and the Civil Engineer Corps, 1940–1946*, vol. 2 (Washington, DC: Government Printing Office, 1947), 163–90.

and muskeg and requisitioning jackets, socks, and boots, as they had not received any winter equipment.[56] Even so, while the 350 men of the battalion took 30 battle casualties, only 40 of the remaining 320 were able to walk five days after landing.[57]

Finally, Army-Navy coordination was in its infancy in May 1943 and neither Service yet truly understood the other. When calling for naval gunfire, Army observers requested destruction of the target, vice neutralization, leading the Navy to expend large numbers of rounds, ammunition it might have needed had the Japanese fleet sortied. The weather, particularly the persistent fog and high winds, inhibited naval gunfire and carrier aviation.[58] While Colonel William O. Eareckson of the U.S. Army Air Forces served as a particularly aggressive air-ground liaison office, borrowing Navy float planes to fly as an airborne forward air controller and on one occasion firing an infantryman's rifle on the Japanese, he could not overcome the weather, which prevented any air support on 11 of the 20 days of the battle.[59] While the Navy and Army Air Forces understood the challenges the weather would bring, the landing force did not, failing to incorporate conditions into its planning, limiting interoperability between the Services.[60]

Even as the battle for Attu raged, the Alaskan Defense Command (ADC) planned for landings on Kiska, which held an even-larger Japanese garrison. Learning from Attu, ADC organized a large force and ensured it was trained and equipped for the conditions. The task force consisted of the 184th Infantry Regiment transferred from Fort Ord, the 17th Infantry Regiment from Attu, ADC's 53d Infantry Regiment, the Canadian 13th Infantry Brigade, the U.S. 87th Mountain Infantry Regiment, and the U.S.-Canadian 1st Special Service Force (FSSF).[61] The latter two units were at the insistence of the U.S. Army chief of staff, General George C. Marshall. Marshall realized the Aleutians campaign was essentially a winter mountain operation given the climate, environment, and topography. When the Kiska task force assembled, Marshall personally gave orders sending the 87th Mountain Infantry Regiment, which had just completed five months of winter mountain training at Camp Hale, Colorado.[62] Marshall also ordered the FSSF, a commando unit organized and trained in Montana to fight on the glaciers of Norway, to join the landings.[63]

Given their winter training, albeit in the high, dry, cold snow of the Rocky

---

[56] Garfield, *The Thousand-Mile War*, 263–64.
[57] *Cold Injury, Ground Type*, 94.
[58] *The Aleutians Campaign, June 1942–August 1943: Combat Narratives* (Washington, DC: Office of Naval Intelligence, U.S. Navy, 1945), 83–84.
[59] Garfield, *The Thousand-Mile War*, 293.
[60] Wesley Frank Craven and James Lea Cate, eds., *The Army Air Forces in World War II*, vol. 4, *The Pacific: Guadalcanal to Saipan, August 1942 to July 1944* (Washington, DC: Government Printing Office, 1950), 386.
[61] *Cold Injury, Ground Type*, 96.
[62] McKay Jenkins, *The Last Ridge: The Epic Story of America's First Mountain Soldiers and the Assault on Hitler's Europe* (New York: Random House, 2003), 122.
[63] Saul David, *The Force: The Legendary Special Ops Unit and WWII's Mission Impossible* (New York: Hachette Books, 2019), 128–29.

Mountains, the 87th Mountain Infantry Regiment and FSSF were better prepared than the 7th Infantry Division had been. The 184th Infantry Regiment, training at Fort Ord, California, had regular contact with the units on Attu, so it too was far better trained.[64] A small handbook, *Soldier's Manual (How to Get Along in the Field)*, distributed to all troops in the landing force, distilled much of the experience on Attu and cold weather doctrine. It included instructions on the care of the feet, especially the use and care of socks, clothing, and cold-weather equipment, the importance of nutrition and hydration, and how to keep fighting positions dry.[65] Finally, the entire task force trained on Adak and Amchitka for several weeks in Aleutian conditions, conducting amphibious landings and marches across the muskeg.[66]

The Kiska landing force's clothing and footgear were also an improvement. Many of the units had longer, hooded parkas to provide better protection. Others kept the Arctic field jacket, but with a wool knit hood or toque to protect the head and neck from the cold.[67] All wore wool-lined trousers and carried full rain gear in their packs. Shoepacs, a boot with a rubber bottom and leather upper, were universally worn.[68] Issued in larger sizes to accommodate two pairs of socks and a felt insole that could be replaced and dried, the shoepacs kept the feet dry but did not provide much support.[69] Conditions on Kiska, due to the timing of the landing in August, were also much better as the snow had melted and the runoff had subsided, so most of the ground was drier than it had been on Attu.[70]

As a result, the 28,450 troops who landed on Kiska only suffered 130 cold casualties, or 1 exposure injury per 219 troops. By comparison, the Attu landing force took 1 cold casualty per 13 men. Of more than 5,000 men of the 87th Mountain Infantry Regiment Combat Team, only 7 experienced trench foot.[71] The inland maneuver on Kiska went unopposed, as the Japanese had evacuated the island two weeks prior to landing, so the force spent a week searching the island, losing 17 Americans and 4 Canadians killed and another 50 wounded to booby traps and friendly fire incidents.[72] But it was the careful preparations for the near-polar conditions on Kiska, whether in the Aleutians or in the mountains in winter, that kept the environment from proving even more dangerous.

FALKLANDS, 1982

On 21 May 1982, Great Britain's 3 Commando Brigade went ashore in San Carlos

---

[64] Garfield, *The Thousand-Mile War*, 376.
[65] *Soldier's Manual (How to Get Along in the Field)* (n.p., 1943); and Garfield, *The Thousand-Mile War*, 377.
[66] Garfield, *The Thousand-Mile War*, 378-79.
[67] *Cold Injury, Ground Type*, 97.
[68] *Cold Injury, Ground Type*.
[69] *Cold Injury, Ground Type*.
[70] *Cold Injury, Ground Type*.
[71] *Cold Injury, Ground Type*, 98.
[72] Garfield, *The Thousand-Mile War*, 380-87.

Sound on the western shore of East Falkland Island. Seven weeks to the day after the Argentine seizure of the islands, the Amphibious Task Force landed three Royal Marine commandos (battalion-size units) and two parachute (para) battalions.[73] While the Falkland Islands lay more than 1,287 km north of the Antarctic Circle, they are sub-Antarctic with a generally cold, wet, and windy climate. Concerns with facing a near-polar winter (June–August in the South Atlantic) was a key factor in Great Britain's speedy dispatch of a task force to retake the Falklands.[74] The Amphibious Task Force quickly came under daylight Argentine air attack, costing most of the landing force's helicopters when the SS *Atlantic Conveyor* (1969) sunk on 25 May, and forcing the logistical offload into the night hours, taking until the 27th.[75]

Faced with the loss of much of its helicopter lift, and under political pressure to engage the enemy, 3 Commando Brigade ordered 45 Commando and 3 Para to walk the nearly 80 km across East Falkland to the main Argentine garrison at Port Stanley.[76] The 2 Para would protect the flank of this foot maneuver by attacking what was thought to be a small Argentinian garrison at the settlement of Goose Green. The garrison proved to be much larger, forcing 2 Para into a 12-hour fight on 28 May before ultimately forcing an Argentine surrender.[77] For three days, 45 Commando yomped and 3 Para tabed across a rocky peatland in the wet and cold of an oncoming winter.[78] Screened by special operations forces, the battalions reached the outer Argentine defenses on 30 May. The next day, 42 Commando helicopter-lifted in to seize a critical height of Mount Kent. And 2 Para, assigned to the just-arrived 5th Infantry Brigade, flew forward on 3 June.[79]

Continuing the buildup of forces, 5th Infantry Brigade landed the 2d Battalion of the Scots Guards from assault ships during the night of 5–6 June at Bluff Cove, to be followed by the 1st Battalion of the Welsh Guards the next night. An Argentine airstrike the morning of 8 June caught the Welsh Guards offloading, hitting one ship, killing 48, and injuring 115.[80] The 1st Battalion, 7th Gurkha Rifles, joined 5th Infantry Brigade and, by 11 June, the two British brigades closed on the outskirts of Port

---

[73] Lawrence Freedman, *The Official History of the Falklands Campaign*, vol. 2, *War and Diplomacy* (New York: Routledge, 2005), 50, 463–92.

[74] Stephen Badsey, "An Overview of the Falklands War: Politics, Strategy and Operations," *NIDS Military History Annual* (2013): 139–66.

[75] Michael Clapp and Ewen Southby-Tailyour, *Amphibious Assault Falklands: The Battle of San Carlos Water* (Barnsley, UK: Pen & Sword Military, 1996), 132–90.

[76] Max Hastings and Simon Jenkins, *The Battle for the Falklands* (New York: W. W. Norton, 1983), 231–53.

[77] Hastings and Jenkins, *The Battle for the Falklands*, 262–74.

[78] The term *yomp* refers to a lengthy and strenuous hike across difficult terrain carrying a loaded pack. The term *tab* refers to World War II-era British slang for a tactical advance to battle (tab) by leaving a drop zone, assembling into designated units, and moving as quickly as possible in full gear to the objective.

[79] Gregory Fremont-Barnes, *The Falklands 1982: Ground Operations in the South Atlantic* (New York: Osprey Publishing, 2012), 43–54.

[80] Robert S. Bolia, "The Bluff Cove Disaster," *Military Review* (November–December 2004): 66–72.

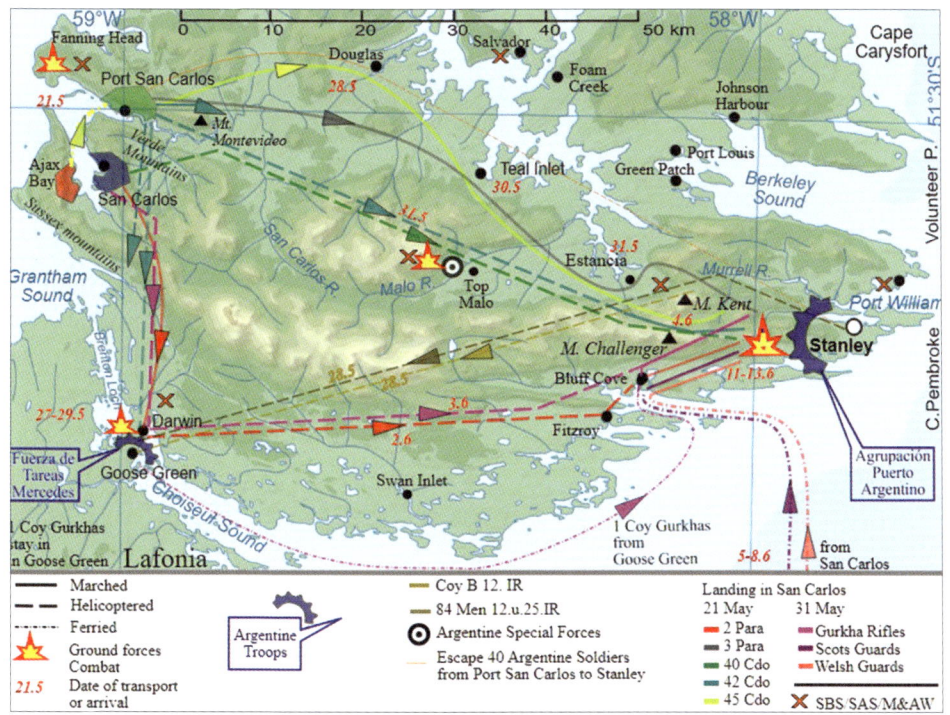

MAP 3
Falkland land operations.
Source: Eric Gaba, adapted by MCUP

Stanley. They launched a series of battalion night attacks, seizing Argentine ridgetop positions: three on the night of 11–12 June and two more on the night of 13–14 June. This precipitated negotiations and then the surrender of the Argentine force on the Falkland Islands on 14 June, ending the campaign.[81]

For their inland maneuver during sub-Antarctic conditions in the face of a looming winter, the British landing force did have a doctrinal base to refer to. The British armed forces emerged from World War II with a set of five military training pamphlets for snow and mountain warfare. These developed from learning the lessons of Narvik, the occupation of Iceland, the training of a mountain division in Scotland, and the experiences of mountain warfare schools in Lebanon and Italy.[82] The pam-

---

[81] Freedman, *The Official History of the Falklands Campaign*, vol. 2, 596–661.
[82] Rob Granger, "British Army Cold Weather and Mountain Warfare Training in the Second World War," *British Journal for Military History* 8, no. 1 (2022): 69–86, https://doi.org/10.25602/GOLD.bjmh.v8i1.1606; and see, for example, *Snow and Mountain Warfare, pt. 4, Whitehead Training and Operations, 1945*, Military Training Pamphlet no. 90 (London: General Staff, War Office, 1945).

Beyond Cold Shores

phlets emphasized the need for forces to be mentally and physically prepared to operate in the cold and mountains. These were updated in the 1970s into two operational manuals on mountainous country and cold climate.[83]

The 3 Commando Brigade made the most use of this doctrine as it had the NATO mission of reinforcing northern Norway. By 1982, the Royal Marines had spent almost a decade developing their mountain and cold weather warfare skills. Between 1973 and 1981, 45 Commando and its attachments deployed every winter to Norway and were based the rest of the time in the comparably cool and damp environs of Scotland.[84] The 3 Commando Brigade, albeit with only 42 Commando due to budgetary constraints, spent January–March 1982 training in Norway, returning just before they deployed to the Falklands.[85] Many of the brigade's officers, noncommissioned officers, and older commandos (both Army and Royal Marine) had experienced five to six Norwegian winters.[86] The two parachute battalions attached to the brigade for the landing, while not having comparable training, were able to take advantage of 3 Commando Brigade's collective experience.[87]

The 5th Infantry Brigade, the other component of the landing force, had little opportunity to apply existing doctrine or do much of anything else, as it had been organized only in January 1982. Intended to operate outside of northwestern Europe, the brigade had two parachute battalions and a Gurkha battalion and had only conducted one exercise by April 1982, which mainly showed the inexperience of the brigade staff.[88] Losing the two parachute battalions to 3 Commando Brigade, 5th Infantry Brigade received two guard battalions just off of public ceremonial duties. These units, while disciplined and well-led, were not formed as light infantry and had not been on cold weather exercises. The brigade trained in Wales at the end of April but focused on platoon and company training evolutions.[89] Setting sail a month after the rest of the task force, the brigade was uncertain if it would be a second landing force, an operational reserve, or a garrison force. Concerns the guardsmen would not be able to walk across East Falkland, and a lack of vehicles or helicopters to move their equipment and supplies forward, led to the decision to move them forward by assault ship, leading to the deadly disaster at Bluff Cove.[90]

---

[83] *Land Operations*, vol. 5, *Operational Techniques under Special Conditions*, pt. 1, *Mountainous Country* (London: Ministry of Defence, 1972); and pt. 4, *Cold Climate* (London: Ministry of Defence, 1977).
[84] Julian Thompson, *No Picnic: 3 Commando Brigade in the South Atlantic, 1982* (London: L. Cooper with Secker & Wargurg, 1985), 8–9.
[85] Nick Vaux, *Take that Hill!: Royal Marines in the Falklands War* (Washington, DC: Pergamon-Brassey's International Defense Publishers, 1986), 11.
[86] Ian Gardiner, *The Yompers: With 45 Commando in the Falklands War* (Havertown, PA: Pen & Sword, 2012), 11, 30–31.
[87] Hugh McManners, *Falkland Commando* (London: William Kimber, 1984), 25.
[88] Hastings and Jenkins, *The Battle for the Falklands*, 267–74.
[89] Freedman, *The Official History of the Falklands Campaign*, vol. 2, 596–601.
[90] Nick van der Bijl and David Aldea, *5th Infantry Brigade in the Falklands* (Barnsley, UK: Pen & Sword, 2014).

As commander, 3 Commando Brigade, Brigadier Julian Thompson, Royal Marines, noted, the landing force had the material for the conditions in the Falklands and, "as important, the knowledge of how to use it properly."[91] Proper use meant dressing in layers—insulating, windproofing, and waterproofing—and altering them based on conditions. For insulation, there was a wide variety of long underwear, civilian jackets, issued sweaters, and combat trousers and jackets, along with cold weather vests, jackets, and trousers of quilted pile. The 3 Commando Brigade wore Arctic windproof hooded smocks and trousers, while 5th Infantry Brigade had Army cold weather hooded parkas and trousers, both made of closely woven cotton gabardine.[92] Initially, 5th Infantry Brigade was only issued 2,000 pairs of trousers and 1,000 parkas, and it took an informal intervention by the House of Lords to complete the issue for all 2,000 men.[93] Waterproof nylon-treated jackets and trousers were issued in the standard disruptive pattern material (DPM) camouflage to 5th Infantry Brigade, but 3 Commando Brigade wore reversible green and white waterproofs.[94] Most of the force wore the regular leather direct molded sole (DMS) boot with short puttees (wraps or leggings), though many replaced those with civilian gaiters. Some Royal Marines wore the dual-purpose leather Arctic ski march boot, while others had civilian hiking boots. The 45 Commando, given its Scottish base and regular forays into Norway, purchased Hawkins Cairngorm hiking boots.[95] A survey after the campaign found 46 different types of boots in the 3 Commando Brigade alone.[96]

The landing force needed this kit and knowledge as the campaign took place during generally cold, wet, and windy conditions. Temperatures hovered around freezing most of the time, dipping down to 10° F on the mountains. There were regular bouts of rain, sleet, and snow. The forced march across the island began with a blizzard on the night of 29 May, and by 5 June, the weather deteriorated even more with wind-driven rain and snow. Winds gusted upward of 45 mph. These conditions, combined with wading ashore from landing craft, meant the force started the campaign wet and stayed wet for 25 days.[97] Unable to dry out their clothing, particularly socks, meant about one-half of the force, particularly in 3 Commando Brigade,

---

[91] Thompson, *No Picnic*, 8.

[92] William Fowler and Michael Chappell, *Battle for the Falklands (1): Land Forces* (London: Osprey, 1982), 32–33.

[93] Freedman, *The Official History of the Falklands Campaign*, vol. 2, 592; and Hastings and Jenkins, *The Battle for the Falklands*, 268.

[94] Rifleman Moore, "Falklands Kit & Uniform-Combats & Windproofs," YouTube video, 16 August 2017; Rifleman Moore, "British Arctic Windproof Combat Smock & Trousers," YouTube video, 5 June 2022; and Rifleman Moore, "Falklands Kit & Uniform-Waterproofs," YouTube video, 14 September 2017.

[95] Rifleman Moore, "British Boots, Ski March," YouTube video, 29 May 2022; and Rifleman Moore, "Hawkins Cairngorm Boots," YouTube video, 29 April 2019.

[96] A. R. Marsh, "A Short but Distant War–The Falklands Campaign," *Journal of the Royal Society of Medicine* 76 (November 1983): 972–82, https://doi.org/10.1177/014107688307601119.

[97] Francis St. Clair Golden et al., "Lessons from History: Morbidity of Cold Injury in the Royal Marines during the Falklands Conflict of 1982," *Extreme Physiology & Medicine* 2 (2013), https:doi.org/10.1186/2046-7648-2-23.

suffered some level of nonfreezing cold injury or trench foot, regardless of the boot worn. Twenty percent of those with trench foot had to seek medical attention and 70 severe cases were transferred to a hospital ship, but the evacuated represented only 14 percent of the battle casualties.[98]

The cold injuries could have been much higher, except for the generally high quality of personnel and leadership. The British landing force was professional, well-trained, and motivated.[99] The enlisted men were volunteers, and most were under the age of 20. Their noncommissioned officers had long service records, and their officers well-versed in their tasks. Officers expected that the men would take care of themselves and their buddies as best as possible, overseen by noncommissioned officers.[100] The Royal Marine units also likely benefited from their Mountain Leaders. These were officers and noncommissioned officers trained by the Mountain and Arctic Warfare Cadre to serve as unit survival, skiing, rock climbing, and mountaineering instructors.[101] While concentrated in the reconnaissance (recce) troops of the three commandos, and the cadre deployed as the brigade recce troop, there were mountain leaders across the entire force, advising and instructing as necessary.[102]

The landing force could have suffered much more from the near-polar conditions had it not made maximum use of the minimum logistical facilities available. One airfield and one stone jetty were the only established facilities available to support the landing, and they were on Ascension Island, 6,365 km north of the Falkland Islands and 6,746 km south of the United Kingdom. After the landing, the amphibious task force struggled, as noted, to build up a brigade maintenance area at Ajax Bay in San Carlos Water. But once inland maneuver began, the landing force had the services of what was then a unique unit: the Commando Logistics Regiment. A decade of experience with limited infrastructure in the mountains of Norway enabled the joint British Army-Navy-Marines regiment to establish and maintain multiple forward support areas to receive helicopter-lifted supplies and move them onward to unit distribution points.[103] To do the latter, the landing force had an unexpected capability in 76 Swedish Bandvagn (BV 202) tracked over-the-snow vehicles assigned to 3 Commando Brigade.[104] However, the commandos ultimately organized porter platoons to carry supplies and ammunition forward to support their attacks.[105] As a

---

[98] Marsh, "A Short but Distant War," 976, table 3.
[99] Freedman, *The Official History of the Falklands Campaign*, vol. 2, 736-37.
[100] Nora Kinzer Stewart, *Mates & Muchachos: Unit Cohesion in the Falklands/Malvinas War* (Washington, DC: Brassey's, 1991).
[101] Mark Bentinck, *Vertical Assault: The Story of the Royal Marines Mountain Leaders' Branch* (Hants, UK: Royal Marines Historical Society, 2008), 57-58.
[102] Rod Boswell, *Mountain Commandos at War in the Falklands: The Royal Marines Mountain and Arctic Warfare Cadre in Action during the 1982 Conflict* (Philadelphia, PA: Pen & Sword Military, 2021).
[103] Kenneth L. Privratsky, *Logistics in the Falklands War: A Case Study in Expeditionary Warfare* (Philadelphia, PA: Pen & Sword, 2014).
[104] Thompson, *No Picnic*, 11.
[105] Thompson, *No Picnic*, 161, 164.

final analysis, it is difficult to argue with Brigadier Thompson's assessment that "it is hard to imagine a brigade more suited to the tasks that lay ahead," including inland maneuver in a sub-Antarctic environment, due to its organization, training, equipping, and experience in mountain winter warfare.[106]

CONCLUSION

These three historical cases unsurprisingly demonstrate that forces prepared for conditions—the Germans at Narvik and the British in the Falklands—were able to conduct successful inland maneuver after polar or near-polar amphibious landings. Forces that were not prepared—the Americans at Attu—suffered greatly from the conditions, potentially putting their maneuver at risk. Prepared forces had supporting doctrine for the extreme conditions. They were organized primarily as light infantry with sufficient support weapons, including artillery. They had suitable material, crucially clothing and boots. The force's leadership and personnel were familiar with the equipment, its use, and how to mitigate the risks of extreme conditions. They also were able to rely on facilities mainly outside of the region, placing a premium on logistical support utilizing multiple means down to the use of porters for the last few kilometers. And successful landing forces were interoperable with their air and naval Services.

These elements of doctrine, organization, training, materiel, leadership and education, personnel, facilities, and interoperability (DOTMLPF-I) were gained historically by preparing for mountain warfare, particularly in winter. In the winter, the combination of cold and wind at elevations above the tree line in the mountains creates analogous polar conditions.[107] Forces trained, prepared, and equipped for winter mountain warfare are thus better prepared to operate in the Arctic and Antarctic. While the challenges of these regions are quite severe, mountain-trained forces will at least arrive in the polar regions with a 70-percent solution for the conditions.[108] Finally, given the limited infrastructure and far distances that define the polar regions, mountain warfare training is a way to prepare forces for polar conditions outside of the poles. Mountain warfare is thus historically demonstrated invaluable preparation for an amphibious force to perform inland maneuver beyond cold shores.

---

[106] Thompson, *No Picnic*, 8.
[107] Raimund Lechner, Thomas Küpper, and Markus Tannheimer, "Challenges of Military Health Service Support in Mountain Warfare," *Wilderness & Environmental Medicine* 29, no. 2 (2018): 266–74, especially table 1, 267, https://doi.org/ 10.1016/j.wem.2018.01.006.
[108] For the challenges of operating in the region, see Capt Nathan Fry, "Survivability, Sustainability, and Maneuverability: The Need for Joint Unity of Effort in Implementing the DOD Arctic Strategy at the Tactical and Operational Levels," *Military Review* 94, no. 6 (November–December 2014): 54–62.

CHAPTER NINE

Soviet Preparations for a Naval Landing against Israel in June 1967 and Their Partial Implementation[1]

Isabella Ginor and Gideon Remez

The Arab-Israeli crisis and war of May–June 1967 exemplified the sea change, literally, both in the USSR's Cold War strategy and in its naval doctrine after the ouster of Nikita Khrushchev less than three years before. In the authors' book, *Foxbats over Dimona*, they demonstrated that far from blundering into this conflict—a belief held by most Western literature, based largely on Kremlin propaganda—the Soviets instigated it deliberately. They prepared a direct military intervention, which was intended to ensure an Arab, and especially Egyptian, victory that would promote Moscow's global and regional interests. The amphibious operations described here were part of this plan. After the scheme's failure in the Six-Day War, naval infantry and other elements of amphibious warfare became a fixture of the peak Soviet presence in Egypt until well after the Yom Kippur War in October 1973.

Khrushchev's ouster from the Soviet leadership freed Admiral Sergey G. Gorshkov, whom he had appointed as navy commander, to pursue his own strategic concept. It aimed to recreate an oceangoing surface force capable of power projection

---

[1] This chapter expands on the relevant passages of Isabella Ginor and Gideon Remez, *Foxbats over Dimona: The Soviets' Nuclear Gamble in the Six-Day War* (New Haven, CT: Yale University Press, 2007); Isabella Ginor and Gideon Remez, "The Six-Day War as a Soviet Initiative: New Evidence and Methodological Issues," *Middle East Review of International Affairs* 12, no. 3 (September 2008); and Isabella Ginor and Gideon Remez, *The Soviet-Israeli War, 1967-1973: The USSR's Military Intervention in the Egyptian-Israeli Conflict* (London: Oxford University Press, 2017), https://doi.org/10.1093/oso/9780190693480.001.0001.

FIGURE 1
A Soviet Marine poses in front of the Suez Canal Company's headquarters, Port Said, Egypt, ca. 1969.
Source: Evgeny Nazarov, "Arab-Israeli Wars," VK.com discussion board, 22 July 2020

worldwide by conventional means, which had been downgraded in favor of Khrushchev's focus on nuclear-missile submarines.[2] Gorshkov had allies in the increasingly powerful Communist Party secretary Leonid Brezhnev and the rising Marshal Andrei A. Grechko, soon to be appointed defense minister. Both had collaborated closely with Gorshkov when the latter commanded a series of successful landing operations against the German invaders on the Black Sea coast during World War II.

The USSR's Naval Infantry (*morskaya pekhota*), the Russian term for marines, who like paratroops are called *desantniki* (descent) or landing troops, were disbanded after that war. Their reestablishment, barely begun in Khrushchev's last years in power, was accelerated after his downfall. Independent marine battalions (*OMBPs*, later expanded to brigades) were attached to each of the navy's fleets. They were initially assembled from land formations, which may account for the marked disparity in professional competence between their role in the operations described here and that of the slapdash landing parties that were raised on the navy's warships, for whom this was a recent and unfamiliar departure.

---

[2] A. B. Shirokorad, *Flot, kotory unichtozhil Khrushchev* [The fleet that Khrushchev destroyed] (Moscow: Vzoi-AST, 2004).

Topping a range of amphibious craft, a new class of large landing ships (*bolshoi desantny korabl'* or BDK, North Atlantic Treaty Organization [NATO] reporting name Alligator and the equivalent of landing ship, tank or LST) was introduced in 1966 as the marines' long-range operational platform. The Black Sea Fleet's first brigade of *desant* ships, the 197th Brigade, was formed in July of that year.[3]

Ships from the Black Sea and Baltic fleets made rotating tours of the Mediterranean for two years, becoming in May 1966 a permanent presence designated temporarily as the combined squadron. Its high-profile formal incorporation as the Fifth *Eskadra* (squadron), reporting directly to navy command in Moscow, was postponed until 14 July 1967, just *after* the Six-Day War, which was to be its first combat test. The *Eskadra*'s main anchorage, at Antikythera west of Crete, lacked shore facilities, and fulfilling the centuries-old Russian aspiration to warm-water bases beyond the bottleneck of the Turkish straits was one of the Soviet goals in the joint planning with Egypt that began in 1965. Despite—or rather, because of—the plan's overall failure, this goal *was* attained for close to 20 years. The Soviet buildup in the Mediterranean thus combined Gorshkov's overarching concept with such regional specifics as halting Israel's nuclear project and intensifying Egyptian dependence on Soviet arms and support as a hedge against U.S. influence.[4]

After sinking as low as a single ship in 1963, the Soviet Navy's Mediterranean average daily presence increased to 10 in 1965 and rose to 24 in 1967.[5] This was accomplished by sending in new flotillas while keeping the previous "watch" in place. By mid-1967, there were more than 30 armed warships in addition to a similar number of auxiliary craft. Submarines were introduced for the first time since 1961, when Albania closed the base it had provided. Nuclear submarines, too big for the Black Sea, were sent in through Gibraltar under the keels of surface vessels.

The Soviets' heightened assertiveness in shadowing the U.S. Sixth Fleet caused increasing concern for the American commanders. Sixth Fleet commander Vice Admiral William I. Martin warned publicly on 17 May 1967 that "a Soviet naval build-up in the Mediterranean is threatening" his fleet. However, his concern was mainly that, "the Fleet [is] now no longer able to devote itself entirely to mounting strike operations against the Soviet Union." If he was aware of the actual Soviet preparations for a Middle East intervention, he did not mention it.[6] Subsequent analyses have ascribed

---

[3] VAdm A. A. Tatarinov et al., ed., *Shtab Rosiyskogo Chernomorskogo Flota: 1831–2001: Istoricheskiy ocherk* (Simferopol: Tavrida, 2002), 77; and Norman Polmar, *Guide to the Soviet Navy*, 3d ed. (Annapolis, MD: Naval Institute Press, 1983), 13, 16.

[4] Vladimir Zaborsky, "Sovetskaya Sredizemnomorskaya Eskadra [Soviet Mediterranean Squadron]," *Nezavisimoye Voennoye Obozreniye* [military supplement of *Nezavisimaya Gazeta*], 13 October 2006.

[5] Gordon H. McCormick, *The Soviet Presence in the Mediterranean* (Santa Monica, CA: Rand, 1987), 7.

[6] Reuters report on Martin's address to the American Club, Rome, 17 May 1967, quoted in Cdr Robert Waring Herrick, *Soviet Naval Strategy: Fifty Years of Theory and Practice* (Annapolis, MD: U.S. Naval Institute, 1968), 154n13.

the overall buildup during 1967 as an effect, rather than a precursor and cause, of the crisis and war that erupted at midyear.[7]

However, the newly recommissioned marines were already in the Mediterranean as an essential part of the naval component in the intervention plan, which also included the other *desantniki*, paratroops.[8] The marines' role in the planned naval operation against Israel was covered up so thoroughly for almost 30 years that it only came to light after the unveiling of another, seemingly much unlikelier, amphibious component. Captain Yuri N. Khripunkov was then the gunnery officer on a brand-new frigate, yet "unchristened" and still known only by its generic appellation, *SKR-6*. His interview in a Ukrainian newspaper in 1994 first revealed this entire operation, which had never been officially disclosed by the USSR or post-Soviet Russia.[9]

In 1966, Aleksandr Kislov was a Middle East correspondent for TASS, the Soviet news agency that played a key intelligence and propaganda function in the crisis. By 2002, Kislov was a professor and head of the Russian Academy of Sciences' Center for Research of Peace Problems. Citing "personal observation," he disputed Khripunkov's claim whereby Moscow's preplanned operation against Israel included improvised landing parties of "volunteer" seamen. Kislov's postscript held that the USSR intended to intervene only "in dire necessity, to stop Israeli aggression." This confirms that a Soviet landing force *was* prepared to strike at Israel. In disputing that claim, he wrote that "*desant* ships with marines who were well-prepared both operationally and psychologically" were present and prepared.[10]

Subsequent references confirmed and detailed these marines' presence and mission. By mid-May a second *BDK* joined its sister ship and two *SDKs* (*sredny desantny korabl'* medium landing craft) that were already attached to the *Eskadra*. Naval histori-

---

[7] McCormick, *The Soviet Presence in the Mediterranean*, 9.
[8] A paratroop division was trained in Crimea (and another in Azerbaijan) for a month before the Six-Day War for a drop in Israel, and they were kept in readiness on the runways for its duration. LtCol Anatoly Isaenko, "Polety na Blizhniy Vostok [Flights to the Middle East]," *NVO* (*Nezavisimoye Voyennoye Obozreniye–Nezavisimaya Gazeta* military supplement), 15 December 2006. Unlike the marines, these units included Jewish conscripts. Two of these later immigrated to Israel and, in interviews with Zeev Katz of the Hebrew University, reported that they spent several days in transport aircraft on the runways prepared for a drop in the Middle East. Professor Katz, personal communication with the authors, June 2000. In a retrospective top-secret assessment, the CIA confirmed reports of this as well as the naval-marine component of the planned Soviet intervention, but in versions that reflect the Soviet propaganda line more than direct knowledge of the actual preparations. "Soviet Policy and the 1967 Arab-Israeli War (Reference Title: Caesar-XXXVIII)," CIA Directorate of Intelligence, 16 March 1969.
[9] The late Capt Khripunkov's account is assembled from a copy he provided of his article, "Khodili my pokhodami . . . : vospominaniya ofitsera flota [The missions we carried out . . . : Memoirs of a naval officer]," *Vecherny Donetsk*, 1994, and subsequent interviews we held with him, as well as his appearance that we arranged in an Israeli documentary: Ilan Ziv, "1967 Six Day War," YouTube video, episode 12, 4 December 2012.
[10] Aleksandr Kislov, "Ne v ladakh s faktami [Incompatible with the facts]," afterword to Isabella Ginor, " 'Shestidnevnaya voyna' 1967 g. i pozitsiya SSSR [The "Six-Day War" and the position of the USSR]," *USA and Canada* (Moscow: Russian Academy of Sciences, USA and Canada Institute, 2002), 76–91.

an Aleksandr Rozin points out that by 10 June *BDK-6* (later named *Krymsky Komsomolets*) was in Port Said with marines of the 309th OMBP on board.[11] This too predates the officially announced postwar entry of a Soviet flotilla into this northern gateway of the Suez Canal, after a renewed Israeli "aggression" on 9 July.[12]

However, an earlier entry confirms a unique but authoritative testimony from the commander of the naval infantry formation's heavy-weapons company, which appeared in 2003 in an online organ of the Belarus Ministry of Defence. Then Lieutenant Colonel Viktor Shevchenko's company was armed with mobile rocket launchers. It would be the first detachment of the new Soviet Naval Infantry to go into combat in the only part of the Soviet intervention plan that is known so far to have produced an actual clash with Israeli forces in the Six-Day War.[13]

Shevchenko was motivated to speak out by a combination of economic hardship and old soldiers' honor, like much of the veterans' literature that by that time was near the end of its "golden age" in the years around the dissolution of the USSR. Though still in uniform as a military academy instructor, he broke the longstanding coverup with a demand for recognition of his troops' battlefield sacrifice, especially those who were killed or injured. As Moscow had never officially acknowledged its failed intervention in 1967, no reference to it was registered in the marines' papers. Neither they nor their survivors received even the small extra allowance for combat veterans over other former servicemember's pensions never mind citations or medals.[14] His protest was therefore short on detail, including even an exact date. But after an initially positive reply from his interviewer when the chapter authors inquired for more information, the entire article was deleted—the copy in the authors' collection may be the only trace—and they were denied access to Shevchenko.

However, once alerted by his startling account, the authors soon discovered corroborating evidence in post-Soviet naval documentation. One such reference dates the dispatch of a *rota* (company) from the 309th OMBP, presumably Shevchenko's, with a number of PT-76 amphibious tanks, to Egypt on 26 May on board its usual BDK operational platform, either the *Krymsky* or the *Voronezhsky Komsomolets*. The

---

[11] Aleksandr Rozin, "Sovetsky VMF v sderzhivanii i prekrashchenii 'chestidnevnoy voiny' v 1967g [The Soviet Navy in Deterrence and Termination of the 'Six-Day War' in 1967]," in A. O. Filonik, ed., *Blizhniy Vostok: Komandirovka na voyn: Sovetskie voennye v Egipte* [Middle East: Mission to War: Soviet Military in Egypt] (Moscow: Academy of Sciences and Moscow State University, 2009), 188; and MajGen Vladimir A. Zolotarev, *Rossiya (SSSR) v lokalnikh voynakh i voyennykh konfliktakh vtoroi poloviny XX veka* [Russia (USSR) in local wars and armed conflicts in the second half of the 20th Century] (Moscow: Russian Federation Institute of Military History, 2000), 185.
[12] Ginor and Remez, *The Soviet-Israeli War, 1967–1973*, 20–29.
[13] Andrei Fyodorov, "Neizvestnaya voyna 'egiptyanina' Shevchenko [The unknown war of 'the Egyptian' Shevchenko]," *Vo slavu rodine*, no. 93, 22 May 2003.
[14] Isabella Ginor and Gideon Remez, "Veterans' Memoirs as a Source for the USSR's Intervention in the Arab-Israeli Conflict: The Fluctuations in Their Appearance and Character with Political Change in Post-Soviet Russia," *Slavic Military Studies* 29, no. 2 (2016): 279–97, https://doi.org/10.1080/13518046.2016.1168136.

latter ship had been attached to the Baltic Fleet since its completion in Kaliningrad in 1966 and is listed as "based in Egyptian ports from June 1967."[15]

But the regular marines' available force was still inadequate for the impending mission. Just before the crisis was sparked by an ostentatious warning from the USSR to Egypt that Israel was massing forces to attack Syria, the deputy commander of the Black Sea Fleet arrived at Antikythera to take command of the "combined" *Eskadra*. Viktor Sysoev's rank, vice admiral, was higher than the squadron's usual chief, indicating preparation for an extraordinary mission. He brought sealed orders for the skippers of the *Eskadra*'s warships that were to be opened after receiving a coded signal. They were to raise landing parties of purported "volunteers" and send them on raids against Israeli coastal targets.[16]

Khripunkov's *SKR-6*, a *Petya II*-class antisubmarine frigate of the fastest and most advanced model in the Soviet Navy (the first to be powered by gas turbines), was a typical component of the *Eskadra*'s buildup. It had just been completed at Kaliningrad's Yantar shipyard and delivered to Baltiysk. On 3 May, well before the overt outbreak of the crisis, it was dispatched to the Mediterranean, along with the *SKR-13*, on their maiden voyage. They were supposedly en route to the Black Sea, "but when we reached the Med, we were told to stay there," Khripunkov recalled.[17]

In interviews held with Khripunkov, the authors were even more astounded to hear the target that he was assigned. His 30-person landing party—one-quarter of his ship's company—was aimed at no less than Haifa port, Israel's main harbor and naval base. Unrealistic as this seemed initially to the authors, it was no less so than to Kislov. Once alerted to it, the authors collected multiple similar testimonies from other ships and officers. They include the published memoir of Ivan Kapitanets, a future admiral of the fleet who was then a destroyer captain. He took on board about 100 naval cadets who were in training on the squadron's flagship, the cruiser *Slava*.[18] Another authoritative source reports that on the submarine tender *Magomed Gadzhiev* (1969), which normally carried a crew of about 450, the landing party of 75 included "every available hand, including medics and even cooks."[19]

On a professional level closer to the marines', a naval commando team was prep-

---

[15] A. B. Morin, "Bol'shye desantnye korabli tipa 'Voronezhsky Komsomolets' pr. 1171 [Large landing ship *Voronezhsky Komsomolets*, project 1171]," *Taifun*, 47 (2005).

[16] O.S. Pevtsov and Yu A. Portnov, "A bylo eto, pomnitsya, tak" [so it was, I remember]," *Podvodnya Flot* [submarine fleet] magazine, no. 9, 2001; and Tatarinov et al., *Shtab Rosiyskogo Chernomorskogo Flota: 1831–2001*, 81.

[17] Capt Yuri N. Khripunkov, telephone interview with authors, August 1999; and Capt Yuri N. Khripunkov, personal interview with authors, October 2006.

[18] Adm Ivan M. Kapitanets (Ret), *Na sluzhbe okeanskomu flotu, 1946–1992: zapiski komandujuschego dvumja flotami* [In the service of the oceanic fleet, 1946–1992: Notes of the commander of two fleets] (Moscow: Andreyevsky Flag, 2000), 174–76; and Ivan M. Kapitanets, telephone interview with authors, 11 January 2003.

[19] Adm V. A. Kravchenko, ed., *Podvodnye sily Chernomorskogo flota* [Submarine forces of the Black Sea Fleet] (Simferopol, Crimea: Tavrida, 2004), 125, 422.

ositioned on a Soviet submarine. Its leader was Gennady Zakharov, a future admiral and the deputy commander of President Boris Yeltsin's guard during the latter's confrontation with the Russian Parliament in 1993. He related a decade later that as a lieutenant in 1967, he commanded a detachment of naval special forces (*spetznaz*): "During the war in the Middle East, we were sitting in a submarine close to shore. Our mission was to destroy Israeli oil terminals and reservoirs," which were located near Haifa. This means they must have been assigned their target and dispatched from their Black Sea base considerably earlier.[20]

Even with inputs like Khripunkov's from 30 ships for a total of about 1,000 mostly untrained and unequipped troops, what could these landings have achieved? All of Israel's able-bodied reservists had been called up and its outnumbered military was stretched to the limit along the borders. A series of such raids might cause serious disruption, damage morale, and drain forces from the front. This might be exacerbated by support for an expected uprising among Israeli Arabs, for which Arabic interpreters attached to the Soviet advisors' *apparat* in Egypt were summoned to the embassy in Cairo, transferred to Alexandria and informed that they would be posted to ships cruising off the Israeli shore. "One of the interpreters . . . said he knew for sure that we would be attached to a *desant* force that would be landing in Haifa or slightly northward," to handle liaison with Israeli Arabs, "who were longing for us."[21]

The orders (Plan Victor) that were issued to Soviet-advised Syrian formations, which were poised to invade Israel from the Golan Heights in the northeast, called for cutting across the country or less than 80 kilometers. They were to link up with an "Egyptian" landing force on the coast north of Haifa, which actually could only have been Soviet.[22]

The entire operation was to be unleashed once Egypt, on signal from Moscow, initiated a series of such provocative moves that Israel would be goaded into a first strike. It was anticipated as a ground offensive, which the Egyptian forces concentrated in Sinai would have to contain until Israel was branded as the aggressor, thus legitimizing a Soviet intervention. When Israel dallied, the Soviets added their own provocation by sending their most advanced aircraft, the still-experimental MiG-25

---

[20] Evgeny Zhirnov, "Rutskogo v Lefortovo ya soprovozhdal sam [I Personally Escorted (Vice President Aleksandr) Rutskoi to Lefortovo; interview with Zakharov]," *Kommersant Vlast'*, 16 April 2002. RAdm Shlomo Erel, who commanded the Israeli Navy in 1967, recalled to the authors a still-mysterious incident on 8 June in which the Israeli destroyer INS *Haifa* (K 38) engaged a submarine 24 kilometers off the naval base of Atlit, south of Haifa. "It was attacked with depth charges . . . oil slicks and debris were spotted, and the engagement was broken off." The submarine's initially assumed identification as Egyptian was later ruled out, but the incident was not investigated further. Shlomo Erel, personal communication with the authors, 7 August 2004; and Ginor and Remez, *Foxbats over Dimona*, 178–79.
[21] Aleksandr Khaldeyev, "Nesostoyavshiisya desant [The landing that did not occur]," *Okna* (Tel Aviv), 14 September 2000.
[22] Syrian documents, some in Russian, reproduced in Yehezkel Hameiri, *Mishnei evrei harama* [On both sides of the heights] (Tel Aviv: Lewin-Epstein, 1970). The rendezvous with a landing force is detailed on p. 58; analysis in Ginor and Remez, *Foxbats over Dimona*, 70–71.

or Foxbat, on two sorties over Israel's nuclear complex at Dimona. This spurred Israel's preemptive air offensive on 5 June, and the Soviets had the pretext they desired.

Khripunkov's captain opened his orders, and ordered the frigate set course for Haifa. The landing party was recruited and received vague orders. Once depth decreased to 15–20 meters, they were to head ashore on the ship's *kater* (cutter, motor launch). Two trips might be needed, which would leave the first 15 troops alone to face whatever awaited them on shore. The personnel were neither trained nor equipped for land warfare, and their leader was given no maps or specific targets. As Khripunkov told the authors, "What were we supposed to accomplish, with my pistol and the sailors' AK-47s? 'Get in there and see,' they told us. 'Throw your RG-42s [depth grenades designed for use against frogmen]. Wipe out the enemy forces'."[23]

Wait for reinforcements, they were told in general terms. Khripunkov was already aware of the marines' presence with the *Eskadra*—another indication that their deployment preceded the war: "there was also a *BDK* with about 40 tanks and maybe a battalion of infantry." But "nothing concrete was said" about the marines' mission. Likewise, "the air force was going to support us." Not that Khripunkov and his men expected much from the promised air support. "How could we contact them? We had nothing ready—no radio gear, no codes, no signal rockets, nothing."[24]

Khripunkov and his troops were thus well aware that they were expendable. "Losing 1,000 men," he remarked at the height of anti-Soviet backlash in newly independent Ukraine, "was nothing for the USSR. They started counting at five million. Each side wanted to demonstrate its dominant role. . . . The United States sends in the [Sixth] Fleet. We bring in our Black Sea Squadron. They send in spy planes. We start preparing a landing in Israel. The Israeli tanks move through Sinai and are ready to skip over the Suez Canal. What then? We land our force and World War III begins?" Still, on board *SKR-6*, only one sailor refused to "volunteer"; he was later transferred to another unit but, as far as Yury Nikolaevich knew, was not otherwise disciplined. "I was a foolish young man then. Today, I too would probably have refused such a mission."[25]

Shevchenko mentioned no such qualms among his marines. The blackout that was reimposed on his account left it unclear whether his outfit's original mission was to engage Israel's frontline forces, as it wound up doing when the Soviet-Egyptian plan backfired spectacularly. Soviet advisors and pilots who were already in Egypt reported not only the near-total devastation of its aircraft but also the destruction

---

[23] Capt Yuri N. Khripunkov, telephone interview with authors, August 1999; and Capt Yuri N. Khripunkov, personal interview with authors, October 2006.
[24] Capt Yuri N. Khripunkov, telephone interview with authors, August 1999; and Capt Yuri N. Khripunkov, personal interview with authors, October 2006.
[25] Capt Yuri N. Khripunkov, telephone interview with authors, August 1999; and Capt Yuri N. Khripunkov, personal interview with authors, October 2006.

of the runways in its bases.²⁶ The Soviet fighter squadrons and strategic bombers that had been readied at the USSR's southernmost bases had nowhere to land and insufficient range for the round trip. While the *Eskadra* already had an amphibious force to match the Sixth Fleet's, it had no aircraft carriers—a situation hardly changed since then. The air component of the Soviet intervention became unfeasible, including the vital air support that had been promised to the landing parties.

As the Egyptians had feared, shorn of any airpower—never mind air superiority—they had no hope of stemming the Israeli ground attack and launching a counteroffensive, even with Soviet support. As remnants of the routed Egyptian Army fled westward across the canal, Shevchenko's *BDK* anchored at Port Said, the marines went ashore, and attempted to cross it eastward. But as Shevchenko related, his company was ravaged by an Israeli air raid, leaving 17 killed and more than 30 injured, including their commander, who 35 years later was still nicknamed "the Egyptian."²⁷

This appears to correspond with a report that, on 8 June, "two battalions of Egyptian artillery which opened fire from the far side of the Canal" at the first Israeli force to arrive "were hit by an Israeli air strike and destroyed." Were these Shevchenko's *Katyushas*? The episode was mentioned only in a "quickie" history of the war by authors who enjoyed privileged access.²⁸ Its disappearance from subsequent versions, including the official Israeli record, appears to reflect Israeli reluctance to highlight direct Soviet involvement and losses, which might provoke retaliation. Together with the Soviets' own censorship, this created a "perfect storm" for obscuring the marines' role. But a coded reference to Shevchenko's engagement appears to be preserved in the *Voronezhsky Komsomolets*' combat record: "Its name was glorified in Port Said during the Arab-Israeli conflict. The ship gave internationalist support to the armed forces of Egypt . . . in repulsing Israeli aggression."²⁹

That Shevchenko's engagement took place no later than 9 June is confirmed by the subsequently published account of another marine officer who was then based in Baltiysk. At 0400 on 10 June, then-lieutenant Valery Mallin relates, the remainder of his 309th OMBP was ordered into combat readiness and was urgently transported overland to the Black Sea. The next day, it sailed for Port Said, evidently in relief of Shevchenko's shattered force; and since their formation's usual *BDK* platform was already in the arena, these marines embarked on a destroyer and a tanker. This ostensibly responsive move—rather than the politically sensitive and therefore undisclosed

---

²⁶ Zolotarev, *Rossiya (SSSR) v lokalnikh voynakh i voyennykh konfliktakh vtoroi poloviny XX veka* [Russia (USSR) in local wars and armed conflicts in the second half of the 20th Century], 183.

²⁷ Fyodorov, "Neizvestnaya voyna 'egiptyanina' Shevchenko [The unknown war of 'the Egyptian' Shevchenko]."

²⁸ Randolph S. Churchill and Winston S. Churchill, *The Six-Day War* (London: Heinemann/Penguin, 1967), 176.

²⁹ Morin, "Bol'shye desantnye korabli tipa 'Voronezhsky Komsomolets' pr. 1171."

earlier involvement in a preplanned offensive—opens Mallin's list of 25 "combat missions in various areas of the world's oceans," in which the Baltic marines formed part of landing forces.[30]

But for lack of air support, any *desant* on the Israeli coast became not merely suicidal but pointless, even though the U.S. Sixth Fleet had been ordered away from the eastern Mediterranean in a display of neutrality. The *Eskadra* had the entire basin virtually to itself, and the landing parties' orders were put on hold, though not entirely rescinded. For the coming five days, their ships cruised up and down the Israeli coast, just outside territorial waters. The head of Israeli signal intelligence at the time told the authors that his stations monitored signals from 42 Soviet vessels, but could not break their code.[31] When, on 8 June, Israeli planes and patrol torpedo (PT) boats attacked the single U.S. Navy ship that was left behind, the signal-gathering vessel USS *Liberty* (AGTR 5), off the Sinai coast, the first ship to approach and offer help was a Soviet destroyer. The improvised naval landing parties continued to train aboard ship. "As an officer, I knew how to use small arms, but the sailors had not fired more than five bullets in target practice, and never had thrown a grenade."[32]

The landing operation in Israel was reactivated when, on 9 June, having overcome both the Egyptian threat and a Jordanian attack, Israel responded to days of Syrian shelling and sent its forces to the Golan Heights. Making its first use of the hotline to Washington, the Soviet leadership threatened "action, including military" if the Israeli advance toward Damascus were not halted. This was not empty bluster; what remained of the Soviet intervention, which had originally been a top-secret operational plan designed to *win* a war, now became an overt deterrent move to *end* it. Khripunkov's frigate was once again turned toward Haifa.[33]

In Lyndon B. Johnson's White House situation room, it was decided to reverse the Sixth Fleet's course and order it back toward the war zone. Defense Secretary Robert S. McNamara and (less plausibly) CIA chief Richard Helms have claimed credit for this decision and thus for deterring Moscow from making good on its threat. In fact, the decision was implemented too late to make any difference. The order to the Sixth Fleet, "reflecting telephoned instructions from McNamara," was transmitted from the Joint Chiefs of Staff at 1522 hours Washington time, that is 2122 Israel time, well after Israel, fearing a direct confrontation with the USSR, accepted and observed a

---

[30] Col Valery Bakirovich Mallin's survey "Boevye Sluzhby Baltiskoy Morskoy Pekhoty [Combat Services of the Baltic Marines]" has appeared in several versions since 1997. The version most accessible at present was posted on *Taifun*, 31 August 2015. An abridged form was included in Filonik, *Blizhniy Vostok*, 143-51.
[31] BGen Yoel Ben-Porat, interview with the authors, 8 March 2002.
[32] James M. Ennes Jr., *Assault on the Liberty: The True Story of the Israeli Attack on an American Intelligence Ship* (New York: Random House, 1979), 116.
[33] Capt Yuri N. Khripunkov, telephone interview with authors, August 1999; and Capt Yuri N. Khripunkov, personal interview with authors, October 2006.

ceasefire.³⁴ *SKR-6* was halted a half hour's sail from its objective. As Zakharov retold it, his submarine-based force "would have carried out [the mission], but the war ended before the final order to act was received."³⁵

This, however, was not the end of the Soviet Marines' involvement. In Moscow, Brezhnev and his allies decided to double down on their commitment to Egypt rather than cut their losses. A massive airlift of materiel to replace the lost hardware was launched while the war was still in progress. A military delegation led by Chief of Staff Matvei Zakharov toured the new front line along the canal to establish its defense, and the Soviet Marines became the first of their country's 50,000 regular troops—distinct from individual advisors—to be stationed in Egypt by 1973, up to 20,000 at a time.³⁶

On 9 July, the Soviets took advantage of the first renewed flareup at Ras el-Ish on the Suez Canal to overtly flaunt the *Eskadra*'s entry into Alexandria and Port Said, where its ships had actually been present since before the war. These ports now effectively became Soviet naval bases, fulfilling one of the USSR's main war aims despite the overall fiasco. Whether Mallin's marines took part in the Ras el-Ish engagement, they now took up positions on the canal's northern sector to hold the line until the Egyptian army could regroup. Their rotating presence in three-month tours of duty became part of the Soviet regulars' combat deployment in Egypt, to be reinforced several times when tensions peaked, as during the latter's War of Attrition with Israel in 1969-70.³⁷

The outbreak of this conflict added special urgency to the otherwise routine dispatch described by then-lieutenant V. I. Dmitriev. His outfit's departure on 15 May 1969 followed the start of massive artillery barrages on 9 March. It included two infantry companies (of which he commanded one), one each of amphibious tanks and of mortars, and a platoon of "shoulder-fired anti-tank missiles"—the earliest known appearance of the *Malyutkas* (Saggers) in Egypt, where they would play a crucial role in the cross-canal offensive of October 1973.³⁸

As in 1967, "the personnel and equipment boarded . . . two destroyers, two minesweepers, a large landing ship and two medium landing ships, and on May 15, 1969 headed for the Mediterranean." On the *BDK*, "the marines were quartered under the tank deck on three-tiered canvas bunks. Their kit was folded into helmets that hung

---

[34] Harriet Dashiell Schwar, ed., *Foreign Relations of the United States, 1964–1968*, vol. 19, *Arab-Israeli Crisis and War, 1967*, doc. 253, "Telegram from the Joint Chiefs of Staff to the Commander-in-Chief European Command (Lemnitzer)," Recorded Date 10 June 1967, 1522Z. Original in National Security File, Country File, Middle East Crisis, vol. 9, 422, Lyndon B. Johnson Library, Austin, TX.

[35] Isabella Ginor and Gideon Remez, "The Six-Day War as a Soviet Initiative: New Evidence and Methodological Issues," *Middle East Review of International Affairs* 12, no. 3 (September 2008).

[36] Ginor and Remez, *The Soviet-Israeli War*, 5–52.

[37] Zolotarev, *Rossiya (SSSR) v lokalnikh voynakh i voyennykh konfliktakh vtoroi poloviny XX veka* [Russia (USSR) in local wars and armed conflicts in the second half of the 20th Century], 185.

[38] V. I. Dmitriev "Zapiski leytenanta morskoy pekhoty [Notes of a lieutenant of the Marine Corps]," in Filonik, *Blizhny Vostok*.

over their heads."³⁹ They arrived on the 19 May in Port Said, where on the next day "we took up combat duty." This included defense "in the second echelon of Egyptian troops" as well as "protection of ships, and evacuation of Soviet military advisers in the event of an [Israeli] breakthrough." The latter was considered so imminent that "on the ships located in Port Said, measures were immediately taken against underwater saboteurs, the sentries threw live grenades over the side of the ship" and divers began daily descents to inspect the hulls. On 9 June, the entire formation carried out an "amphibious landing" at Port Fuad, the Egyptians' only remaining foothold east of the canal facing the Israeli positions. "The uniform of the landing party was 'tropical' and consisted of a cap, shorts and a short-sleeved tunic, all blue. For action in combat conditions there was an army fatigue uniform, and for ceremonial purposes a black field uniform."⁴⁰

On 20 July, the Egyptians' initial, massive numerical advantage in firepower was reversed when Israel launched its air force into action as "flying artillery." The relative immunity that Port Said had enjoyed thanks to the Soviet naval presence was ended. Two days into the Israeli air offensive, Dmitriev witnessed an attack on Egyptian missile boats, even though they had "nestled up to [his] *BDK* for shelter. One burst of aircraft [cannon] fire perforated a UAZ-452 ambulance that was on the upper deck."⁴¹

The same day, British sources reported "Soviet marines sighted in Port Said" from among the "thousands of marines" on Soviet warships and landing craft in that harbor. This was an exceptional item, as Western media had little remaining presence in Egypt and none in the canal zone. Both "British and US officials . . . said that positioning Soviet naval commando units outside the USSR is an innovation for Russia and if true, this fact might be a very significant event on the way to dangerous confrontation in areas of tension worldwide." But on the record, the reports were downplayed as usual: the State Department held that the department "cannot even verify whether there indeed *were* marines on board" the Soviet ships.⁴²

On their return voyage in August, Dmitriev relates that his outfit took part in the Fifth *Eskadra*'s "first joint maneuvers" with the Egyptian and Syrian navies. The Israeli incursion did not materialize, but the marines' rotating presence remained a fixture at Port Said even after the bulk of Soviet regulars were withdrawn from Egypt once the ceasefire of August 1970 accomplished most of their mission. Much of this

---

³⁹ Dmitriev "Zapiski leytenanta morskoy pekhoty [Notes of a lieutenant of the Marine Corps]," in Filonik, *Blizhny Vostok*, 22–26.

⁴⁰ Dmitriev "Zapiski leytenanta morskoy pekhoty [Notes of a lieutenant of the Marine Corps]," in Filonik, *Blizhny Vostok*.

⁴¹ Yaacov Bar-Siman-Tov, *The Israeli-Egyptian War of Attrition, 1969–70: A Case Study of Limited Local War* (New York: Columbia University Press, 1980), 57–58; G. V. Karpov, "Vospominaniya sovetskogo voennogo sovetnika v Egipte," in Filonik, *Blizhniy Vostok*, 105; and Dmitriev "Zapiski leytenanta morskoy pekhoty [Notes of a lieutenant of the Marine Corps]," in Filonik, *Blizhny Vostok*, 22–26.

⁴² "Ein ishur layedi'ot al nehatim sovietim bate'alah [No confirmation for reports of Soviet marines on the canal]," *Ma'ariv*, 21 July 1967, 1, quoting State Department spokesman Robert McCloskey and "U.S. and British officials."

withdrawal was disguised as a unilateral "expulsion" of the Soviet "advisors," as the result of a fictitious rift with Moscow; however, little pretense was made to conceal the marines' continuing presence.[43] After a talk with the Soviet military attaché and GRU (military intelligence) *rezident*, a British counterpart reported that Port Said still "provided a haven for Soviet commando units as a counter to marine forces of the Sixth Fleet." The Briton noted that his Soviet interlocutor, Rear Admiral Nikolay Ivliev, had been very forthcoming on all subjects except the matter of Port Said, on which he seemed uncomfortable and confirmed in effect that the main change was in visibility: "Soviet ratings are still forbidden to go ashore in uniform." In London, it was correctly assumed "that there is something peculiar about the use to which the Russians put Port Said (we have always suspected this)." Ivliev's aim was understood "to convince us that the Soviet naval presence . . . was smaller than it really is."[44]

This was borne out in less than a year as one of several indications that the USSR was privy to Egyptian preparations for the cross-canal offensive that would be launched on 6 October. As in 1969, the impending war called for reinforcing the marines' presence beyond the usual rota of available naval-infantry units. The procedure that followed was remarkably reminiscent of the previous instances, going back to 1967, that included makeshift landing parties of seamen.

As early as 28 September—a week before the Egyptian-Syrian surprise offensive against Israel—the Soviet Baltic Fleet's marine force was once again put on alert. Some of its complement was already deployed in West Africa; the remainder, under Lieutenant Colonel V. I. Gorokhov, was flown in transport planes to Sevastopol with personal arms only. There it was loaded, with full battle gear and weapons borrowed from the Black Sea Fleet's counterpart formation, onto a *BDK*. Additional units followed the same day by train, to embark on two *SDK*s; all of them set sail for the Mediterranean. Another reinforced marine battalion followed on the same day directly from Baltiysk on the Baltic Fleet's own *BDK*, *Krasnaya Presnya*. The urgency and mode of the additional *desantniki*'s dispatch indicate preparation for a highly extraordinary mission: "forming an amphibious assault for operations in the conflict area." According to Captain Vladimir Zaborsky, on 17 October (after the Israeli canal crossing farther south), "preliminary plans for a limited 'demonstration' landing of Soviet naval infantry on the west bank of the canal were drafted. . . . One large and six medium landing ships were already in the region but they were all being used for equipment transport"—that is, the resupply sealift for Egypt that was already in full swing. So,

---

[43] Isabella Ginor and Gideon Remez, "The Origins of a Misnomer: The 'Expulsion of Soviet Advisers' from Egypt in 1972," in Nigel J. Ashton, ed., *The Cold War in the Middle East: Regional Conflict and the Superpowers 1967-73* (London: Routledge, 2007), 136-63, https://doi.org/10.4324/9780203945803.

[44] Amb Richard Beaumont, Cairo, to Foreign and Commonwealth Office (FCO), 13 November 1972; naval attaché J. P. Marriott, Cairo, to Ministry of Defense, "Call on Admiral Ivliev," 14 November 1972; D. A. S. Gladstone to A. J. M. Craig, Near East and North Africa Department, FCO, "Soviet/Egyptian Relations in the Military Field," 21 November 1972, all in file FCO 39/1265, National Archive (Public Records Office), London.

incredibly, once again "Gorshkov ordered . . . a landing force to be assembled of 'volunteers' from the crews of all combatant and auxiliary ships."[45]

This time, although an Israeli advance on Port Fuad again failed to materialize, it was not an imaginary scenario; such a move had first been proposed by the military command and supported by Defense Minister Moshe Dayan as early as 7 October, on the mistaken assumption that "the Soviets have evacuated Port Said." It was repeatedly postponed and finally ruled out only on the 22 October.[46] Whether the Soviets were aware of this, they were finally ordered to carry out the mission on 24 October. Until then, "some thousand men had signed up" for the landing parties. Captain Evgeny Semenov, then the *Eskadra*'s chief of staff, wrote in his journal: "Seems we're going to save Port Said from the Israelis." Their landing was called off "at the last minute"; Lyle Goldstein and Yury Zhukov, quoting from Semenov's unpublished manuscript, conclude that "this resort to volunteers is a sign that the *Eskadra* was to some extent in over its head." But as in 1967, the "volunteer" character of the force was risible, and the concept had, remarkably, remained an operational option. Semenov wrote that only after a "very difficult combat service," the force made a friendly landing at Tartus on 7 December.[47]

Mallin, by then a captain, commanded another such deployment in Syria via Tartus in August–December 1975. The last marine operation on his list was from February to August 1989 in Angola. Writing in 2015, he ended his survey on a doleful note: "Changes in Soviet policy that took place in the second half of the 1980s, the abandonment by the USSR of its interests in many regions, led to the loss of gains that had been made over decades" such as the Soviet role as "one of the deterrents in the permanent Arab-Israeli confrontation . . . the naval infantry ceased to perform combat service."[48]

But this was soon to come full circle. The very same year, Russia's intervention in the Syrian Civil War restored Russian naval presence in the Eastern Mediterranean, including amphibious capability, close to the Soviet peak in the 1960s and 1970s, with Tartus and Latakia replacing Alexandria and Port Said. Some of the 16 original

---

[45] Zaborsky, "Sovetskaya Sredizemnomorskaya Eskadra [Soviet Mediterranean Squadron]"; Zaborsky, "Zapiski o neizvestnoy voyne [Notes on an unknown war]"; Popov, "Desantnye korabli osvaivayut Sredizemnoye more [Landing ships in the Mediterranean]"; and an unpublished journal by Semenov, all quoted in Goldstein and Zhukov, "A Tale of Two Fleets," 27–63.

[46] Shimon Golan, *Kabbalat Hahlatot ba-Pikkud ha-Elyon be-Milhemet Yom Kippur* [Decision Making in the High Command in the Yom Kippur War] (Tel Aviv: Ma'arakhot [Israel Defense Forces publishing] and Modan, 2013), 436, 453–54, 1146, 1150.

[47] This account of the *Eskadra*'s moves in 1973 is based on Vladimir Zaborsky, "Sovetskaya Sredizemnomorskaya Eskadra [Soviet Mediterranean Squadron]," *NVO*, 13 October 2006; Vladimir Zaborsky, "Zapiski o neizvestnoy voyne [Notes on an unknown war]," *Morskoy sbornik* 3 (March 1999); V. I. Popov, "Desantnye korabli osvaivayut Sredizemnoye more [Landing ships in the Mediterranean]," *Taifun*, February 2002; and an unpublished journal by Semenov, all quoted in Lyle J. Goldstein and Yury M. Zhukov, "A Tale of Two Fleets: A Russian Perspective on the 1973 Naval Standoff in the Mediterranean," *Naval War College Review* 57, no. 2 (Spring 2004): 27–63.

[48] Mallin's remark is in Russian as quoted from Filonik, *Blizhniy Vostok*.

*BDK*s were reactivated as troop and materiel transports, after being mothballed in the 1990s. Besides exemplifying the continuity from Soviet to Russian strategy, these alligators are among the oldest warships still in service with a major navy. *Plus ca change . . .*

CHAPTER TEN

Operation Husky

The Challenges of Joint Amphibious Operations

Darren Johnson

In January 1943, leaders from the United States and Great Britain met in Casablanca, Morocco, to solidify Allied strategy in the Mediterranean theater of operations (MTO) during World War II. After the successful landings in French Morocco and Algeria (Operation Torch) on 8 November 1942 by Allied forces, President Franklin D. Roosevelt and Prime Minister Winston S. Churchill agreed on a strategy that would focus on eliminating the Axis's military presence in North Africa, securing lines of communication in the Mediterranean, and relieving military pressure on the Soviet Union by forcing the Germans to shift forces from the eastern front to the MTO.[1]

Cooperation between the United States and Great Britain was not a smooth process. In the summer of 1940, the groundwork for United States and Great Brit-

---

[1] In a joint Anglo-American letter to Soviet premier Joseph Stalin, President Roosevelt and Prime Minister Churchill said of Operation Husky, "We have made the decision to launch large-scale amphibious operations in the Mediterranean at the earliest possible moment. Preparation for these operations is now under way and will involve a considerable concentration of forces, including landing craft and shipping in Egyptian and North African ports." President Roosevelt and Prime Minister Churchill were conscious of the need to support the Soviet Union not just materially through Lend-Lease but also through direct military action against the German armed forces. Iskander Magadeyev and Olga Kucherenko, "Casablanca: A Table Just for Two (November 1942 to January 1943)," in *The Kremlin Letters: Stalin's Wartime Correspondence with Churchill and Roosevelt*, ed., David Reynolds and Vladimir Pechatnov (London: Yale University Press, 2018), 169–203, https://doi.org/10.2307/j.ctv7cjvz5.14.

ain's military collaboration was laid by U.S. naval observer, Rear Admiral Robert L. Ghormley, whose mission was to establish naval cooperation between the two nations should the United States get drawn into the war.[2] After the Japanese attack on Pearl Harbor on 7 December 1941, the United States and Great Britain developed a strategic framework at the Arcadia Conference in Washington, DC.[3] Foundational to the Allied strategy was to focus their efforts on the defeat of Germany first and then defeat the Japanese in the Pacific.[4] Beyond that decision, no substantive agreements were made by the two Allies at the conference. In the United States' viewpoint, defeating Germany required immediate action, a drive straight to Berlin from an invasion along the northwestern coastline of France. In a note on 22 January 1942, future Supreme Allied Commander in Europe, Dwight D. Eisenhower wrote, "We've got to go to Europe and fight. And we've got to quit wasting resources all over the world—and still worse—wasting time."[5] In contrast to the American sentiment, Churchill warned that a defeat on the French coast was "the only way in which we could possibly lose this war."[6] The British approach to defeating Germany lay in fighting on the periphery, or what they called the "soft-underbelly" of Europe.[7] While the United States and Great Britain differed on the strategy to defeat Germany, the need to militarily engage with Germany in the near term became the priority. Operation Torch, the invasion of Axis occupied North Africa, was a compromise between the two nations. The Americans were able to finally engage Germany in combat operations on land and British desires for a peripheral strategy were placated. This study argues that the challenges in conducting amphibious operations during Operation Husky in Sicily necessitated an increased level of cooperation between the United States and Great Britain for the remainder of the war.

Lieutenant General Eisenhower led Operation Torch, with British officers serving as his chief deputies.[8] A collective force of 125,000 soldiers, sailors, and airmen from the United States and Great Britain conducted simultaneous amphibious inva-

---

[2] Gordon A. Harrison, *Cross-Channel Attack: The European Theater of Operations*, U.S. Army in World War II (Washington, DC: U.S. Army Center of Military History, 1993), 1.
[3] The Arcadia Conference was a series of 12 meetings held by American and British leaders in Washington, DC. *Proceedings of the American-British Joint Chiefs of Staff Conferences*, 2 pts. (Washington, DC: Joint Chiefs of Staff, 1941).
[4] LtCol Albert N. Garland and Howard McGaw Smyth, *Sicily and the Surrender of Italy: The Mediterranean Theater of Operations*, U.S. Army in World War II (Washington, DC: U.S. Army Center of Military History, 1993), 2.
[5] Rick Atkinson, *An Army at Dawn: The War in North Africa, 1942–1943*, vol. 1 (New York: Owl Book, an imprint of Henry Holt, 2003), 11.
[6] Atkinson, *An Army at Dawn*, 13.
[7] The British experience at Dunkirk, France, and the failed Dieppe amphibious raid all contributed to the British strategy of delaying a direct assault against German forces in Western Europe. To many American leaders, the Mediterranean strategy was only perpetuating British imperial ambitions in the region. Atkinson, *An Army at Dawn*, 14.
[8] This would be a normal occurrence during the war, ensuring cooperation and partnership between the Allies in planning and execution of major operations.

sions at Casablanca, Oran, and Algiers on 8 November 1942. Operation Torch planners were unsure if the Axis aligned Vichy French forces would fight against an Allied invasion. Intelligence reports indicated that if the Allies encountered stiff resistance, their advance toward Tunisia may be delayed by up to three months.[9] The complexity of the Operation Torch landings was already high as three separate task forces descended on nine different landing sites from embarkation points in the United States and Great Britain. Navigational errors, delays in landing, darkness, weather, and sea currents all had an impact on the various amphibious landing locations. Sporadic Vichy French resistance delayed the Allied advance inland but was isolated and not well coordinated.[10] A more determined enemy may have capitalized on the friction the Allies experienced during the execution of Operation Torch.[11] Despite the various challenges the Allies endured during Operation Torch, the experience provided them, especially the Americans, a blueprint for conducting amphibious operations in a joint environment.

Inter-Allied disputes came to the forefront once again at the Casablanca Conference in January 1943. The American delegation continued to advocate for a cross-channel invasion from England in 1943, with U.S. Army Chief of Staff general George C. Marshall being the most prominent supporter. Given the amount of personnel and resources already in North Africa after the Operation Torch and Tunisia campaigns, the prudent choice was to use these forces against German and Italian strongholds in the Mediterranean. With reluctance, the American military leadership, with President Roosevelt's approval, agreed to the British peripheral strategy, and began planning for what would eventually be Operation Husky, the invasion of Sicily in July 1943.[12] The United States gave their support to the invasion of Sicily in 1943 for British assurances (but not guarantees) of a cross-channel invasion of northwest France in 1944.[13]

The Combined Chiefs of Staff (CCS), along with their staffs, developed plans for

---

[9] Atkinson, *An Army at Dawn*, 23.

[10] Charles A. Anderson, *Algeria-French Morocco: The U.S. Army Campaigns of World War II* (Washington, DC: U.S. Army Center of Military History, 2003).

[11] Anderson, *Algeria-French Morocco*, 30. Anderson highlights that landing ships for infantry could avoid many sandbars along the landing sites that the heavier landing ships for vehicles could not. This resulted in many vehicles being off-loaded in water that disabled their electrical systems.

[12] The Soviet premier, Joseph Stalin, expressed his displeasure that the "Anglo-American alliance" was not conducting a cross-channel invasion as was promised in earlier communications. Stalin highlights that due to the Anglo-American failure to invade Western Europe resulted in the transfer of 36 German divisions to the eastern front, putting additional pressure on Soviet forces. "Operations: Operation Husky: Stalin to Prime Minister Husky Cannot Replace Second Front in France," 15 March 1943, vol. 17, folio 326, FO 954/17B/326, National Archives, Kew.

[13] Robert M. Citino, "Smashing the Axis: Operation Husky and the Sicilian Campaign," in *The Wehrmacht Retreats: Fighting a Lost War, 1943* (Lawrence: University Press of Kansas, 2012), 163-97. Despite the assurance of a cross-channel invasion in 1944, a long-term and unified strategy was yet to be determined by the Allies at this time in the war. The "next step" for actions in the Mediterranean had yet to be planned.

Operation Husky, all while fighting in North Africa continued.[14] General Dwight D. Eisenhower was named as Supreme Allied Commander, with three British officers as subordinates: General Harold Alexander as commander of ground forces, Admiral Andrew B. Cunningham in command of the naval forces, and Air Chief Marshal Arthur W. Tedder commanding the air forces. Placing British military leaders in direct command of the invasion forces enabled the British to keep a "watchful" eye on their American allies.[15] The combined Anglo-American staffs generated more than eight plans to invade Sicily, none of which receiving full support from the various officers charged with executing the operation. Much of the difficulty in planning Operation Husky was due to the geographic separation of the Allied staffs. With the North African campaign still on-going, planning cells were scattered across the battlefield with commanders still focused on defeating the Germans and Italians in Tunisia. Allied planners relied on imperfect intelligence to determine the composition and disposition of the German and Italian forces that occupied Sicily.[16] Moreover, Allied leaders could only guess as to the morale of Axis forces on Sicily. Heavy Allied bombardment, coal and food shortages, and logistical constraints were expected to demoralize Axis soldiers' morale and their commitment to fight. Regardless of the enemy efforts on the battlefield, Operation Husky planners identified ports, airfields, and major population centers as key terrain that needed to be seized and secured to enable a successful amphibious assault.[17]

Initially, planners focused on dispersing the Allied landing sites on Sicily. This was done to avoid concentrating too many naval assets in one area as well as to maximize the seizure of airfields for use by Allied aircraft.[18] At the behest of British General Bernard L. Montgomery, who commanded the British Eighth Army on Sicily, subsequent invasion plans brought the proposed American landing sites closer to the British invasion area. By consolidating the entire invasion force along a 160-kilometer stretch of beach, each amphibious landing force offered mutual support should a determined German and Italian counterattack occur.[19] German forces fought tena-

---

[14] The Combined Chiefs of Staff (CCS) included the senior staffs of the United States and Great Britain that, with head of state approval, established military policy decisions during World War II.

[15] Citino, "Smashing the Axis," 167. In the Tunisian campaign, British Gen Alexander distrusted the American fighting ability and relegated the American Army's II Corps to a minor role of flank security for the British, which was a similar role American forces would have on Sicily. Carlo D'Este, *Bitter Victory: The Battle for Sicily, July–August 1943* (New York: Harper Collins, 1988), 66.

[16] Estimates of Axis forces on Sicily ranged from 300,000 to 365,000, with 40,000–62,000 of them being German. D'Este, *Bitter Victory*, 606.

[17] D'Este, *Bitter Victory*, 145.

[18] Darren Johnson and Claudio Innocenti, *The West Point Guide to the Campaigns of World War II: Sicily* (New York: Rowan Technology Solutions, 2022); and D'Este, *Bitter Victory*, 113. Considered a double envelopment, the British Eighth Army would land along the southeastern coast of Sicily between Augusta and Gela with the American Seventh Army conducting two distinct assaults: one in the southwest of Sicily in the Sciacca-Mazara region and another in northwest Sicily near the Castellammare, Capaci, and Trappeto areas.

[19] Garland, *Sicily and the Surrender of Italy*, 58.

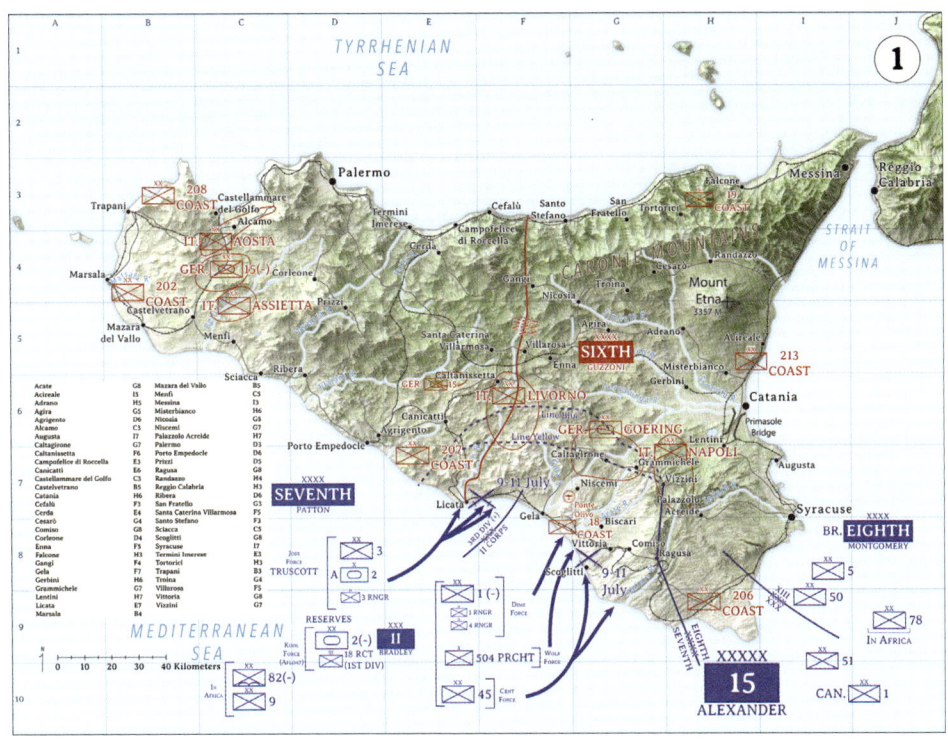

MAP 1
Operation Husky invasion plan.
Source: courtesy of the Department of History, United States Military Academy

ciously in Tunisia and General Montgomery, along with General Eisenhower, feared a more determined Italian defense of their home soil.[20] The final plan called for seven Allied divisions landing at more than 26 locations in the southeastern portion of Sicily, between Licata as the most western amphibious landing and Syracuse as the eastern.[21]

Moving inland under fire, securing the beachhead, and advancing to secure key terrain was not going to be a simple task for Allied forces on Sicily.[22] Prior to the outbreak of World War II, the United States was ill-prepared for amphibious op-

---

[20] Andrew J. Birtle, *Sicily, 9 July–17 August 1943* (Washington, DC: U.S. Army Center of Military History, 2021), 10.

[21] Johnson and Innocenti, *The West Point Guide to the Campaigns of World War II*; and Keys and Cummings, "Report of Operations: Initial Plan, 1 October 1943." In addition, elements of two Allied airborne divisions would land behind the amphibious landings to facilitate securing the beachhead. Birtle, *Sicily, 9 July–17 August 1943*, 10.

[22] Rick Atkinson, *The Day of Battle: The War in Sicily and Italy, 1943-1944*, vol. 2 (New York: Henry Holt, 2007), 52.

Operation Husky

201

erations at the magnitude of Operation Husky. Amphibious operations had been limited to river crossings or raids on enemy-held terrain.[23] Between World War I and World War II, the United States military sought to develop its amphibious landing doctrine.[24] The U.S. Marine Corps and Army each developed amphibious warfare doctrine that placed an increased emphasis on the decentralization of decision-making after landing.[25] In June 1940, after the collapse of France, American military planners recognized that large scale amphibious operations would be necessary to fight a European conflict. Much of the Marine Corps and Navy doctrine that was developed in the 1930s was eventually adopted by the Army in preparation for an increased role in amphibious operations.[26] The Army established the Amphibious Training Center (ATC) at Camp Edwards, Massachusetts, in May 1942 to develop its own doctrine and experience in ship-to-shore and shore-to-shore amphibious operations from embarkation to expansion of the beachhead.[27] Once the Army developed their own training centers and doctrine, coordination between the Army and Navy was almost nonexistent, though the Army was reliant on the Navy for future amphibious operations.[28]

There was also a lack of coordination between the Allied ground and air forces in the months preceding Operation Husky. The Allies unified their air forces in the Mediterranean theater under British Air Marshal Arthur Tedder in February 1943. The Mediterranean Air Command adhered to the doctrine of theater airpower,

---

[23] Capt Marshall O. Becker, *The Army Ground Forces*, Amphibious Training Center Study no. 22 (Washington, DC: Historical Section, Army Ground Forces, 1946), 1.

[24] Milan Vego, "On Major Naval Operations," *Naval War College Review* 60, no. 2 (2007): 101. The U.S. Marine Corps published the *Tentative Manual for Landing Operations* in 1934. The manual used lessons learned from the failed Gallipoli landings by the entente powers in World War I and contrasted it with the successful amphibious landing (Operation Albion) by the German Army and Navy in World War I. *Tentative Manual for Landing Operations*, 1934, HAF 39, COLL/3634, MCHD, Quantico, VA

[25] Bruce Gudmundsson, "Ambiguous Application: The Study of Amphibious Warfare at the Marine Corps Schools, 1920–1933," in *On Contested Shores: The Evolving Role of Amphibious Operations in the History of Warfare*, ed., Timothy Heck and B. A. Friedman (Quantico, VA: Marine Corps University Press, 2020), 184, https://doi.org/10.56686/9781732003149.

[26] John T. Greenwood, "The U.S. Army and Amphibious Warfare during World War II," *Army History*, no. 27 (Summer 1993): 3.

[27] Becker, *The Army Ground Forces*, 5. In addition, the Army was directed to train 12 divisions by February 1943 to be capable of conducting amphibious operations. Unfortunately, no plan was developed with the Navy to produce landing craft to facilitate this training. The ATC, from its inception, suffered from a lack of trained personnel, available landing craft, and proper facilities to support the training of an Army division. American military leaders disbanded the ATC and made their facilities available for the Navy by June 1943, ending the short-lived Army-centric ATC. Becker, *The Army Ground Forces*, 17.

[28] *Field Service Regulations: Operations, May 22, 1941*, Field Manual 100-5 (Fort Leavenworth, KS: U.S. Army Command and General Staff College Press, 1992). *Field Service Regulations: Operations* and *Joint Action of the Army and the Navy* (Washington, DC: Government Printing Office, 1927) were the predominant doctrine of the United States when conducting combined arms or Joint operations. The Army referenced *Landing Operations on Hostile Shores*, FM 31-5 (Washington, DC: Government Printing Office, 1941) when developing their own methods in conducting amphibious operations.

which placed emphasis on "air superiority missions, battlefield interdiction tasks, and close air support, in that order of importance."[29] Securing the Sicilian airfields within days of the landings was of paramount importance. If Sicilian airfields remained in Axis control, they posed a significant risk to the success of Allied amphibious landing. Axis air elements would be able to operate over the landing locations for 45 minutes every hour, while the Allies could only provide 15 minutes of air coverage.[30] Allied ground commanders expected the air contingent of Operation Husky to focus on close air support, as opposed to "deep strike" targets in the Axis rear elements in Sicily and on mainland Italy.

The Army's *Field Service Regulations: Larger Units*, Field Manual (FM) 100-15, from 1942, outlines the need for air forces to perform both the "deep strike" and close air support functions.[31] U.S. Army ground and air leaders interpreted *Larger Units* to fit their own idea of air support, resulting in frustration and inefficiency in the planning and execution phases of Operation Husky. Historian Alexander Fitzgerald-Black, argues that Tedder's focus on gaining air superiority over Axis air forces was militarily sound, but his lack of clear details on how the Allied air forces would provide direct air support of the amphibious landings was a failure.[32] Tedder's emphasis on gaining air superiority was valid because it would, inevitably, aid the landings by preventing Axis air and ground movement to and from the beachhead that would enable greater Allied freedom of maneuver from the beachhead. However, Army planners had difficulty understanding when and where they should expect air support in the tenuous beginning stages of the invasion. The lack of detail and specifics in Tedder's plan led to distrust between the Allied Service components that remained during the execution phase of Operation Husky.

In April 1942, prior to Operation Torch, the U.S. Army published, *Basic Field Manual: Aviation in Support of Ground Forces*, FM 31-35.[33] Using *Aviation in Support of Ground Forces* as a guide, the Army ground forces developed a nine-week air-ground coordination training exercise at Fort Benning (now Fort Moore), Georgia, that was intended to further integrate the ground and air elements of the U.S. Army. The comprehensive training included observation, bombing and strafing, communications,

---

[29] Alexander Fitzgerald-Black, *Eagles over Husky: The Allied Air Forces in the Sicilian Campaign, 14 May to 17 August 1943* (Solihull, UK: Helion, 2018), xxi.
[30] Fitzgerald-Black, *Eagles over Husky*, 35.
[31] *Field Service Regulations: Larger Units*, FM 100-15 (Washington, DC: Government Printing Office, 1942). *Larger Units* was used as a guide for commanders and staffs of air forces, corps, armies, or groups of armies.
[32] Fitzgerald-Black, *Eagles over Husky*, 37.
[33] *Basic Field Manual: Aviation in Support of Ground Forces*, FM 31-35 (Washington, DC: Government Printing Office, 1942).

identification, control, defense, and exploitation.[34] While comprehensive in nature, as with the ATC, the actual execution of the air-ground coordination exercises lacked realism due to a variety of factors. The most significant limitation was the pressing need for qualified air crews and functioning aircraft to support the war effort overseas. Air-ground coordination inefficiencies were not solely related to wartime necessity but rather mistrust and skepticism. In a statement to General George Marshall in December 1942, Major General Lesley J. McNair, the commanding general of Army Ground Forces said, "We have made little progress in air-ground cooperation, in spite of our efforts, if we view frankly the conditions that must obtain in order to secure effective results in combat . . . [and] the trouble is that the air side of the setup has been too sketchy to permit effective training. I say this without criticism of the air forces."[35] McNair's comments provide a glimpse into the mentality of some officers with respect to the effectiveness of the Army Air Corps and the potential for integration with ground elements overseas.

For the United States during the North African campaign, lack of experience and coordination with air support hindered the Army in achieving mission success. In March 1943, in the latter stages of the North African campaign, Major General John P. Lucas, a former corps commander serving as a deputy under Eisenhower in the Mediterranean, was tasked by the Army Ground Forces to provide a report on his observations and to provide recommendations for doctrinal changes. Lucas specifically identified weaknesses in air-ground integration and aerial reconnaissance during operations in North Africa. He recommended liaison officers be placed within the echelons of command to improve Joint coordination.[36] Throughout the planning process, fractures in coordination between Allies and Services was evident. The success of Operation Husky necessitated the Allies and Services have unity of effort, especially during the initial stages of the amphibious invasion.

The Combined Chiefs of Staff approved the final plan for Operation Husky on 13 May 1943.[37] To be completed in five phases, American, British, and Canadian forces would:

1 – Gain naval and air supremacy around Sicily. 2 – Airborne and glider elements

---

[34] Kent R. Greenfield, *Army Ground Forces and the Air-Ground Battle Team Including Organic Light Aviation*, Forces Study no. 35 (Fort Monroe, VA: Historical Section, Army Ground Forces, 1948), 9. The Army Ground Forces were the precursor to the modern-day Army Forces Command and Army Training and Doctrine Command. They were created in March 1942 under the command of MajGen Lesley McNair.
[35] Greenfield, *Army Ground Forces and the Air-Ground Battle Team Including Organic Light Aviation*, 18.
[36] Johnson and Innocenti, *The West Point Guide to the Campaigns of World War II: Sicily*; and John P. Lucas, "Report of Visit to the North African Theater of Operations, 28 April 1943," in *Observer Report [Army Ground Forces]* (Carlisle, PA: Army War College, 1943), 4. In addition, MajGen Lucas mentioned that the U.S. Army was significantly less capable than the Germans in air-ground integration.
[37] The Combined Chiefs of Staff were American and British military leaders, with approval from President Roosevelt and Prime Minister Churchill, set the major policy decisions for the two nations during World War II.

would land on 9/10 July to disrupt Axis movement and communications inland near airfields while Allied forces conduct amphibious assaults on the southeastern coast of Sicily. 3 – Establish a secure lodgment for future operations. 4 – Capture the ports of Augusta and Catania and the Gerbini airfields. 5 – Secure Sicily through the reduction of Axis forces.[38]

For General Montgomery, Eighth Army's initial objectives were to seize and secure the port of Syracuse and the area near Pachino. Once the beachheads were secured, British and Canadian forces were to establish a general front in the Syracuse-Pozzallo-Ragusa region, make contact with the U.S. Seventh Army, and rapidly move forward to capture the Catania plain and the Gerbini airfields.[39] The U.S. Seventh Army, under the command of Lieutenant General George S. Patton, were initially tasked with securing the port and airfield at Licata along with the airfields of Ponte Olivo, Biscari, and Comiso. Once those tasks were complete, the Seventh Army was expected to make contact with the British Eighth Army, secure the airfields, and protect the British left flank from Axis interference.[40] For the Allies, the initial objectives culminated in reaching the "Yellow Line," a notional location roughly 32 kilometers inland that would deny Axis forces from using indirect fires on seized ports and airfields.[41]

Planners anticipated the most dangerous phase for the Allies during Operation Husky was the initial landing. As such, they developed a comprehensive plan to integrate the air, naval, and field artillery fires plan to secure the beachhead and expand the lodgment. As noted earlier, the air support plan for Operation Husky focused on three primary tasks: 1) negate the enemy air forces' ability to seriously influence shipping, projected landing locations, and subsequent operations; 2) impede the enemy's freedom of maneuver on land and at sea; and 3) provide the maximum support to Allied land forces in the assault and subsequent phases of the operation. For the Allied air forces, destroying the enemy air forces had priority over all other tasks.[42] Naval gunfire was to provide support down to the division level, which could assign priority of fires to their subordinate units. Naval gunfire was to eliminate shore batteries to protect shipping and the landings as well as to support the advance inland. Field artillery support was also at the division level and could be assigned as needed by the division commanders.[43]

---

[38] D'Este, *Bitter Victory*, 144–45.
[39] D'Este, *Bitter Victory*, 148.
[40] D'Este, *Bitter Victory*, 150–51.
[41] Atkinson, *Day of Battle*, 69.
[42] Johnson and Innocenti, *The West Point Guide to the Campaigns of World War II*; and Fitzgerald-Black, *Eagles over Husky*, 36–37.
[43] Johnson and Innocenti, *The West Point Guide to the Campaigns of World War II*; and Hewitt, "Annex #8 to Field Order #8: Air Support Plan, 23 June 1943," 2.

In the constantly changing environment of combined arms warfare and the complexity of amphibious operations, having a "method of marking" enemy and friendly forces was paramount. In planning documents issued to II Corps, under the command of Lieutenant General Omar N. Bradley, the air support method of marking emphasized that ground teams would use pyrotechnics, large alphabetic symbols, and landmarks to identify enemy and friendly locations in the event of radio failure.[44] While not as comprehensive in detail as the air support plan, naval gunfire used forward observers, pyrotechnics, and landmarks to shift their fires in the event of friendly forces in the area.[45]

To minimize confusion during the invasion, training centers were established in North Africa for ship-to-shore and shore-to-shore amphibious rehearsals. Soldiers were trained on loading and unloading personnel and vehicles from Landing Ship, Tanks (LSTs), Landing Craft, Tanks (LCTs), Landing Craft, Infantry (LCIs), and Landing Craft, Vehicle, Personnel (LCVPs).[46] The U.S. Army's 1st Infantry Division and 45th Infantry Division conducted full rehearsals during 23–25 June 1943 that would replicate conditions they would face during the amphibious invasion of Sicily. Navigational errors, delays in timelines, and general operational friction during rehearsals provided these units experience with what they could expect during Operation Husky.[47] As with many actions during World War II, Joint amphibious operations were a relatively novel endeavor. Doctrine for conducting amphibious operations was new and untested. Major General Lucian K. Truscott, the 3d Infantry Division commander, remarked that Operation Husky would be "the first real test of shore-to-shore operations under actual conditions of war with adequate equipment."[48]

The Allies faced an enemy coalition on Sicily that was strained, not only under the weight of the Allied bombardment, but by distrust and resentment. In the first half of 1943, tensions between Germany and Italy increased as the Allies captured Tunis and expelled the Axis forces from the African continent. Adolf Hitler, along with many German military leaders, viewed the Italians as the weak link in the Axis coalition.[49] On Sicily, disagreements between the Italian Sixth Army commander Alfredo Guzzoni and German field marshal Albert Kesselring on how to defend the

---

[44] Johnson and Innocenti, *The West Point Guide to the Campaigns of World War II*; and Hewitt, "Annex #3 to Field Order #8: Air Support Plan," 2.

[45] Johnson and Innocenti, *The West Point Guide to the Campaigns of World War II*; and Hewitt, "Annex #8 to Field Order #8: Air Support Plan," 2.

[46] "War Cabinet and Cabinet: Chiefs of Staff Committee: Minutes, 30 January 1943," CAB 79-59-8, National Archives, Kew. Notes from a secret British document pertaining to the need to transport a variety of additional landing craft to the Middle East to facilitate training multiple brigade-size elements in preparation for Operation Husky.

[47] Johnson and Innocenti, *The West Point Guide to the Campaigns of World War II*; and Atkinson, *Day of Battle*, 40.

[48] Barbara Brooks Tomblin, "Gearing up for Operation Husky," in *With Utmost Spirit: Allied Naval Operations in the Mediterranean, 1942–1945* (Lexington: University Press of Kentucky, 2004), 132.

[49] D'Este, *Bitter Victory*, 193.

966 kilometers of Sicilian coastline further fractured Axis unity. Guzzoni believed the Allies would attack on the southeastern coastline of Sicily and sought to position German armored units near the coastline once the Allies invaded. Kesselring agreed on the templated location of the Allied attack but argued to maintain a mobile reserve of German forces that could counterattack the Allies in the event of multiple amphibious landings.[50]

On paper, the estimated 300,000–350,000 Axis soldiers scattered along the Sicilian coastline and key areas inland appeared formidable, but many suffered from a lack of training, poor morale, and indiscipline.[51] Carlo D'Este, in *Bitter Victory: The Battle for Sicily, 1943*, reports that, "during surprise visits guards were found asleep at their posts, telephones inoperable and at one battalion headquarters the duty telephonist was found sleeping soundly."[52] In the first days after the amphibious invasion, intelligence reports from the U.S. 3d Infantry Division emphasized the Italian soldiers low morale as mass numbers willingly surrendered to American forces after the landing.[53] Much of the demoralization among the Axis forces was due to the aerial bombardment by the Allies in the weeks leading up to Operation Husky. Italian and German prisoners of war, when interviewed by Allied intelligence officers, complained of the seemingly constant bombing they endured day and night, which contributed to their willingness to surrender.[54] In an interview after the war, former commanding general of the German Air Force in Italy, General Maximillian von Pohl spoke of the operational changes Axis forces had to make due to the Allies bombing efforts in Italy. General von Pohl remarks that the evacuation of Sicily "was caused by the air force attacks on railroads in southern Italy and the sea area off Messina, which effectively delayed the arrival of German reserves and supplies."[55] Allied bombing prevented Axis forces on Sicily from being resupplied from mainland Italy, effectively isolating them prior to the amphibious invasion.

The Allies made concerted efforts to deceive the Germans and Italians about where subsequent operations would take place at the conclusion of the North African campaign. In preparation for Husky, Operations Barclay and Mincemeat were

---

[50] D'Este, *Bitter Victory*, 196–98.
[51] Walter Fries, General Der Panzertruppen, *The Battle for Sicily*, U.S., WWII Foreign Military Studies, 1945-1954, Record Group 338, National Archives, 8–9.
[52] D'Este, *Bitter Victory*, 194–95.
[53] Johnson and Innocenti, *The West Point Guide to the Campaigns of World War II*; and Atkinson, *Day of Battle*, 40.
[54] On 12 July, an Italian officer offered the following ditty to a U.S. 3d Infantry Division intelligence officer: "It certainly would be a treat, when Hitler and Mussolini meet, in the armored train at Brenner Pass, their lair, to find a bomb awaiting them there, what would the outcome be? Why, of course, the salvation of humanity!" Johnson and Innocenti, *The West Point Guide to the Campaigns of World War II*; and Walter, 3d Infantry Division G2, "Enemy Situation at End of Period," 6–7.
[55] Headquarters, MAAF, Intelligence Section, *Mediterranean Allied Air Forces: Air Surrender Documents*, pt. 3,. World War II Operational Documents, Combined Arms Research Library, 21.

launched.⁵⁶ Both operations were used to influence Germany to shift forces away from Sicily and the eastern front, with Western Europe, the Balkans, Greece, and Crete being the ruse amphibious invasion locations.⁵⁷

Admiral Andrew B. Cunningham, the commander of Allied naval forces during Operation Husky, split his force into an Eastern Task Force (British and Canadian) and Western Task Force (American). Led by British Admiral Bertram H. Ramsey, the Eastern Task Force was organized into three assault forces designated as A, B, and V, which carried British and Canadian Army units to the various beaches in the British sector.⁵⁸ American Vice Admiral Henry Kent Hewitt's Western Task Force was similarly organized into three attack forces; Cent, Dime, and Joss, which brought the American forces to Sicily in both ship-to-shore and shore-to-shore capable landing craft.⁵⁹ The Allied invasion fleet needed cooperative weather as they all navigated to their assigned beaches. Weather studies were conducted in the months preceding the invasion that examined cloud cover, precipitation, winds, and seas and surf that would impact the invasion. Until 9 July, the weather in the western Mediterranean was typical for the season, but hours before the invasion fleet approached the Sicilian coastline, winds averaged 31 knots with gusts up to 37 knots.⁶⁰

A Western Task Force "Action Report" from August 1943 described the weather as being "most unfavorable for craft convoys" to maintain their formations and timelines for the invasion.⁶¹ Despite the difficult weather, the Dime, Cent, and Joss attack forces reached their rendezvous locations, generally, at their prescribed times.⁶² Much of the credit was attributed to the use of "beacon submarines" that acted as navigational guides for the attack forces as well as reconnaissance in the event of Axis naval forces in the region.⁶³

H-hour for the amphibious invasion was planned for 0245 on 10 July, for both the

---

⁵⁶ D'Este, *Bitter Victory*, 181–91.
⁵⁷ Maj Donald J. Bacon, *Second World War Deception: Lessons Learned for Today's Joint Planner*, Wright Flyer Paper no. 5 (Maxwell Air Force Base, AL: Air Command and Staff College, 1998), 2–3.
⁵⁸ Assault Force A consisted of the British 5th Infantry and 50th Divisions; Assault Force B included the British 51st Division and 231st Infantry Brigade; and Assault Force V was made up of the 1st Canadian Division. A submarine force would also support the task forces by potentially intercepting Axis warships and guiding the assault forces to the correct beaches for invasion. D'Este, *Bitter Victory*, 153.
⁵⁹ D'Este, *Bitter Victory*, 153. Named the DUKW, this shore-to-shore landing craft carried troops and equipment from the departure to embarkation to debarkation points in an amphibious invasion. Attack Force Cent consisted of the 45th Infantry Division; Dime included the 1st Infantry Division; and Joss was made up of the 3d Infantry Division. D'Este, *Bitter Victory*, 151.
⁶⁰ *Aerology and Amphibious Warfare: The Invasion of Sicily*, NAVAER 50-30T-1 (Washington, DC: Aerology Section, Chief of Naval Operations, 1944).
⁶¹ VAdm H. K. Hewitt, *Action Report: Western Naval Task Force—The Sicilian Campaign: Operation "Husky" July–August 1943*, Combined Arms Research Library, 36, hereafter *The Sicilian Campaign*.
⁶² While the weather did not severely impact the amphibious invasion, Atkinson in *The Day of Battle*, 86, highlights multiple incidents where soldiers and sailors lost their lives due to severe winds and waves during the process of ship-shore operations. The high seas did cause difficulties as supplies and vehicles were being transported to the beaches after the invasion forces landed.
⁶³ Hewitt, *The Sicilian Campaign*, 37–38.

Western and Eastern Task Forces, slight delays in landing was followed by sporadic Axis resistance, generally concentrated in the Dime beach landing. Preassault naval gunfire on designated targets enabled the relative ease in assault by the Cent force consisting primarily of 45th Infantry Division soldiers.[64] British seaborne forces from Assault Force A landed around 0400 on 10 July near their assigned beaches. Soldiers there found many of the defensive positions unmanned and came under very little organized resistance as they moved inland from the landing. Most of the defenders were Italians who were eager to surrender, not seeking to fight.[65] In terms of planning for Operation Husky, the success of the amphibious invasion hinged on the neutralization of beach defenses. The primary means to accomplish this task fell on naval and air bombardment. As seaborne soldiers moved inland to secure the beachhead, the risk of friendly fire by naval or air bombardment to support soldiers was too high. The task of "softening" the beach defenses fell on the airborne contingent of Operation Husky.

While the amphibious invasion forces were scheduled to reach their assigned landing sites by 0245 on 10 July 1943, Allied paratroop and glider forces reached Sicilian air space a few hours prior to the seaborne forces.[66] For the British glider elements, problems arose immediately when the 1st Airlanding Brigade of the British Army prematurely released the gliders from the towplanes, resulting in 47 gliders crashing into the Mediterranean Sea.[67] Elements of the U.S. Army's 504th Parachute Infantry Regiment of the 82d Airborne Division successfully landed on Sicily in the late evening hours of 9 July, causing confusion among Axis forces and disrupting their efforts to reach the beaches as Allied seaborne forces waded ashore.[68] The relatively successful preamphibious invasion airborne assault by Allied forces was followed by another airborne assault by further elements of the 504th Parachute Infantry Regiment in the late evening hours of 11 July.

Major General Matthew B. Ridgway, commanding general of the 82d Airborne Division, warned of the potential for friendly fire in the weeks preceding Operation Husky. Ridgway did not receive confirmation for an air corridor for the 144 Douglas C-47 Skytrain transport aircraft by the Navy until 5 July, just days before the planned invasion.[69] Delays in the dissemination of the planned airborne drop resulted in some Army and Navy units not knowing of the operation until hours before it began, and

---

[64] Hewitt, *The Sicilian Campaign*, 40–41.
[65] Tomblin, "Gearing Up for Operation Husky," 152.
[66] Operation Ladbroke (British glider landing on Syracuse) and Operation Fustian (British airborne insertion at the Primosole Bridge) were supplemental operations that preceded the larger airborne insertion of Operation Husky.
[67] Tomblin, "Gearing Up for Operation Husky," 149. The weather and enemy fire have both been blamed for the premature release of the gliders.
[68] "United States Army 82nd Airborne Division narratives from operations in Sicily, Italy, Normandy, Holland, Ardennes, and Central Europe, August 1942–May 1945," D78 Item nos. 2000-2019, Maneuver Center of Excellence, Donovan Research Library, Fort Moore, GA, 33–38.
[69] Atkinson, *The Day of Battle*, 108.

some not at all."⁷⁰ At 2240 on 11 July, the first group of C-47s entered Sicilian airspace through their prescribed air corridors and dropped each of their 12 paratroopers on their assigned drop zones. Soon after, red tracers from friendly naval antiaircraft batteries lit up the sky and wreaked havoc on the transport planes. The severity of the antiaircraft fire broke up aerial formations as each plane sought to avoid the deadly fire. Having endured repeated Axis air attacks in the daytime, naval gunners were ready to repel what they believed to be further attacks on the tenuously held Allied beachhead. There was no safe refuge for the pilots and paratroopers as "men died in their planes, men died descending in their parachutes, and at least four were shot dead on the ground by comrades convinced they were Germans."⁷¹ The resulting friendly fire incident destroyed 23 planes and severely damaged another 37. Investigators estimated the casualties to be 410, although that number has not been confirmed. No one was found personally culpable for the tragedy on 11 July. Patton described the incident as "an unavoidable incident of combat."⁷² This thought process minimizes the responsibility of leaders to mitigate the risk to the mission and the force.

Major General Matthew Ridgway, Commander of the 82d Airborne Division, wrote a memorandum on 27 November 1943 outlining the use of airborne units in operations. The memorandum reads as a brief after action report of Operation Husky from the airborne perspective and highlights recommendations for future operations involving airborne forces. Emphasizing concerns that many leaders in Operation Husky held, Ridgway believes that "there must be continuous detailed coordination between airborne, air, ground, and sea forces throughout the entire planning and operational stages of an operation."⁷³ American Seventh Army after action reports outline that during amphibious operations, or any operation, the failure to coordinate between services "results in confusion, inefficiencies, and unwarranted delay."⁷⁴ With the case of the 504th Parachute Infantry Regiment, the lack of coordination between the Services resulted in the death of American soldiers, which hindered the success of the operation.

By the conclusion of the initial 48 hours of Operation Husky, Allied beachheads were established from Licata in the American sector to Syracuse in the British sector.⁷⁵ The situation was not all satisfactory, however. Fighting remained fierce in the

---

⁷⁰ Atkinson, *The Day of Battle*, 107. Patton signed the final approved order at 0845 on 11 July but delays in the signal room resulted in the order not being disseminated until close to 1620, much too late to ensure all antiaircraft batteries both on sea and on land were informed of the operation.

⁷¹ Atkinson, *The Day of Battle*, 109.

⁷² Atkinson, *The Day of Battle*, 109, 112.

⁷³ LtCol John T. Ellis Jr., *The Army Ground Forces*, Airborne Command and Center Study no. 25. (Washington, DC: Historical Section, Army Ground Forces, 1946), 136.

⁷⁴ Johnson and Innocenti, *The West Point Guide to the Campaigns of World War II*; and Keys and Cummings, "Operations," 1 October 1943, in Seventh Army Sicily Source Packet, Report of Operations, author's collection, 1.

⁷⁵ Kent Roberts Greenfield, *The War against Germany and Italy: Mediterranean and Adjacent Areas*, U.S. Army in World War II, Pictorial Record (Washington, DC: U.S. Army Center of Military History, 1988).

American 1st and 45th Infantry Divisions' sectors and German aircraft remained a constant harassment over the Allied positions in within the first 36 hours of the invasion. German and Italian forces counterattacked in force against the American 1st Infantry Division near Gela and the Ponte Olivo airfield. Elements of the Hermann Goering *Panzer Division* penetrated the 1st Infantry Division lodgment and threatened the beachhead with dozens of heavy tanks on 11 July.[76] The Navy, having been marginalized in the bombardment of known or suspected enemy targets to neutralize beach defenses, were called on to provide indirect fire support to halt the advancing German tank formations near the Gela (Dime) beachhead, which they did with devastating results on their German counterparts.[77] In a similar situation, Paul A. Disney, then a reconnaissance battalion commander during Operation Husky, later commented how Navy observers with vehicle-mounted radios provided supporting fires from two cruisers to dislodge enemy tank formations that were threatening his position on 18 July.[78] These accounts provide clear examples of how Joint coordination and cooperation between the Services in the initial stages of the operation was successful to accomplishing the mission and sustaining the force on the ground.

Despite not being the first amphibious operation for the Allies during World War II, Operation Husky was a crucible of learning that necessitated greater unified effort in the planning and execution phases of future Joint amphibious assaults. Multiple shortfalls occurred in the planning and execution of Operation Husky, including the lack of unified command. Allied planners were scattered across the North African landscape and planned in a series of relative vacuums that did not involve in-depth planning and coordination between the various Services and Allied partners. Liaison officers, within the command structure of Operation Husky, may have prevented some of the operational and tactical failures that were experienced. A British liaison that was embedded with the U.S. Seventh Army in the planning stages of Operation Husky highlighted that "officers employed on such duties must be qualified by ample operational experience, should already have the confidence of one Army Commander, and should be capable of rapidly gaining the confidence of the other."[79] The liaison officer can keep their "parent" element informed and provide substantial benefit to the gaining organization, freeing up commanders to make decisions in a rapidly developing operational environment. Specifically for the Western Task Force, despite repeated requests, no air representative attended any of the Joint

---

[76] D'Este, *Bitter Victory*, 295–97.
[77] Hewitt, *The Sicilian Campaign*, 44–45.
[78] Paul A. Disney, *Operations of the 82nd Armored Reconnaissance Battalion in Sicilian Campaign, July 10–22, 1943 (Personal Experience of a Battalion Commander)* (Fort Leavenworth, KS: U.S. Army Command and General Staff College, 1947), 12.
[79] Johnson and Innocenti, *The West Point Guide to the Campaigns of World War II*; and Combined Operations Headquarters, "Notes on Planning and Assault Phases of the Sicilian Campaign," 1 October 1943, COHQ Bulletin No. Y/1, Combined Arms Research Library, 2.

planning board that the Joss Force commanders established.[80] By not having an air liaison officer embedded within the Joss force, uncertainty about the air aspect of the operation resulted in mistrust and confusion. A Seventh Army after action report from Operation Husky identified the liaison shortfall and emphasized the importance of having a liaison officer "available and function at the inception of planning" in all facets of the operation.[81]

The Allied air forces were blamed for much of the lack of coordination during the invasion due to their insistence on focusing on the deep targets inland as opposed to supporting the shipping, beachhead, and subsequent objectives inland.[82] British Air Marshal Arthur Tedder emphasized the importance of destroying enemy airfields and aircraft while Allied ground commanders desired greater close air support. The Allied air forces were determined to carve out their own strategic role rather than serve as support to the ground and sea elements of Operation Husky.[83] From the Joss Force perspective, the naval leadership "left North Africa with very little idea of what part our air forces were to play in the initial assault."[84] The lack of clarity of the air support plan created an unnecessary level of friction between the Services that degraded the mission effectiveness of all levels of command.

From specifically the naval perspective, the decision to marginalize the Navy during the initial invasion was under the belief that surprise on the assault beaches should be achieved and the use of naval gunfire would violate that. Hewitt's action report highlighted that the "old-fashioned military concept that naval guns are unsuitable for shore bombardment needs revision."[85] The employment of naval gunfire to neutralize enemy defenses during the initial invasion would bring greater firepower to bear than all of the organic artillery capabilities that the assault forces could bring ashore.[86] For future operations, mobile naval guns can be used to overwhelm the opposing force to facilitate a rapid seizure of the beachhead and assault objectives inland.

Operation Husky offers a learning experience for the Allies in Joint amphibious operations. The lessons from Sicily were carried forward to subsequent amphibious operations on mainland Italy and eventually in Normandy in June 1944. The failings during Operation Husky were substantial and were suffered at tremendous cost, but the lessons of coordination, unity of command, and trust between partner nations and services were further solidified as a result.

---

[80] "Notes on Planning and Assault Phases of the Sicilian Campaign," 2.

[81] Johnson and Innocenti, *The West Point Guide to the Campaigns of World War II*, 1.

[82] David Jablonsky, Donald Kagan, and Frederick Kagan, "Unity in Practice: Sicily and Italy, May–December 1943," in *War by Land, Sea, and Air: Dwight Eisenhower and the Concept of Unified Command* (New Haven, CT: Yale University Press, 2010), 96–97.

[83] Tomblin, "Gearing Up for Operation Husky," 138.

[84] "Notes on Planning and Assault Phases of the Sicilian Campaign," 2.

[85] Hewitt, *The Sicilian Campaign*, 44.

[86] Hewitt, *The Sicilian Campaign*, 44.

CHAPTER ELEVEN

A New Zealand-led "Commando Raid" in the South Pacific

The Green Islands, 30–31 January 1944[1]

Shaun Mawdsley

The Green Islands "Commando Raid" has been called "the largest and most complex New Zealand-led special operations mission of the Second World War."[2] The mission serves as a classic example of the utility of amphibious raids, with a unique international flavor, and aligns with raiding characteristics promulgated in *Amphibious Operations*, Joint Publication 3-02.[3] Conducted in late-January 1944, the raid was the only one of its kind involving U.S. and New Zealand forces. It originated from a need for accurate intelligence on the Japanese-held Green Islands, located about 63 kilometers northwest of Bougainville in the northern Solomon Islands, which were then the target of an amphibious assault, Operation Squarepeg, set for mid-February 1944. However, unlike other amphibious operations in the Solomons, which benefited from an established intelligence gathering network of coastwatchers, Allied planners lacked basic information on the islands, their inhabitants, and the waters surrounding them. Even aerial reconnaissance proved inadequate ow-

---

[1] Parts of this chapter appear in Shaun Mawdsley, " 'With the Utmost Precision and Team Play': The 3rd New Zealand Division and Operation 'Squarepeg' " (MA thesis, Massey University, 2013).
[2] Rhys Ball and Shaun Mawdsley, "Australasian Special Operations in the Second World War," in *The Routledge History of the Second World War*, ed., Paul Bartrop (Oxon, UK: Routledge, 2022), 616.
[3] *Amphibious Operations*, Joint Publication 3-02 (Washington, DC: Joint Chiefs of Staffs, 2019).

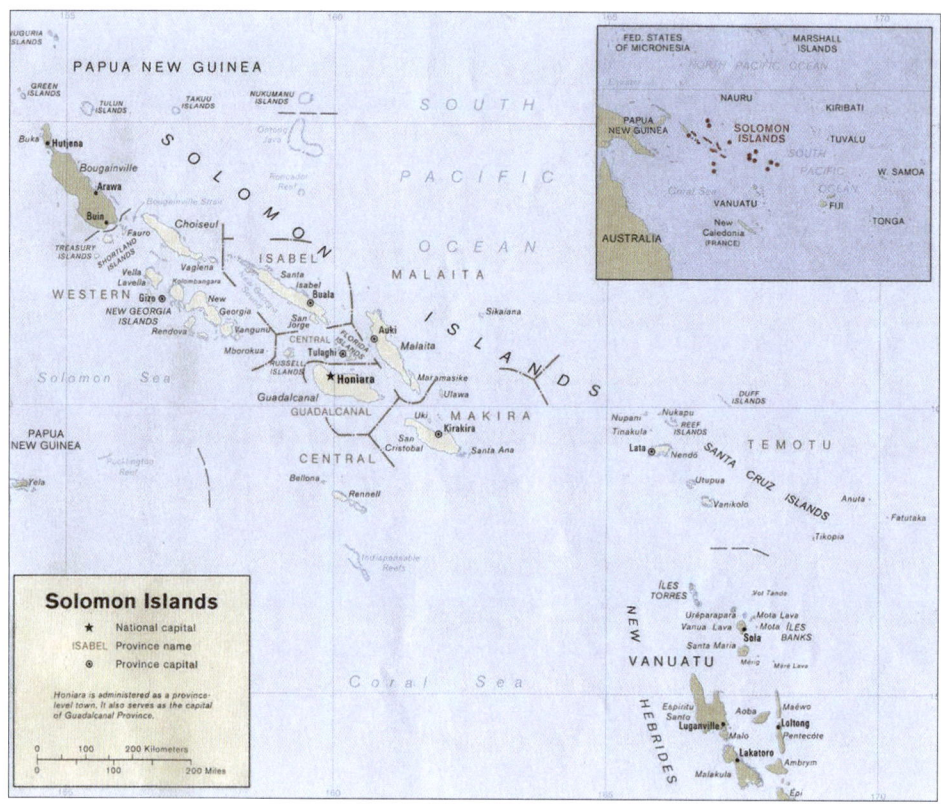

MAP 1
Postwar map of Bougainville, the Solomon, Santa Cruz, and New Hebrides Islands, and the Green Islands (top left corner).
Source: U.S. Central Intelligence Agency, Perry-Castañeda Library Map Collection: Solomon Islands Maps, University of Texas at Austin, University of Texas Libraries

ing to the density of the vegetation.[4] Subsequently, no accurate threat estimation could be provided from which to base the amphibious planning; an unacceptable scenario for a type of operation that required intricate and detailed planning procedures. There were also questions around the allegiance of the local islanders, with the

---

[4] "Photo Intelligence Unit, 12th AAF Photo Intelligence Detachment, USAFISPA-COMSOPAC, APO 502, Green Island: Photo-Interpretation Study," 30 December 1943, Headquarters 3d Division–Office records–Squarepeg Operations, DAZ 121/9/A50/4/2, 1512, WAII1, 18886, ADQZ, Archives New Zealand (ANZ); Douglas Ford, "US Assessments of Japanese Ground Warfare Tactics and the Army's Campaigns in the Pacific Theatres, 1943-1945: Lessons Learned and Methods Applied," *War in History* 16, no. 3 (2009): 330, https://doi.org/10.1177/0968344509104; Matthew Wright, *Pacific War: New Zealand and Japan, 1941-45* (Auckland, NZ: Reed Publishing, 2003), 123-24; and Oliver A. Gillespie, *The Pacific*, Official History of New Zealand in the Second World War, 1939-45 (Wellington: War History Branch, Department of Internal Affairs, 1952), 170.

general assumption that they were hostile toward the Allies. With no knowledge of the Japanese garrison, potentially hostile islanders, and inaccurate naval and marine charts, the Green Islands were shrouded in mystery.

In late 1943, shipping limitations created by the United States Pacific Fleet requirements in Micronesia risked imposing an operational downturn to Admiral William F. Halsey's (commander, South Pacific area) forces in the Solomons until mid-1944.[5] Fearing a loss of initiative, Halsey and his staff—Navy rear admiral Robert B. Carney, Navy commander H. Douglas Moulton, Marine Corps colonel William E. Riley, and Navy captain W. F. Riggs Jr.—consulted with Army general Douglas MacArthur and his staff at Port Moresby, New Guinea, on 20 December 1943 for possible intermediate operations.[6] After reaching an understanding, Halsey sent personal messages to Rear Admiral T. S. Wilkinson (commander, Third Amphibious Force, Task Force 31), Vice Admiral Aubrey W. Fitch (commander, aircraft in the South Pacific), and Major General Roy S. Geiger (I Marine Amphibious Corps) on 22 December, advising them of his desire for an interim operation aimed at the Green Islands by employing elements of the 3d New Zealand (NZ) Division.[7] On 24 December 1943, Wilkinson, Fitch, Geiger, and members of their respective staffs met to discuss Halsey's proposal.[8]

After thoroughly deliberating, they remained unconvinced of Halsey's suggestion and instead recommended investigation of other islands. For the next four days, the staff wrestled with the options set before them. During that time, Geiger visited the headquarters 3d NZ Division and its commander, Major General Harold E. Barrowclough, on Vella Lavella, an island about 435 kilometers west-northwest of Guadalcanal. Geiger remained close-lipped about the possible future operation and revealed nothing to Barrowclough, his visit likely an information-gathering activity to check on the state of the New Zealanders.[9] With Geiger's opinion satisfied, on 28 December, Colonel Riley, Halsey's operations officer, wrote a memorandum noting the

---

[5] Samuel Eliot Morison, *History of United States Naval Operations in World War II*, vol. 6, *Breaking the Bismarcks Barrier: 22 July 1942–1 May 1944* (Boston, MA: Little, Brown, 1989), 413, hereafter *Breaking the Bismarcks Barrier*.
[6] John Miller Jr., *Cartwheel: The Reduction of Rabaul*, U.S. Army in World War II: The War in the Pacific (Washington, DC: Office of the Chief of Military History, Department of the Army, 1959), 313; and Maj John N. Rentz, *Bougainville and the Northern Solomons* (Washington, DC: Historical Section, Division of Public Information, Headquarters Marine Corps, 1948), 114–15.
[7] Halsey to Wilkinson, 22 December 1943, Appendix A, "Memorandum for Commander South Pacific," 28 December 1943, Folder 8, Box 9, Wilkinson Papers, Library of Congress; and Henry I. Shaw Jr. and Maj Douglas T. Kane, *History of the U.S. Marine Corps in World War II*, vol. 2, *Isolation of Rabaul* (Washington, DC: Historical Branch, G-3 Division, Headquarters Marine Corps, 1963), 178, 507, hereafter *Isolation of Rabaul*.
[8] "Memorandum on Conference at COMAIRSOPAC on December 24," 25 December 1943, Folder 8, Box 9, Wilkinson Papers, Library of Congress.
[9] Official War Diary of Gen Barrowclough, 20–27 December 1943, Acc. No. 1998.834, Kippenberger Military Archive, hereafter Barrowclough diary, date.

A New Zealand-led "Commando Raid"

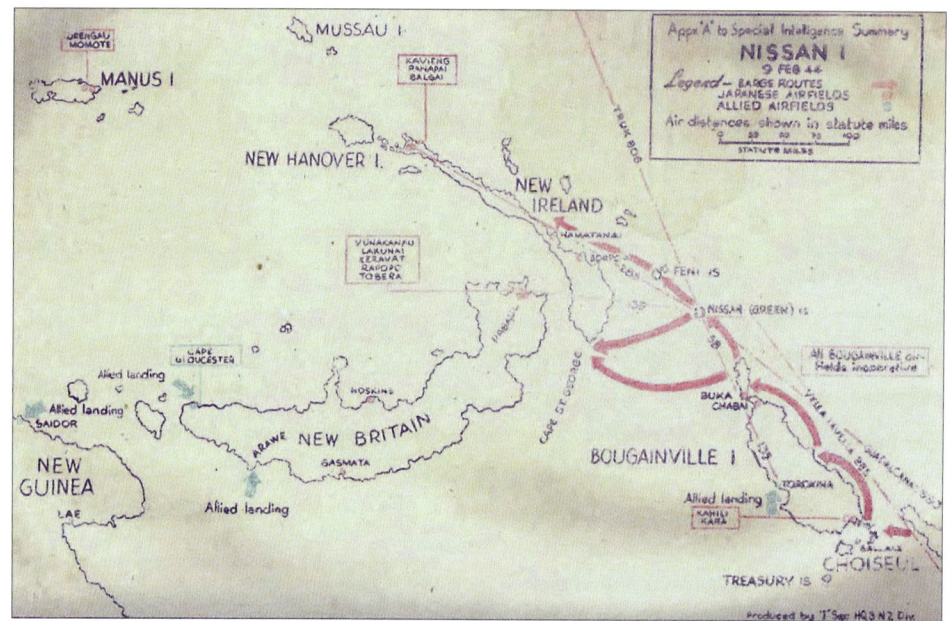

MAP 2
Map of the northern Solomons and the Bismarck Archipelago, with the Green Islands located roughly halfway between Bougainville and New Ireland.
Source: Headquarters 3d Division G Branch, War Diary, February 1944, DAZ 121.1/1/13, 1092, WAII1, 18886, ADQZ, ANZ

alternative operations as too resource-intensive (Borpop Harbor) or offering lesser opportunities for future operations (Boang Island) and therefore endorsing Halsey's preference for the Green Islands.[10]

The next day, Barrowclough received a signal to report to Wilkinson's headquarters on Guadalcanal. On 30 December, Barrowclough, his general staff officer 1 (senior operations officer), and his assistant adjutant and quartermaster general, the division's senior logistics officer, departed Vella Lavella. After their arrival on Guadalcanal, they were informed that Wilkinson was in New Caledonia, an indication of the challenges of commanding dispersed forces in the South Pacific as well as Wilkinson's confidence in his staff's planning abilities.[11] On New Year's Eve, with Wilkinson still away, the New Zealanders met Lieutenant General Millard F. Harmon (commander of U.S. Army forces in the South Pacific area), Rear Admiral George H. Fort (commander of amphibious craft in the South Pacific area), and the rest of headquarters

---

[10] "Memorandum for Commander South Pacific. Subject: Intermediate Operations to Precede Forearm or its Equivalent," 28 December 1943, Folder 8, Box 9, Wilkinson Papers, Library of Congress; Gillespie, *The Pacific*, 169; and Shaw and Kane, *Isolation of Rabaul*, 507.
[11] Barrowclough diary, 30 December 1943.

Task Force 31 staff to discuss "the nature of the proposed operation," particularly as it pertained to the construction of airfields.[12] This conference became an impromptu mission analysis with details presented in a preliminary manner, including objectives, the criteria for the end state ashore, and an (initially proposed) invasion date of 25 January.

According to Barrowclough, it was Harmon, not Fort, who "desired to send a reconnaissance party to report" on suitable landing beaches, airfield sites, and enemy dispositions.[13] For security and planning reasons, Barrowclough insisted that the reconnaissance mission and the main landing (then still set for 25 January) occur as close together as possible so as not to provide too much forewarning to the Japanese yet still allow sufficient time to incorporate new information into the operational plan. However, it was clear within the first week of January that all the components required for the main landing could not be gathered by the original date and the main operation was twice postponed (apparently by MacArthur).[14] Of course, this complicated matters, as any postponements to the main landing required the main amphibious force components be notified, in addition to the reconnaissance force elements, which had to be stood-down. Evidently, the New Zealanders harbored some frustrations at this time as no definite confirmation was released to them until after 10 January, despite having already relocated sections of their divisional headquarters to Guadalcanal to assist planning with Task Force 31.[15] In the meantime, the New Zealanders continued with their own preparations.

The U.S. Navy's 1938 *Landing Operations Doctrine* (FTP 167)—the doctrine to which the New Zealanders adhered—emphasized the importance of intelligence collection ahead of the main landing.[16] The Americans in particular were mindful of this requirement, being anxious to avoid a repeat of Tarawa in November 1943, when failure to conduct adequate hydrographic reconnaissance contributed to excessive casualties

---

[12] Fort was well-known to the New Zealanders, having commanded the first echelon of the task force that invaded the Treasury Islands in October 1943 for Operation Goodtime. Miller, *Cartwheel: The Reduction of Rabaul*, 69; Barrowclough diary, 31 December 1943; and Morison, *Breaking the Bismarcks Barrier*, 294.
[13] Barrowclough diary, 31 December 1943.
[14] Letter from Barrowclough to Puttick, 6 January 1944, Official Papers kept by General Barrowclough, Acc. No. 1998.835, Kippenberger Military Archive; Gillespie, *The Pacific*, 170–71; Barrowclough diary, 1–9 January 1944; and FlAdm William F. Halsey and LtCdr J. Bryan III, *Admiral Halsey's Story* (New York: McGraw-Hill, 1947), 188.
[15] These included the chief royal artillery, assistant director medical services, chief royal engineers, assistant adjutant and quartermaster general, GSO 3 (intelligence), GSO 2, commander signals, deputy director medical service, typists, miscellaneous staff, and the 14th Brigade liaison officer. Headquarters 3d NZ Division, G Branch-War Diary, 1–7 January 1944, DAZ 121.1/1/15, 1089, WAII1, 18886, ADQZ, ANZ.
[16] *Landing Operations Doctrine*, FTP 167 (Washington, DC: Office of Naval Operations, Division of Fleet Training, 1938), 6; and letter from RAdm R. K. Turner to Barrowclough, 16 May 1943, Folder 14: Correspondence, Box 1, Series 1, Papers of Adm Richmond Kelly Turner, USN, Operational Archives Branch, Naval Historical Center.

A New Zealand-led "Commando Raid"

among the assaulting Marines.[17] Such failure was enhanced by earlier operational experiences that had emphasized the importance of intelligence collection in creating the necessary conditions for successful mission execution, a good example being the lessons of the 1942 Makin Island raid controversy.[18] The New Zealanders were also aware of these bitter experiences and the 3d NZ Division could not afford excessive casualties as New Zealand did not have the reserves of manpower, or the political willpower, to withstand heavy losses in the Pacific. Conducting a thorough reconnaissance mission was therefore a high priority.

As mentioned previously, a key intelligence shortcoming was accurate data on the Green Islands, especially hydrographic information. Preliminary reports advised no landings should be attempted on the exterior of the main atoll, owing to extensive reefs and rugged cliffs 60 feet high. Many of these contained caves, which could have formed natural defensive positions for the Japanese; however, insufficient information was available on alterative landing sites.[19] The dearth of information forced intelligence personnel to cast a wide net, and they resorted to interviewing any known visitors to the islands, with questions pertaining to tides and water depth being high on the agenda.[20] Most charts described the Green Islands as consisting of four densely forested islands 14.5 kilometers in length and 8 kilometers wide, which formed an oval shape with a central lagoon, where the largest island, Nissan, served as the site of prewar plantations and thus was probably best suited for the construction of an airfield.[21] Importantly, there was no source of fresh water, which had to be taken into account by the reconnaissance force.[22]

The personnel who would comprise that force were decided when Barrowclough met with Brigadier Leslie Potter, commanding officer of 14th NZ Brigade, who immediately nominated his 30th Battalion for the mission.[23] It was at this time that security

---

[17] "Memorandum for Commander South Pacific. Subject: Intermediate Operations to Precede Forearm or Its Equivalent"; Col Joseph H. Alexander, *Utmost Savagery: The Three Days of Tarawa* (Annapolis, MD: Naval Institute Press, 1995), 73, 76–78; Reg Newell, *Operation Squarepeg: The Allied Invasion of the Green Islands, February 1944* (Jefferson, NC: McFarland, 2017), 19; and Kenneth Macksey, *Commando Strike: The Story of Amphibious Raiding in World War II* (London: Guild Publishing, 1985), 173, 198.

[18] VAdm George C. Dyer, *The Amphibians Came to Conquer: The Story of Admiral Richmond Kelly Turner*, vol. 2 (Washington, DC: Government Printing Office, 1972), 681.

[19] "Photo Intelligence Unit, 12th AAF Photo Intelligence Detachment, USAFISPA-COMSOPAC, APO 502, Green Island: Photo-Interpretation Study," 30 December 1943.

[20] Commander Third Amphibious Force, Intelligence Section: "Objective Data-Green (Nissan) Island, 9 January 1944"; "Interview with Capt Fairfax Ross, AIF (8 January 1944)"; South Pacific Force of the United States Pacific Fleet, Headquarters of the Commander, 14 January 1944, Nissan (Green) Island Group-Objective Data on: "Report of Interview with Capt W. A. Forman, AIF"; "Report of Interview with Lt A. C. Medlrum, RANVR(s)"; "Report of Interview with Bishop Wade"; and "Report of Interview with Cdr Robert Crookshank, RN (Ret)," all in DAZ 121/9/A50/4/2, 1512, WAII1, 18886, ADQZ, ANZ.

[21] Gillespie, *The Pacific*, 168.

[22] Letter from Barrowclough to Puttick, 6 January 1944, Official Papers kept by General Barrowclough, Acc. No. 1998.835, Kippenberger Military Archive.

[23] Barrowclough diary, 1 January 1944.

concerns convinced Barrowclough to call the reconnaissance a "commando raid" in the hopes of deceiving Japanese intelligence as to its true purpose. The mission was designed to be interpreted by the Japanese as a raid: the reconnaissance force was meant to be discovered, hence them planting fake operation orders to substantiate the presence of a "raiding" force. Moreover, the troops were to imitate raider-type actions while the specialists conducted their surveys. If a small force was used to conduct the mission, the likelihood of it being destroyed by the enemy garrison was unacceptably high. Indeed, as the locals were believed sympathetic to the Japanese, Barrowclough could not hope to land a small team without it being noticed and if this occurred the force would require a certain degree of firepower for its defense.[24]

The 30th Battalion was an odd choice for the raid as it was the only infantry unit in the division without combat experience; however, Barrowclough was eager to give the battalion an opportunity to prove itself before the formation was disbanded due to New Zealand's manpower pressures.[25] Only 308 troops of the battalion were selected, and they readily embraced the mantle of commandos, helping to foster a sense of pride for an otherwise green unit. They were accompanied by mortar, signals, intelligence, reconnaissance, medical personnel, engineers, artillery specialists, hydrographers, photographers, native scouts, and radar technicians, bringing the total force to 362, including 51 officers.[26] The types of personnel selected illustrated the broad nature of the tasks required, even an Australian officer was attached for his local knowledge and expertise in pidgin.[27] Any forewarning this large force may have provided the Japanese once ashore was outweighed by the crucial information it could collect.[28]

Heading this force was Lieutenant Colonel Frederick C. Cornwall, commanding officer 30th Battalion. At 52 years old, Cornwall was well over the usual age of commando raid leaders. Moreover, although a decorated Great War veteran, Cornwall's last combat experience was in 1917, and his last hostile landing was Gallipoli in 1915.[29] Ostensibly, he held no special distinguishable characteristics that would have qualified him for such a mission. Fortunately, Cornwall's relative inexperience was offset by the presence of Navy commander J. McDonald Smith (Landing Craft, Infantry Flotilla 5), who controlled the naval units, and Navy captain Ralph Earle (commander

---

[24] Barrowclough diary, 1 January 1944.
[25] Barrowclough diary, 1 January 1944; and Letter from Barrowclough to Puttick, 6 January 1944, 5, 1, PUTTICK5, 8477, ACGR, ANZ.
[26] "30 NZ BN 'Commando' Force OO No. 1, 24 January 1944," 30th Battalion-War Diary, February 1944, DAZ 156/1/40, 1154, WAII1, 18886, ADQZ, ANZ. The figure usually reported is 360 but see amendments in Appendix A, to "C.O. 30 Bn 'Commando Raid,' 24 January 1944," Appendix VI, Headquarters 14th NZ Brigade-War Diary, January 1944, DAZ 155/1/25, 1151, WAII1, 18886, ADQZ, ANZ.
[27] The term *pidgin* refers to the combination of several languages to simplify communication between people who do not share a common language.
[28] "Report on Operations-3 N.Z. Division. 1 Jan. 44 to 30 June 44," MajGen Barrowclough (Personal), March 1944–August 1944, S1, WAII9, 18907, ADQZ, ANZ.
[29] Newell, *Operation Squarepeg*, 23.

Destroyer Squadron 45), who held overall command of the raid, with the three levels of command illustrating the inherent complexities of amphibious operations.[30]

Even with such experienced U.S. Navy commanders, planning could not move forward without preliminary assessments of the area. Subsequently, two patrol-torpedo (PT) boats conducted soundings of the main channel on 10-11 January to assess its depth for the raid's landing craft. Satisfied, Wilkinson signaled for preparations to continue.[31] On 12 January, after further discussions between Barrowclough and U.S. commanders and staff, an operations memorandum was dispatched to 14th NZ Brigade advising the date of the raid as 30 January.[32] This was followed by another 10 days of intense meetings and conferences involving personnel from all Services and at all levels of command from battalion to theater task force. A photographic mosaic (and later a sand model) of the Greens Islands was shown to 30th Battalion officers to assist their planning.[33] Evidently, Cornwall was keen to get things rolling and, in a reversal of the usual planning process, submitted his operation order on 22 January, preempting Divisional Headquarters' Operation Instruction No. 53 by two days.[34]

These set out the raiders' main tasks as "(a) Recce Green I[sland]. with a view to est[ablish] an Air Base and P.T. Base; (b) Recce [reconnaissance of] landing facilities for craft and ships; (c) Make general terrain and hydrographical recce as may be practicable under the circumstances."[35] The raiding force was to make its way through the main channel at night, turn to starboard, land, and establish a defensive position in Pokonian Planation. There they were to wait until sunrise before separating into three groups: one remaining at Pokonian to conduct base reconnaissance, the second moved to Barahun Island to identify suitable landing sites, while the last proceeded across the lagoon to reconnoiter Tangalan Planation and the possible airfield location. All the while, fighting patrols were to destroy enemy equipment and stores without becoming heavily engaged. With the tests completed, the detachments were to regroup at Pokonian before reembarking their landing craft to rendezvous with

---

[30] "Seizure and Occupation of Green Is Report of Third Amphibious Force," 16 April 1944, Appendix 1, DAZ 121.1/1/15, 1089, WAII1, 18886, ADQZ, ANZ; and Newell, *Operation Squarepeg*, 24.

[31] Capt Robert J. Bulkley Jr., *At Close Quarters: PT Boats in the United States Navy* (Washington, DC: Naval History Division, 1962), 147-48.

[32] "Operations Memorandum 26," 12 January 1944, Appendix 9, DAZ 121.1/1/15, 1089, WAII1, 18886, ADQZ, ANZ.

[33] Headquarters 14th NZ Brigade-War Diary, 10 and 17 January 1944, DAZ 155/1/25, 1151, WAII1, 18886, ADQZ, ANZ; and Clive B. Sage, *Pacific Pioneers: The Story of the Engineers of the New Zealand Expeditionary Force in the Pacific* (Wellington: A. H. & A. W. Reed, 1947), 99.

[34] "30 NZ BN 'Commando' Force OO No. 1," 22 January 1944, Headquarters 14th NZ Brigade-Office records-Commando Raid-30 Battalion Commando Forces Raid on Nissan, DAZ 155/9/1, 1551, WAII1, 18886, ADQZ, ANZ; and "3 NZ Div Op Instn No. 53 Op 'Squarepeg'-Orders for Commando Raid," 24 January 1944, Appendix 26, DAZ 121.1/1/15, 1089, WAII1, 18886, ADQZ, ANZ.

[35] "C.O. 30 Bn 'Commando Raid," 24 January 1944, Appendix VI, DAZ 155/1/25, 1151, WAII1, 18886, ADQZ, ANZ.

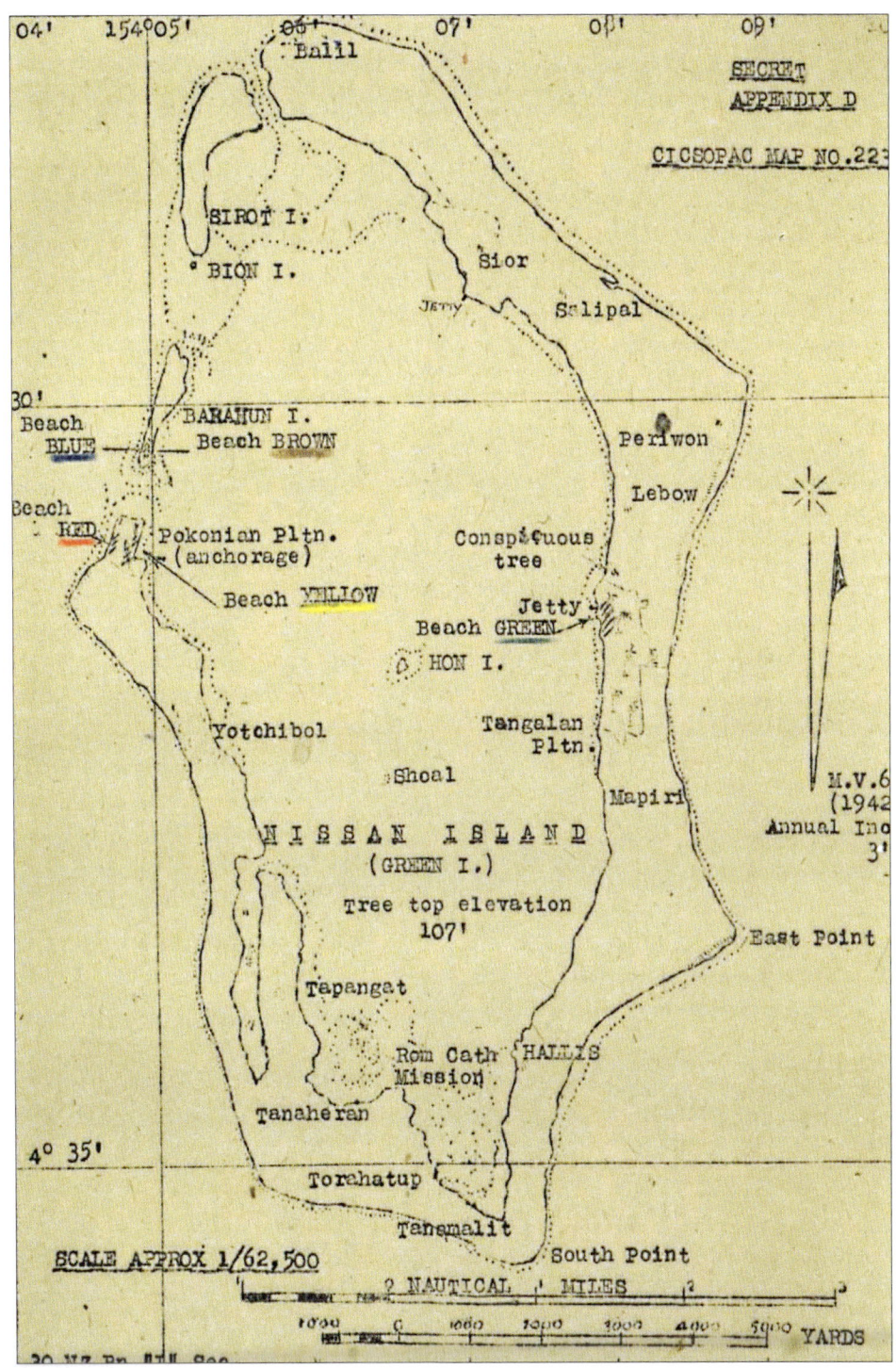

MAP 3
Green Islands depicting the raiders' landing beaches. Beaches blue and red were not used.
Source: Headquarters 14th NZ Brigade-Office Records-Commando Raid-30 Battalion Commando Forces Raid on Nissan, DAZ 155/9/1, 1551, WAII1, 18886, ADQZ, ANZ

A New Zealand-led "Commando Raid"

the awaiting destroyers.³⁶ Prisoners were to be taken if possible but not if it required excessive effort, which demonstrated the planners' low expectations in actually capturing Japanese soldiers. Additionally, placing greater emphasis on taking prisoners may have unnecessarily jeopardized the mission and could have resulted in enemy reinforcements being sent to the islands. Understandably, therefore, defended localities were also to be bypassed where possible. To foster good relations with native islanders, and to prevent an accidental confrontation before the main landing, the local population was to be left alone "unless definitely hostile."³⁷ The mission was to last no longer than 24 hours.

Although the 30th Battalion was without combat experience, it had undergone further jungle warfare and amphibious training that equipped it for such a mission.³⁸ Drawing on New Zealand's recent lessons on Vella Lavella, and spurred on by headquarters 14th NZ Brigade, training directed rehearsing for landings on hostile beaches, establishment of beachhead and perimeter defense at night, silent digging techniques, information collection, and beach reconnaissance.³⁹ With these efforts, and once established in a "bush line," each man knew the exact position of their platoon members, and most importantly, their commander.⁴⁰ Orientation was also assisted through the use of the sand table map in relief shown to every person before departure.⁴¹ Despite the additional training, the soldiers had not undergone any specialist selection in the traditional sense, and were thus very much ordinary soldiers designated to conduct an extraordinary task.

The infantry platoons were reduced to 25 soldiers to accommodate the attached technicians and specialists, which was not appreciated as many were "Left Out of Battle."⁴² As was the standard for jungle warfare, grenades and automatic weapons ammunition were a priority with 525 Bren machine gun, 500 Thompson submachine gun rounds, and 62 grenades distributed per section, alongside 100 rifle rounds per person. Each carried 48 hours of rations and full water bottles, a further two two-gallon tins of water per section was issued, along with emergency rations, which were held aboard the landing craft, everything else was kept to a minimum.⁴³ Five Wire-

---

[36] "30 NZ BN 'Commando' Force OO No. 1," 24 January 1944, DAZ 156/1/40, 1154, WAII1, 18886, ADQZ, ANZ.

[37] "30 NZ BN 'Commando' Force OO No. 1," 22 January 1944, DAZ 155/9/1, 1551, WAII1, 18886, ADQZ, ANZ.

[38] Letter from Barrowclough to Puttick, 6 January 1944; and *Officers' Book 14th Brigade New Zealand Expeditionary Force in Pacific* (n.d.), Kippenberger Military Archive.

[39] Frank Rennie, *Regular Soldier: A Life in the New Zealand Army* (Auckland: Endeavour Press, 1986), 50; "Training Memorandum No. 2," 14 January 1944, Appendix IV; and "Training Directives-Island Patrols," 5 January 1944, Appendix I, DAZ 155/1/25, 1151, WAII1, 18886, ADQZ, ANZ.

[40] H. L. Bioletti, *Pacific Kiwis: Being the Story of the Service in the Pacific of the 30th Battalion, Third Division, Second New Zealand Expeditionary Force* (Wellington: A. H. & A. W. Reed, 1947), 88.

[41] Bioletti, *Pacific Kiwis*, 89.

[42] Gillespie, *The Pacific*, 170–71.

[43] "30 NZ BN 'Commando' Force OO No. 1," 22 January 1944.

less Set No. 48s were carried, and a divisional signals detachment was also assigned to maintain long-range communications with Task Force 31 and coordinate a quick withdrawal should it be required.[44]

Discussions between 30th Battalion, 3d NZ Division, and Task Force 31 contributed to the issue of specially designed topography questionnaires to the raiding force, which assisted in noting observations and recording data.[45] Strict security measures were enacted while preparations were underway; however, reports indicated that many officers and enlisted breached security by revealing details of the upcoming raid to other personnel. This was quite a serious matter considering the operational risks and was an indication of a general lack of security awareness among members of the 14th NZ Brigade.[46]

On 25 January, Cornwall briefed his troops, and four days later on 29 January, the mission began with the assembly of three auxiliary personnel destroyers (APDs) (converted destroyers modified to carry around 185 personnel) and four escorting destroyers. The presence of APDs indicated that speed during the movement phase was of the utmost importance, as APDs were faster and more seaworthy than the larger landing craft specifically designed for amphibious landings. The Landing Craft, Infantry (LCI) could carry the same number of personnel but were notoriously prone to excessive yawing and rolling in even moderate seas, as well as being about 25 percent slower than APDs.[47] Once aboard the APDs, U.S. Navy and New Zealand commanders held a final conference for the rehearsal later that night. However, despite their best efforts, the (first) rehearsal landing was abandoned as the original beach could not be identified in the darkness, and the troops were forced to land on another beach—further evidence of the necessity for alternate plans and the requirement for adequate communications to enact them.[48] The near failure of the rehearsal phase went unmentioned within the action report of the commander for Task Group 31.8; no doubt an attempt by Captain Earle to brush off responsibility for the mishap, but also one that was adequately rectified.[49]

After the rehearsal, the task force sequenced its movement north, escorted by Consolidated PBY Catalinas (flying boats), and rendezvoused with two PT boats—

---

[44] Gillespie, *The Pacific*, 174; and "30 NZ BN 'Commando' Force OO No. 1," 24 January 1944.

[45] "30th NZ Battalion, Report on Operations Jan/Feb 1944-Green Island Group," n.d., Headquarters 14th NZ Brigade-Office Records-Unit Reports Squarepeg Operation Including Signals Report, DAZ 155/9/4, 1551, WAII1, 18886, ADQZ, ANZ.

[46] "Breaches of Security," 25 January 1944, Appendix V, DAZ 155/1/25, 1151, WAII1, ADQZ, ANZ.

[47] "Characteristics of Landing Craft Likely to Be Used for Move to Forward Area," 8 September 1943, Headquarters 14th NZ Brigade-Office Records-Amphibious, DAZ 155/9/2, 1551, WAII1, 18886, ADQZ, ANZ.

[48] Bioletti, *Pacific Kiwis*, 88–89.

[49] "Narrative of APD Activities during Raid and Reconnaissance in Force-Green Islands, B.S.I.," 4 February 1944, Commander Transport Division 12, in "Action Reports Covering Operations of Task Force 31 from 28 January 1944 to 17 February 1944," 24 March 1944, Serial 00177, Box 126, Record Group 38, National Archives and Records Administration (NARA).

the same boats that had conducted the preliminary soundings and were thus able to provide navigational marks for the larger landing craft.[50] The passage was uneventful except for the unexpected rescue of one U.S. Marine Corps reservist, Lieutenant Ranegan, from a rubber boat, after his Vought F4U Corsair had "been forced down by engine trouble."[51] On arrival off the Green Islands, the troops descended into the lowered landing craft and proceeded to the rendezvous area a few hundred yards offshore. Some of the landing craft had difficulty forming up, leaving a number to continue with the scheduled timetable without them. It was decided that the landing craft would be towed through the main channel by a PT boat to minimize the noise of multiple engines. After gathering speed, and when the tide was right, the boat cut its engines on approaching the main channel entrance and used the inward current to drift through almost silently before executing a near perfect landing.[52] There was considerable angst during the movement through the narrow channel, as testified by an officer, who commented that "it would have been disastrous if we had been fired upon . . . as the 12 barges went through the gap."[53] Within 30 minutes of boarding the landing craft, the first "commandos" were ashore. Their training kept them in good stead as they established a defensive perimeter without a detectable sound, which was quite a feat on a moonless night in the jungle and with many suffering from seasickness. Their success and speed can be attributed to the insistence on training for night amphibious landings, something regular American forces did not ordinarily conduct.[54]

The soldiers and specialists dug-in and waited four hours until sunrise before setting off on their tasks, by which time inquisitive islanders had infiltrated the perimeter happy to engage in conversation. The U.S. Navy hydrographic team investigated the two channels leading into the lagoon for depth and ran sounding lines along the shore to assess landing sites for the larger landing craft and vessels.[55] Some troops protected the specialists while others imitated raider tactics to deceive Japanese eyes. One group set out across the lagoon to the site of the potential airfield, where they were buzzed by a New Zealand aircraft dropping a roll of toilet paper. The commandos, not impressed with what they thought a poor joke, were unaware

---

[50] Morison, *Breaking the Bismarcks Barrier*, 414.
[51] "Action Report, covering operations of Task Group 31.8 from January 28, 1944 to February 1, 1944," Commander Destroyer Squadron 45 (Commander Task Group 31.8), 10 February 1944, Serial 0048, in "Action Reports Covering Operations of Task Force 31 from 28 January 1944, to 17 February 1944," 24 March 1944, Serial 00177, Box 126, Record Group 38, NARA.
[52] "Narrative of APD Activities during Raid and Reconnaissance in Force-Green Islands, B.S.I.," 4 February 1944.
[53] Rennie, *Regular Soldier*, 50. Contemporary documents use the terms *barge* and *landing craft* interchangeably.
[54] Gordon L. Rottman, *US World War II Amphibious Tactics: Army & Marine Corps, Pacific Theater* (Oxfordshire, UK: Osprey Publishing, 2005), 6.
[55] "The WWII Recollections of Captain Junius T. Jarman, USC&GS of the Wartime Experiences of the USS Pathfinder Forward," in *Pathfinder: Recollections of Those Who Served, 1942-1971* (Silver Spring, MD: Office of National Oceanic and Atmospheric Administration Corps Operations, 1994).

of the message tucked inside, alerting them to the presence of Japanese barges on the opposite shore.[56]

In addition, the battalion reconnaissance party journeyed along the western edge of the lagoon in three landing craft, searching for suitable landing areas, and in the process discovered suspicious silhouettes near the waterline. After observing the objects through binoculars, and seeing no movement, they decided to investigate.[57] Unfortunately, the landing craft pilots initiated a frontal approach and on nearing the shore, they came under accurate Japanese fire at close range, killing three and wounding four of the raiders including one of the craft pilots. The decision to investigate the suspicious objects was sound, but in retrospect the frontal approach was risky, and it was fortunate that the craft withdrew without further casualties. This was a serious, yet simple, error by Commander Smith and New Zealand lieutenant Patrick O'Dowd who had controlled the landing craft.[58]

The area was later engaged with mortar fire and a counterattack launched, but not before Cornwall ordered Smith to stand down and await the completion of reconnaissance activities, indicating the two commanders' very different levels of aggression. In late afternoon, two landing craft with one infantry platoon each sailed toward the enemy positions, while four other landing craft engaged the area with automatic fire. Unfortunately, just after depositing the platoons ashore, the landing craft were strafed by Japanese aircraft, demonstrating the precarious position of assaulting amphibious troops during the ship-to-shore or shore-to-shore phases of a landing. The Japanese air retaliation was serious enough for the small force to break radio silence and send an uncoded message: "Being heavily strafed. Request air support."[59] The attack shook the New Zealanders' confidence, and soon after they disembarked at the locality, the troops were recalled due to fears of further enemy aerial attacks.[60] For some unknown reason, the enemy aircraft failed to make a second pass. It was fortuitous. Had they done so, casualties could have been severe.

As night fell on 31 January, and with reconnaissance tasks completed, the troops prepared to reembark for rendezvous with the returning ships. Quite astutely, the decision had been taken to place Wilkinson's chief of staff aboard one of the APDs on this night. Senior New Zealand officers had also taken the opportunity to observe conditions first-hand, with Potter and three of his staff officers watching from a destroyer.[61] Their presence provided additional observation of operating conditions that

---

[56] Newell, *Operation Squarepeg*, 32-33.
[57] The objects were actually two well-camouflaged Japanese landing craft.
[58] Smith redeemed himself by extracting his landing craft from the kill zone while under fire, but O'Dowd died of his wounds two hours later. Gillespie, *The Pacific*, 174-76; and Bioletti, *Pacific Kiwis*, 91-94.
[59] Headquarters 14th NZ Brigade-War Diary, 31 January 1944, DAZ 155/1/25, 1151, WAII1, 18886, ADQZ, ANZ.
[60] Rennie, *Regular Soldier*, 54-55; and Gillespie, *The Pacific*, 176.
[61] Headquarters 14th NZ Brigade-War Diary, 1 February 1944, DAZ 155/1/26, 1151, WAII1, 18886, ADQZ, ANZ.

may have affected the main landing. One final drama occurred when, as the landing craft returned to the APDs, they encountered heavy seas that impeded the recovery of the craft.[62] This experience, in conjunction with the rough surf encountered on 10 January, further indicated the difficulty of landing on the beaches of the outer coastline.

Once the raiding force returned to Vella Lavella, Barrowclough reported that "the whole operation was daringly conceived and splendidly carried out."[63] Indeed, the raid had the desired effect by quickly enabling U.S. and New Zealand forces to draft operational orders for the main amphibious landing.[64] In particular, it verified the viability of the key objectives for the main operation, namely securing a suitable area for the construction of an airfield and a PT boat base. This, along with the beach analysis, identified the operation's decisive points around the main channel and the main landing beaches at Pokonian and Tangalan Plantations. It also evidenced the smooth interoperability between U.S. and New Zealand forces at the planning and tactical stages, demonstrating a common grasp of doctrine and staff work, which was quite a feat considering the New Zealanders hailed from the British system, with its different staff designations, relationships, cognitive approaches, and traditions. That New Zealand and U.S. personnel operated effectively together despite having been awake for the better part of two days was testament to their fitness and training and the experience of the U.S. Navy crews.

From the moment Barrowclough returned to Vella Lavella and informed his staff of the mission on 1 January until the issuance of Wilkinson's operation order on 25 January, it had taken 25 days to plan the raid. Considering proper naval planning could not begin before 11 January, when the channel was found accessible to landing craft, the coordination and gathering of resources was impressive. The value of the raid could also be measured by the acquisition of information and data "of inestimable value in planning the main operation" at a cost of four killed, six wounded and three injured during the process of embarking and disembarking landing craft.[65] The mission's contribution to the main landing was aptly demonstrated on the morning of 15 February, when elements of the 3d New Zealand Division and various U.S. units quickly established themselves on the islands. The occupation finally severed Japanese lines of communication between Rabaul and Bougainville and brought potential landings zones in the Bismarck Archipelago within range of Allied air power. This led Admiral Halsey to declare that "the entire Green [Islands] operation was thoroughly

---

[62] "Narrative of APD Activities during Raid and Reconnaissance in Force–GREEN ISLANDS, B.S.I.," 4 February 1944.
[63] "Letter from Major General Barrowclough to the Prime Minister, 5 August 1944: Report on Operations–3d New Zealand Division, 1 January 1944 to 30 June 1944," in *Documents Relating to New Zealand's Participation in the Second World War, 1939–45*, vol. 3 (Wellington: Department of Internal Affairs, War History Branch, 1963), 447.
[64] Headquarters 14th NZ Brigade-War Diary, 4–5 February 1944, DAZ 155/1/26, 1151, WAII1, 18886, ADQZ, ANZ.
[65] "Letter from Major General Barrowclough to the Prime Minister," 5 August 1944, 447.

planned and was executed with the utmost precision and team play."[66] The mission vindicated the benchmark cost effectiveness of Second World War amphibious raids in that it required limited time in which to train and prepare personnel, the low casualty rates justified the information obtained, and it did not prove a drain to supporting services.[67] Lastly, the raid was unique by employing regular soldiers on a special operations-type mission. To be sure, they did not face severe resistance, but some of the U.S. Navy's finest commanders sought to assign such specialized tasks to these troops, which signaled significant confidence in the New Zealanders' capabilities. In the end, with sufficient training, these regular soldiers adapted their normal mission skillsets to suit operational requirements, demonstrating that regular forces held more flexibility than ordinarily presumed.

---

[66] Commander, South Pacific to Commander in Chief, United States Fleet, "Seizure and Occupation of GREEN ISLANDS, 15 February to 15 March 1944," S14, 1, WAII9, 18907, ADQZ, ANZ.
[67] Maj Peter Evans, RM, "The Value of Amphibious Raiding in the Twentieth Century: A Historical Perspective," *Defence Studies* 1, no. 3 (Autumn 2001): 103, https://doi.org/10.1080/714000047.

A New Zealand-led "Commando Raid"

CHAPTER TWELVE

PLA Amphibious Campaigns and the Origins of the Joint Island Landing Campaign

Xiaobing Li

In August 2022, the world witnessed the fourth Taiwan Strait Crisis as the People's Republic of China (PRC) launched one of its largest military exercises by sending more than 100 warplanes, deploying 10 destroyers, and firing 12 missiles around Taiwan (the Republic of China, ROC) after two U.S. congressional delegations visited the island.[1] The PRC reaction prompted a greater focus on how the United States would respond if the Chinese People's Liberation Army (PLA) launched an assault, particularly an amphibious invasion of Taiwan.[2] After consolidating power at the 20th National Congress of the Chinese Communist Party (CCP), PRC President Xi Jinping has adopted many of Mao Zedong's strategies as his own, including those concerning Taiwan. For example, Mao developed a strategy to use Taiwan to deal with America by putting more pressure on Washington. Moreover, Mao designed a local war (or limited war) doctrine in the Taiwan Strait by concentrating a large landing force, controlling air and sea, and attacking one island at the time without American intervention. In the 1950s, the PLA seized 32 Taiwan-controlled islands during Mao Zedong's regime. These historical actions can shed light on the current crisis.

The questions this chapter examines include: How did the PLA plan, orches-

---

[1] For more on the crisis, see "Taiwan," in *2022 Report to Congress of the U.S.-China Economic and Security Review Commission*, 117th Cong., 2d Sess. (November 2022), chap. 4.
[2] In the People's Republic of China, the PLA is the term for the army as well as the entire defense force; so, the navy is known as the PLA Navy, and the air force is the PLA Air Force.

trate, and execute amphibious landings on Taiwan's offshore islands? Why were the Chinese Communist offenses not thwarted by American armed forces? What lessons do Chinese strategists and tacticians derive and inherit from their past encounters in the Taiwan Strait? This chapter focuses on the PLA's Yijiangshan (1955), Hainan (1950), and Quemoy (1949) landing campaigns using official Chinese documents, military writings, and interviews of both PLA and Taiwanese generals.[3] No matter how antiquated, the PLA's real experience in the immediate post-World War II period is the service's only meaningful reference for its Cold War amphibious capabilities. The PLA's experience facing the Taiwanese and U.S. forces in the Taiwan Strait necessitated the Chinese military restructuring and reforms. To match their opponents and capably improve, Chinese generals shifted their concept of classical amphibious warfare, demanded improvements in naval and air support, and enhanced their firepower, transportation, and logistics. While PLA modernization efforts have improved, this chapter argues for continuity and adaptation in the Chinese joint island landing campaign concept. Xi Jinping adopted Mao Zedong's island attack doctrine like other Chinese leaders, and this was already evident in former PLA general chief of staff Deng Xiaoping's amphibious battle against Vietnamese forces on the Paracel Islands in the South China Sea in 1974, the invasion of the Spratly Islands in 1988, and former president Jiang Zemin's Taiwan Strait missile crisis in 1995-96. Deng launched the "limited" landing campaigns in the South China Sea after the U.S. armed force left South Vietnam.[4] Jiang Zemin step down from his military threats on Taiwan in 1996 after the Clinton administration sent U.S. aircraft carrier battle groups to the Taiwan Strait.[5]

PLA AMPHIBIOUS OPERATION GUIDELINES: LESSONS LEARNED

Soon after the Chinese Nationalist forces left the mainland, Nationalist president Chiang Kai-shek deployed 60,000 Kuomintang (KMT) troops on Quemoy, 100,000 men on Hainan, 120,000 along the Zhoushan Island group, and 200,000 on Taiwan after the PLA overtook the mainland in October 1949.[6] Although taking the small islands should have been a simple part of the PLA's attempt to control the strait, the PLA's 1949 landing on Quemoy island was a disaster since the PLA had very little experience

---

[3] Also referred to as Jinmen, Quemoy, or Kinmen in some sources.
[4] Liu Huaqing, "Carry on Deng Xiaoping's New Thoughts to Build a Strong, Modern Military," in *Liu Huaqing junshi wenxuan* [Selected Military Works of Liu Huaqing], vol. 2 (Beijing: PLA Press, 2008), 546-47. Adm Liu was the commander the PLA Navy from 1982 to 1988.
[5] Zhang Wannian, *Zhang Wannian zizhuan* [Autobiography of Zhang Wannian], vol. 2 (Beijing: PLA Press, 2011), 433-35. Gen Zhang was the CMC vice chairman and the commander of the 1996 PLA joint landing exercise along the eastern coast.
[6] To prevent confusion, the more common naming/spelling conventions for people and places will be used throughout. *Guojun houqin shi* [Logistics History of the GMD Armed Forces], vol. 6 (Taipei: Bureau of History and Political Records, Defense Ministry, 1992), 199-200.

in amphibious campaigns during World War II or the Chinese Civil War (map 1).⁷

In 1949, the small island group of Quemoy, lying just three kilometers from the mainland, had a population of 40,000 civilians. The island is not in the open ocean but lies in Xiamen (Amoy) harbor, the southeast mainland's largest seaport (map 2). By 17 October, Xiamen's nearby mainland KMT garrison was swiftly overtaken by the PLA's 10th Army Group. Unfortunately, PLA officers did not consider an amphibious landing much different than a ground assault when the army group commander, General Ye Fei, ordered the 28th Army to attack Quemoy. Without updated intelligence, naval assistance, or air support, the 28th Army's commander positioned 10,000 troops, in three regiments, in a disconcerted first-wave attack on the evening of 24 October 1949. The commanders felt the landing troops' perceived element of surprise would allow for a quick penetration in depth resulting in the defeat of the nationalist garrison on the island. However, successful mainland tactics relied on during the civil war were ineffective and disastrous on Quemoy. First, the 28th Army did not have adequate landing craft and used 200 fishing boats that had been gathered from around Xiamen. The fishing boats were promptly destroyed by KMT naval and air forces on Quemoy the next morning.⁸ Second, the KMT island garrison counterattacked using armor forces to separate the landing troops into several pockets, inflicting heavy casualties on the PLA forces. The 150,000-strong PLA 10th Army Group left without transportation, could only listen helplessly to their comrades' pleas for reinforcement over the radio. The remaining PLA landing troops were surrounded on the second day in a small village, Guningtou, near the landing zone, and three days later the landing party was decimated by the KMT defenders, having lost only 1,000 casualties, and the PLA losing 9,086 PLA attackers and more than 3,000 prisoners.⁹

Mao Zedong was shocked when news reached Beijing on 28 October regarding the 10th Army Group's losses. The army, which was one of the 3d Field Army's best units, lost three regiments on the beaches of Quemoy. A circular drafted by Mao warned all PLA commanders, "especially those high-level commanders at army level and above," that they "must learn a good lesson from the Jinmen [Quemoy] failure."¹⁰

---

⁷ Toshi Yoshihara, *Chinese Lessons from the Pacific War: Implications for PLA Warfighting* (Washington, DC: Center for Strategic and Budgetary Assessments, 2023).

⁸ Gen Hau Pei-tsun (Ret), interview with the author, Taibei, Taiwan, May 1994. Hau was ROC Army commander on the offshore islands during the PLA attack on Quemoy in 1949; he served as the defense minister of Taiwan in the 1980s.

⁹ *A History of the Republic of China*, vol. 2 (Taipei: Modern China Press, 1981), 297. The ROC Army officially claimed PLA casualties of about 20,000 troops, including 7,200 prisoners. According to the author's interviews both in Taiwan and China, 10,000 PLA casualties seem most acceptable.

¹⁰ "Circular on the Setback of Jinmen Battle, 29 October 1949," Central Military Commission (CMC), Beijing. This document was sealed and issued by the CMC. In 1987, the Archives and Research Division of the CCP Central Committee found that Mao drafted the original document. The division reprinted it from Mao's manuscript and included it in *Jianguo yilai Mao Zedong wengao, 1949–1976* [Mao Zedong's Manuscripts since the Founding of the State, 1949–1976], vol. 1 (Beijing: CCP Central Archival and Manuscript Press, 1993), 100–1, hereafter *Mao's Manuscripts since 1949*.

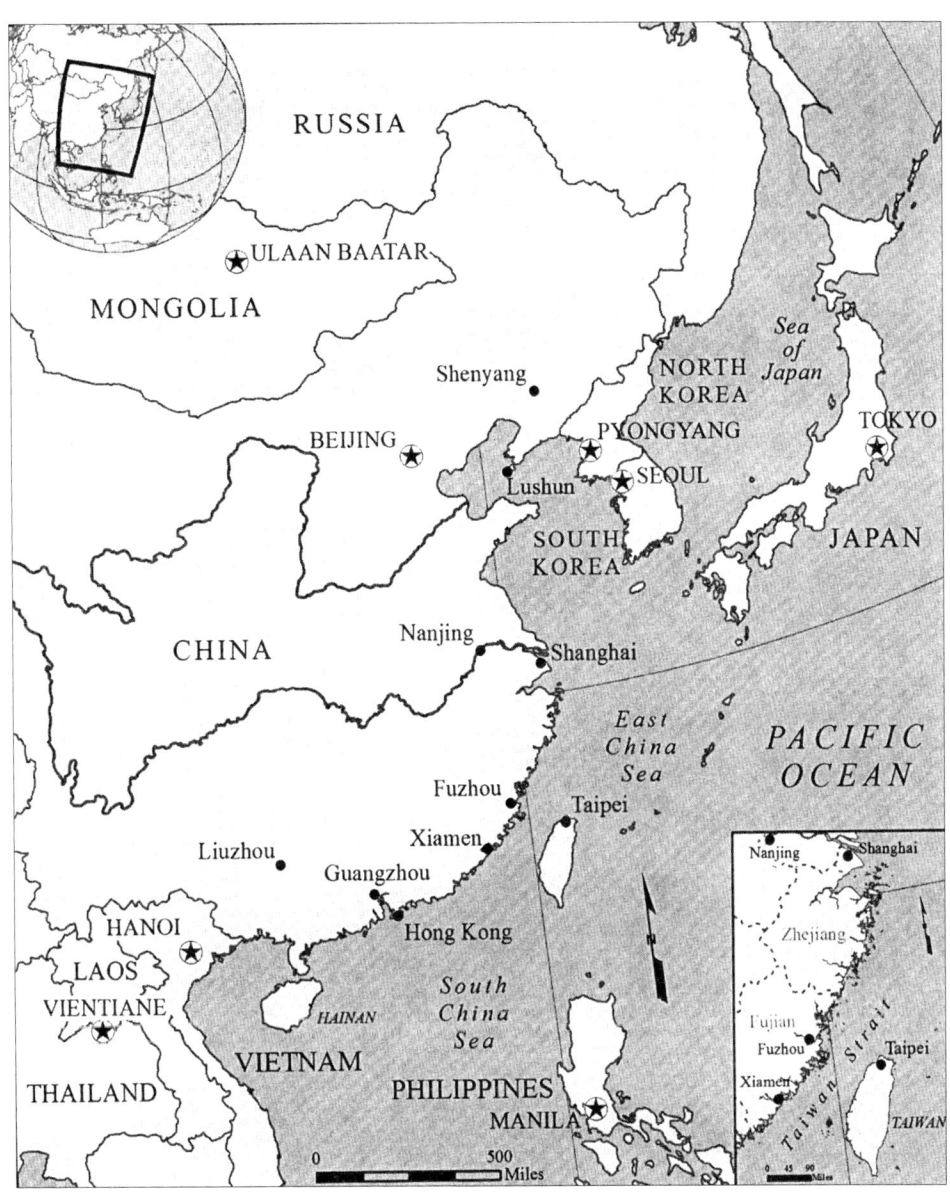

MAP 1
China and the Taiwan Strait.
Source: Xiaobing Li, The Cold War in East Asia (New York: Routledge, 2018), 14, prepared by Brad Watkins

Mao also ordered 4th Field Army commander Lin Biao to halt all amphibious operations on the South China Sea coast on 31 October, and telegraphed the 3d Field Army's deputy commander Su Yu in early November to postpone any East China Sea

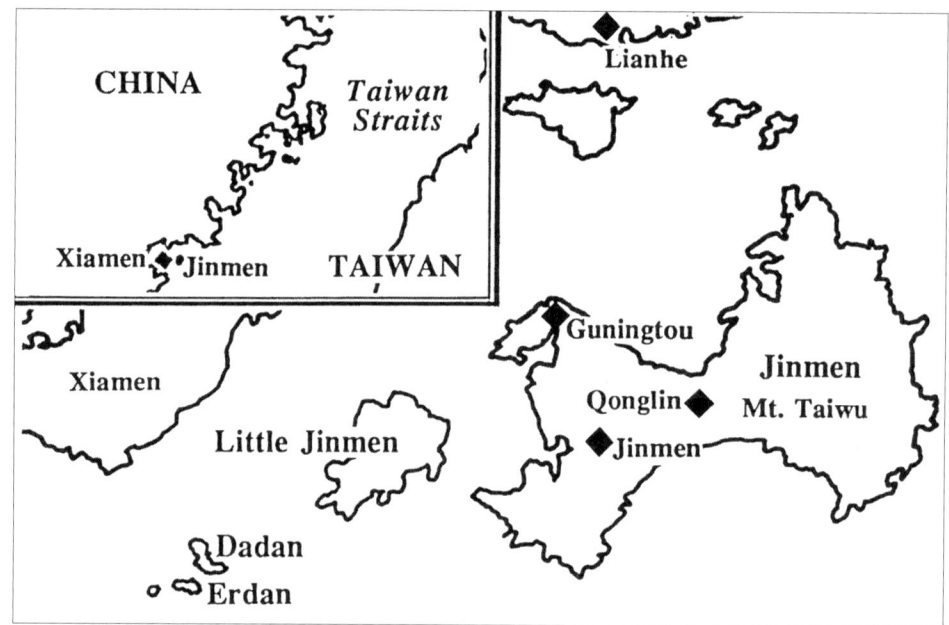

MAP 2
The Quemoy Islands.
*Source: Xiaobing Li, A History of the Modern Chinese Army
(Lexington: University Press of Kentucky, 2007), 181*

island assaults.[11] Mao did not want another disaster that might affect the morale of his forces or provide confidence to the nationalists or their allies.

By 14 November 1949, only the field army headquarters could authorize amphibious landing operations as Su relayed orders to the army group commanders for the 7th, 9th, and 10th Army Groups.[12] In demonstrating extreme caution after Quemoy's failure, Mao stressed preparedness and calculation to Su for future amphibious operational training. In November, Mao telegraphed the field army commanders again that the "cross-strait campaign is totally different from all experience our army had

---

[11] Mao, "Telegram to Lin Biao: My Suggestions on Your Troops Disposition and Battle Array, October 31, 1949." In his telegram, Mao alerted Lin: "Do not attack the Leizhou Peninsula, much less a chance to attack the Hainan Island." *Mao's Manuscripts since 1949*, vol. 1, 107. Two of these CMC telegrams were drafted by Mao to Su Yu. The first one is the "Telegram for the Operation Plan of the Dinghai Campaign, November 4, 1949," and the second is the "Telegram: The Disposition of the Dinghai Campaign, November 14, 1949." The latter reads, "In view of the military failure on Jinmen, you must check out closely and seriously all problems, such as boat transportation, troop reinforcement, and attack opportunity on the Dinghai Landing. If it is not well prepared, we could rather postpone the attack than feel sorry about it later." *Mao's Manuscripts since 1949*, vol. 1, 118, 120, 137.

[12] He Di, "The Last Campaign to Unify China: The CCP's Unrealized Plan to Liberate Taiwan, 1949-1950," in *Chinese Warfighting: The PLA Experience since 1949*, eds. Mark A. Ryan, David M. Finkelstein, and Michael A. McDevitt (Armonk, NY: M. E. Sharpe, 2003), 88.

in the past." Mao warned commanders to "guard against arrogance, avoid underestimating the enemy, and be well prepared."[13] In training the 3d Field Army, Su also advised the high command that it would be "extremely difficult to operate a large-scale cross-ocean amphibious landing operation without air and sea control."[14] Amphibious operations, according to Su, presented "a new warfare" or "modern warfare, different from all the wars we have fought before."[15] Of course, most of these lessons had been explored by Allied forces in the Pacific in the 1940s.

The KMT account of the PLA's loss on Quemoy, nevertheless, differed. According to Taiwanese generals, first, the PLA attackers, after overpowering Xiamen, were conceited and arrogant in their underestimation of Quemoy's defensibility. PLA landing troops were unprepared for substantial resistance and saw the mere landing on Quemoy as a success. Second, the KMT garrison received reinforcement from the 18th and 19th Armies before the PLA's landing on Quemoy, something of which the latter was unaware. Third, instead of varying landing zones and times, the PLA 28th Army chose only one landing zone and a detrimental landing time, thereby allowing for concentrated KMT firepower on the landing site. Finally, the PLA's first wave landed without antitank guns and supporting fire and reserved no boats for potentially reinforcing the island thereafter. During the author's interview with KMT General Chiang Wei-kuo, the general recalled that the Battle of Quemoy boosted his father Chiang Kai-shek's notion that the KMT could build up a strong defense against PLA amphibious threats, survive on the islands, and continue as leader of the ROC.[16]

Mao quickly realized that to successfully execute any significant amphibious operations, the PLA required air and naval support. Thus, the PLA's Air Force (PLAAF) was established on 11 November 1949. According to Xiaoming Zhang, "Chinese Communist concepts for the development of airpower derived primarily from Mao Zedong's plan for the invasion of Taiwan in 1949."[17] Furthermore, the PLA Navy (PLAN) headquarters was formed from the 4th Field Army's 12th Army Group in December with Admiral Xiao Jinguang as the naval commander.[18] Mao desperately needed to equip the new naval and air forces, and he visited Moscow on 16 December to broker

---

[13] CMC document, drafted by Mao, "Circular on the Lesson of Jinmen Battle, October 29, 1949," in *Mao's Manuscripts since 1949*, vol. 1, 101.

[14] Gen Ye Fei, *Ye Fei huiyilu* [Memoirs of Ye Fei] (Beijing: PLA Press, 1988), 608. The author's interview of the staff member of the 10th Army Group headquarters at Hangzhou, Zhejiang, 6 July 2006, hereafter Ye 2006 interview. Ye was the commander of the 10th Army Group in 1949-55. *Xinghuo liaoyuan* [Composition Department], *Zhongguo renmin jiefangjun jiangshuai minglu* [Marshals and Generals of the PLA], vol. 1 (Beijing: PLA Press, 1992), 58–59.

[15] Ye, *Ye Fei huiyilu* [Memoirs of Ye Fei], 608; Ye 2006 interview; and *Xinghuo liaoyuan* [Composition Department], *Zhongguo renmin jiefangjun jiangshuai minglu* [Marshals and Generals of the PLA], vol. 1, 58–59.

[16] Gen Chiang Wei-kuo, ROC Army, (Ret), interview with the author, Rongzong Hospital, Taipei, Taiwan, 26 May 1994.

[17] Xiaoming Zhang, *Red Wings over the Yalu: China, the Soviet Union, and the Air War in Korea* (College Station: Taxes A&M University Press, 2002), 6.

[18] RAdm Yang Guoyu, *Dangdai Zhongguo haijun* [Contemporary Chinese Navy] (Beijing: China's Social Science Press, 1987), 17. Yang was deputy commander and chief of the staff of the PLAN in 1978-85.

an alliance between the PRC and USSR. Soviet leader Joseph Stalin eventually agreed to loan warships and equipment, totaling $150 million (USD, 1950 value), after Mao's two-month stay in the Soviet Union.[19] Later, Mao augmented the new air force when he ordered 340 Russian warplanes for 1.2 billion rubles ($380 million USD).[20] On 11 February 1950, Mao wrote to Stalin and purchased an additional 628 airplanes from Moscow.[21] However, neither the newly created PLAN nor PLAAF were able to support the 1950 landing campaigns as purchased Soviet warships and planes arrived too late to see combat.

The first statement on PLA amphibious operations was a landing campaign checklist composed by Mao Zedong in mid-December 1949.[22] Mao warned the Chinese generals that they "must study the lesson [of Quemoy]."[23] The checklist stated that cross-strait attacks should, first, establish a centralized chain of command; second, the invading force must be superior in number over the defense garrison; third, the invading force should receive proper training, transportation, and supplies; and fourth, the operation must have air and naval support. From 1949 to 1962, the PLA continued amphibious campaign preparation and improved its island intelligence, near-sea communication, and offshore combat effectiveness. By the early 1960s, the balance of power favored the PLA in the Taiwan Strait and the Chinese generals met Mao's cross-strait attack guidelines, launching amphibious landing campaigns and seizing Taiwanese-held islands one by one.

LANDING CAMPAIGNS AND U.S. INVOLVEMENT

With the disaster of Quemoy still fresh, the PLA's 4th Field Army prepared for an amphibious landing at Hainan in late 1949. The PLA commanders had implemented most of Mao's new doctrines, except air and naval support. First, Commander Lin Biao deployed his 15th Army Group to conduct a three-month landing preparation from December 1949 to March 1950 for the Hainan campaign. Deng Hua, commander of the 15th Army Group, organized beachhead assault training, antiship attacks, and landing coordination and communication. Deng also employed 6,000 boat crews and 2,100 fishing boats for cross-strait transportation. With better training and transpor-

---

[19] Yang, *Dangdai Zhongguo haijun* [Contemporary Chinese Navy], 48, 52.

[20] LtGen Han Huaizhi, *Dangdai zhongguo jundui de junshi gongzuo* [Military Affairs of Contemporary China's Armed Forces], vol. 2 (Beijing: China's Social Science Press, 1989), 161.

[21] Mao's telegrams to Stalin on 11, 15, and 25 February 1950, were quoted in Chu Feng, "20 Shiji 50 niandai zhongsu junshi guanxi yanjiu" [The Sino-Soviet Military Relations in the 1950s] (PhD diss., Party University of the CCP Central Committee, Beijing, 2006), 45, 59.

[22] Military History Research Division, PLA Academy of Military Sciences (AMS), *Zhongguo renmin jiefangjun zhanshi* [War History of the Chinese People's Liberation Army], vol. 3 (Beijing: Military Science Press, 1987), 359.

[23] Mao Zedong telegram to Lin Biao, 18 December 1949, responding Lin's telegram on 10 December about the 4th Field Army's campaign proposal, including landing campaign on Hainan Island. As quoted in *Mao's Military Manuscripts since 1949*, 104–6.

tation options, the 4th Field Army's Hainan operation was approved by Mao after his return from Moscow.[24]

To ensure an overwhelming landing force, the 15th Army Group commander Deng Hua assigned his 40th and 43d Armies, three artillery regiments, and combat engineering troops, totaling 100,000 fighters, for the Hainan landing campaign. To support the 15th Army Group's landing efforts, the PLA could also count on the strength of about 20,000 guerrilla soldiers already operating on Hainan.[25]

In addition to the training and development of the force, the PLA established a centralized chain of command to ensure that the CCP's Central Military Commission (CMC), 4th Field Army, and the 15th Army Group worked closely together. Mao instructed the party and CMC on 10 January 1950 "to make an effort to solve the problem of the Hainan Island in the spring and summer seasons."[26] The CCP Central China Bureau held a party/army Hainan campaign conference on 1 February, where the party and the military leaders decided to combine large- and small-scale amphibious crossings to counter the KMT's air and naval superiority in the Ch'iongchou Strait (about 32 kilometers wide). Mao reiterated to Lin Biao on 12 February, "[You] must confirm the guaranty of landing transportation and preparation before you launch the attack. Avoid push and rush, avoid mistake and loss."[27]

After training for three months, the 15th Army Group was ready for its amphibious campaign against Hainan Island. Deng Hua opened the campaign with guerrilla tactics, which had been successful tactics during the Civil War, and small battalion-size landings from 5–10 March. These troops successfully overtook the KMT garrison and joined up with local guerrilla forces. Then on 26–31 March, the 43d Army launched regiment-size landings with artillery pieces to establish large landing zones, secure two small harbors, and prepare for the 15th Army Group's arrival.[28]

On 16 April, the first major landing wave of 50,000 troops from the 15th Army Group on 350 boats embarked at 1930 that evening. After the fleet left the shore, the KMT air patrol soon observed the landing forces resulting in six warships attacking the PLA landing forces in transit but failed to stop the offensive. During the battles, the PLA boats sank one KMT ship and damaged two. The 40th Army's 118th Division landed at Hainan by 0600 the next morning, followed by the 119th Division, which

---

[24] Mao, "Approval of the Plan to Attack Dinghai First, Jinmen Second, 8 March 1950," in *Mao's Manuscripts since 1949*, vol. 1, 282.

[25] Gen Zhang Aiping, *Zhongguo renmin jiefangjun* [The Chinese People's Liberation Army], vol. 1 (Beijing: Contemporary China Press, 1994), 75–76.

[26] "Mao's Telegram to Lin Biao on the Issues of the Battle of Hainan Island, 10 January 1950," in *Mao's Manuscripts since 1949*, vol. 1, 77–78.

[27] "Mao's Telegram to Lin Biao, Agree on the 43rd Army's Landing Plan on Hainan, 12 January 1950," in *Mao's Military Manuscripts since 1949*, vol. 1, 123.

[28] RAdm Zhang Hancheng, "The Logistics during the Naval Operations in the Early Years," in *Hanjun huiyi shiliao* [The Navy: Memoirs and History Records] (Classified), ed., Navy Compilation Committee, PLA Historical Documents and Collections Series, vol. 2 (Beijing: Ocean Wave Publishing House, 1994), 890–92.

penetrated the KMT defense that afternoon. On 18 April, the 43d Army's 128th Division advanced deeper and attacked the KMT 252d Division, the main defensive force of Hainan. KMT defenses collapsed by 22 April after the 252d Division's destruction. The PLA's second landing wave of 50,000 troops left the mainland on 23 April and arrived at Hainan the next morning. By 1 May, the Battle of Hainan Island ended in a PLA victory.[29] The PLA's successful execution of amphibious landings on Hainan inspired further preparation for the invasion of Taiwan, scheduled for the summer of 1950.

The PLA's high command convened with Mao after his return from Moscow on 4 March 1950. Mao ordered the acting chief of the General Staff, Nie Rongzhen, and Su Yu to plan an attack on Taiwan, and Mao recommended mobilizing additional divisions and training airborne forces for the attack.[30] Su visited PLAN Commander Xiao Jinguang on 11 March and assigned naval units for the Taiwan invasion. The CMC approved the Su-Xiao plan in April, and later that spring, the PLA's 3d Field Army commenced amphibious training exercises. Half a million troops from the 3d Field Army, comprising the 7th, 8th, and 9th Army Groups, and the navy, mobilized for the invasion.[31] The 4th Field Army also participated in the Taiwan operations by deploying its 13th Army Group as a landing reserve force and 19th Army Group as a mobile force along the coast. All told, the Taiwan invasion force consisted of nearly 800,000 PLA troops.[32] The Taiwan invasion plan codified Mao's guidelines for a PLA amphibious landing campaign as a continuation of the Chinese Civil War military doctrine.

In preparation for the invasion, the 3d Field Army's 9th Army Group routed 120,000 KMT troops on the Zhoushan island group and occupied it by May 1950. The 9th Army Group landed on 18 islands of the Dongshan and Wanshan island groups in the East China Sea and overcame KMT defensive forces in early June. The 3d Field Army's success in the East China Sea bolstered morale in the PLA and encouraged final preparation for Taiwan's invasion.[33] Meanwhile, Mao announced that the party's priority was the "liberation" of Taiwan at the CCP's Seventh National Congress during its Third Plenary Session on 6–9 June in Beijing. After Su briefed the party leaders

---

[29] Han, *Dangdai zhongguo jundui de junshi gongzuo* [Military Affairs of Contemporary China's Armed Forces], vol. 1, 150.
[30] "Mao's telegram to Liu Shaoqi, Approval of Disposing Four Divisions for Landing Campaign, 10 February 1950"; "Mao's Comments on the Proposal of Attacking Dinghai First, Jinmen Second, 28 March 1950"; and "Mao to Su Yu, Instructions on Paratroops Training," *Mao's Manuscripts since 1949*, vol. 1, 256–57, 282.
[31] Gen Xiao Jinguang, *Xiao Jinguang huiyilu* [Memoirs of Xiao Jinguang], vol. 2 (Beijing: PLA Press, 1988), 8, 26.
[32] He, "The Last Campaign to Unify China," 82–83.
[33] Jiang Weiguo, interview with the author, Rongzong (Glory General) Hospital, Taibei, Taiwan, 23 May 1994. Gen Jiang recalled that his father, Chiang Kai-shek, and ROC intelligence had the information on the PLA landing preparation in the spring of 1950.

on PLA preparations for Taiwan's invasion, the CCP approved the plan.[34] However, Mao's priority was involuntarily altered and the CCP was forced to shift its objectives after the outbreak of the Korean War on 25 June 1950.[35]

Mao was blindsided by the invasion of South Korea, as neither the Soviets nor the North Koreans informed Chinese leadership of the planned 25 June attack on South Korea.[36] U.S. policy toward Taiwan also shifted as Washington abruptly and unexpectedly switched from "hands off" to "hands-on" regarding all things Asian.[37] As a deterrent against potential Chinese Communist attacks on ROC-held Taiwan, President Harry S. Truman deployed the U.S. Seventh Fleet to the Taiwan Strait two days after North Korea's invasion. By the end of 1950, Truman's stance prevented the PLA's planned invasion and secured the ROC with continued Seventh Fleet patrols in the Taiwan Strait, which marked a major obstacle in the cross-strait war plan. Direct American involvement in the Taiwan Strait presented the PLA with a challenge that they were not equipped politically or militarily to counter.[38] Before June, the PLA's primary task was liberating Taiwan from nationalist forces. But, as reflected in one of Mao's speeches, after June 1950, "The American armed forces have occupied Taiwan, invaded Korea, and reached the boundary of Northeast China. Now we must fight against the American forces in both Korea and Taiwan."[39] What had been a civil war on the Korean Peninsula quickly transformed into an international conflict and Communist leaders faced a new challenge. Any decision on a PLA amphibious invasion of Taiwan would require consideration of American military options after the outbreak of the Korean War in 1950. The window for the attack was closing.

---

[34] CCP Party History Research Division, *Zhongguo gongchandang lishi dashiji, 1919-1987* [Major Historical Events of the CCP, 1919-1987] (Beijing: People's Press, 1989), 191-92.

[35] Ye Fei, interview with the author, Hangzhou, Zhejiang, July 1996. Gen Ye was the commander of the 10th Army Group, 3d Field Army, of the PLA in 1949-51.

[36] Mao was very dissatisfied with this and later confided, "They [North Koreans] are our next door neighbor, but they did not consult with us about the outbreak of the war." As quoted in Li Haiwen, "When Did the CCP Central Committee Decide to Send the Volunteers to Fight Abroad?," *Dang de Wenxian* [Party Literature and Archives] vol. 5 (1993), 85, from Shen Zhihua, "China Sends Troops to Korea: Beijing's Policy-making Process," in *China and the United States; A New Cold War History*, eds., Xiaobing Li and Hongshan Li (Lanham, MD: University Press of America, 1998), 20.

[37] Xiaobing Li, "Truman and Taiwan: A U.S. Policy Change from Face to Faith," in *Northeast Asia and the Legacy of Harry S. Truman: Japan, China, and the Two Koreas*, ed., James I. Matray (Kirksville, MO: Truman State University Press, 2012), 127-28.

[38] Hau Pei-stun, interviews with the author, Taipei, Taiwan, 23-24 May 1994. Hau, as the commander of the front artillery force on Quemoy Island, felt relieved when he was informed of the U.S. Seventh Fleet's patrol in the Taiwan Strait in June 1950. See also Xiao, *Xiao Jinguang huiyilu* [Memoirs of Xiao Jinguang], vol. 2, 26.

[39] Mao, "The Great Achievements of the Three Glorious Movements" (speech, Third Plenary Session of the First National Committee of the Chinese People's Political Consultative Conference, 23 October 1951), as quoted in *Mao's Manuscripts since 1949*, vol. 2, 481-86.

JOINT OPERATIONS AND CHINA'S "LOCAL WARS"

The advantage of avoiding a full-scale war against the United States was a primary lesson that Chinese leaders learned from the Korean War. Much like the West, to mitigate economic expenses and human losses, the PLA would adopt a policy of focusing on limited or "local wars," rather than a major conflict. To the Chinese, avoiding total war with Washington and making limited, calculated attacks in the Taiwan Strait promoted Beijing's interests with the least amount of risk. From the mid-1950s, the nature of the strait crisis transformed from a civil war between China and Taiwan to a Cold War–style international conflict among Beijing, Taipei, and the United States. The PLA high command had to include America's response in their planning, preparation, and execution of their joint amphibious island landing campaigns after the Korean War ended in 1953. The PLA's new joint island landing doctrine emphasized the limited scale, remote location, and quick operation to prevent possible U.S. intervention. During PLA amphibious island landing campaigns from 1954 to 1965, their assaults remained small scale on distant small islands for quick landing victories to avoid countering the advantages of because of the significant technological gaps between Chinese and U.S. air and naval forces.

The PLA resumed its focus on conquering the ROC offshore islands and planned amphibious campaigns in 1954. Beginning with the smaller, northernmost Tachen Islands, which lay more than 322 kilometers away from Taiwan and 160 kilometers away from the U.S. Seventh Fleet headquarters in Yokosuka, Japan. The East China Military Region's (ECMR) Zhejiang commander Zhang Aiping proposed a "piecemeal" offense for taking the islands one by one (map 3). Since the PLA then possessed no antiaccess/area-denial (A2/AD) weapons, Zhang's "piecemeal" proposal leveraged the geographical positioning of PLA forces to avoid U.S. forces. After the initial operation, the PLA would then move south to take the larger islands, one at a time.[40] Zhang's amphibious campaign included a three-step joint air, naval, and land campaign. The first step was to engage the Taiwanese Air Force in the East China Sea and conduct air raids on the islands to establish PLA air domination over the Tachens. The second step would isolate the ROC garrisoned on the Tachens by gaining sea control through naval engagements. The third step would be the landing assaults by the 24th Army on the individual islands.[41] Zhang's plan was unique in that it marked the first implementation of joint operations between the PLA's various branches, even though this type of combined operation was normal

---

[40] LtGen Xu Changyou, interview with the author, Shanghai, April 2000. Xu served as Gen Zhang Aiping's aide and then the deputy secretary general of the CMC. He was vice commissar of the PLAN East Sea Fleet at the time of the interview.

[41] Dong Fanghe, *Zhang Aiping zhuan* [Biography of Zhang Aiping], vol. 2 (Beijing: People's Press, 2000), 663–64; and Xiaobing Li, "PLA Attacks and Amphibious Operations during the Taiwan Straits Crises of 1954–55 and 1958," in *Chinese Warfighting*, 146.

MAP 3
Offshore islands in the East China Sea.
*Source: Xiaobing Li, A History of the Modern Chinese Army
(Lexington: University Press of Kentucky, 2007), 132*

in most other nations. The high command approved Zhang's three-phased plan for capturing the islands.

Per the plan, the PLAAF 2d Division engaged the Taiwanese Air Force over the Tachen area in the spring of 1954. While Taiwan had American-made Republic F-84 Thunderjets, the Taiwanese failed to deploy them in time. The PLAAF lost only two fighters during six air engagements, whereas six ROC fighters were shot down. PLAAF pilots, in Soviet-made MiG-15 jet fighters and experienced from the Korean War, quickly outmatched ROC pilots, and the PLA controlled the skies north of the Tachens by May.[42] Taiwan's President Chiang Kai-shek personally visited the garrisons on the Tachens on 6–7 May, where he pressed his troops, as the situation seemed unfavorable, to remain calm and avoid panic. Rumors about evacuation were quelled

---

[42] The air force bases in east coast cities like Shanghai, Hangzhou, and Ningbo were also used by Zhang's jets in the air campaigns.

PLA Amphibious Campaigns

and the Tachens received more supplies and reinforcement after this tour, and ROC troop morale was raised.[43]

In contrast to the PLAAF, however, PLAN exhibited poor performance during the second stage of the Tanchen campaign. The East Sea Fleet (ESF) of the ECMR had 12 engagements against the ROC Navy from 18 March to 20 May, resulting in the damaging of several ROC ships but losing the PLAN warship *Ruijin* during the battle. The only ESF success was an attack on the Sanmen Wan bay north of the Tachens, sinking one ROC warship and damaging another.[44] Zhongtian Han contends the PLA was successful at the strategic adaptation of joint operation but failed at the operational level with an uneven performance between the air force and navy.[45] To prepare for the Tachen landing campaign, the ECMR established the 1st Marine Regiment and an amphibious tank regiment in Shanghai in April 1953. The next year, the PLA established ECMR's 1st Marine Division in December 1954 with greater landing combat and coastal defense capabilities.

The Dongji, a group of small islands north of Tachens, became the PLA's next landing target in early May 1954. Zhang Aiping deployed PLAN ships to the water around the Dongji and isolated the ROC garrison on the islands. On 15 May, Zhang landed PLA troops on the islands and defeated the ROC forces, capturing 60 prisoners.[46] Because PLA commanders adjusted to joint operations of "local war" conditions, PLA amphibious operations evolved rapidly. On the heels of Zhang's success, the CMC ordered the ECMR in July to launch similar amphibious offenses on the much larger island groups along the Zhejiang coast.[47] The ECMR instructed Zhang Aiping and his command to prepare a landing campaign on the Tachens in September.[48]

For the Tachen campaign, Zhang Aiping established the joint Zhejiang Front Command (ZFC) at Ningbo. The ZPF housed commanders from the army, navy, and air force comprised of the tripartite command headquarters. On 31 August, PLA commanders met to examine Zhang's meticulous new plan for the invasion of the Tachens. Zhang sent infantry liaison officers to the air force and navy units to en-

---

[43] Chiang Wei-kuo, interview with the author, Rongzong Hospital, Taipei, Taiwan, 25-27 May 1994. Gen Chiang, when asked during the interview about his father's secret visit to the Tachens, pointed out that his father recognized the strategic importance of these islands after the Korean War. Chiang Kai-shek made his trip to these offshore islands without informing any ROC officials or American representatives in Taiwan other than his naval commanders.

[44] Adm Hu Yanlin, *Weizheng haijing: renmin haijun zhengzhan jishi* [Shocking the Sea: Records of the People's Navy's Battles] (Beijing: PLA National Defense University Press, 1996), 59. Hu was PLAN political commissar in 2003-8 and served as an admiral in 2004.

[45] Zhongtian Han, "The PRC's Naval-Air Campaign in the East China Sea, 1954-1955" (conference paper presented at the annual meeting of Chinese Military History Society, via Zoom, 16 April 2020).

[46] Zhang, *Zhongguo renmin jiefangjun* [The Chinese People's Liberation Army], vol. 1, 189-90.

[47] Dong, *Zhang Aiping zhuan* [Biography of Zhang Aiping], vol. 2, 664-65; and Li, "PLA Attacks and Amphibious Operations during the Taiwan Straits Crises of 1954-55 and 1958," 148.

[48] Hu, *Weizheng haijing* [Shocking the Sea], 209-10.

hance cooperation between the PLA's various branches. At a ZFC meeting in September, Zhang and Nie Fengzhi, ZFC air force commander, decided to launch the attack at Yijiangshan Island as the first target of the Tachen campaign.[49]

Yijiangshan, a half-square mile islet, 11 kilometers north of Tachen Islands, was defended by a garrison of more than 1,200 ROC troops. In October, although the ZFC was ready to launch the attack on Yijiangshan, the PLA high command instructed Zhang Aiping to conduct more preparation, landing training, and naval and air attacks to isolate the Yijiangshan garrison. To carry out the high command's order, Zhang and Nie Fengzhi launched three air raids against the island. On 1 November, the PLAAF commenced its first heavy bombing when bombers and fighters flew more than 100 sorties, dropping more than 1,000 bombs on Yijiangshan and the Tachens in four days, thereby ensuring ZFC air dominance.[50] Then, Nie conducted the second bombing between 21 December 1954 and 10 January 1955. The ZFC air force launched 28 bomber and 116 fighter sorties in five raids against the islands. A third raid on 14–15 January targeted the Tachen harbor, sinking one ROC tank landing ship and damaging four others.[51]

Weather permitting, a joint attack was scheduled to destroy ROC forces on 18 January. The attack began at 0800 as 54 bombers and 18 fighters attacked Yijiangshan and Tachen, destroying key ROC defense works, artillery sites, and headquarters. The PLA bombers dropped 127 tons of ordinance within four hours. Then, at 1220, a two-hour artillery shelling of Yijiangshan, from coastal guns at Toumenshan began. During these two hours, the island was barraged by 40,000 shells from 4 artillery battalions and 12 artillery companies. Finally, between 1318 and 1415, the island's defense positions were also fired on by two gunboats and four escort ships. By the end of the day, nearly all Yijiangshan's beach positions, bunkers, and communications were eliminated during the prelanding bombardment (map 4). The PLA's heavy bombing and shelling also neutralized supporting fire that could reach Yijiangshan from the Tachen.[52] The PLA had successfully prepared the battlefield through massive aerial, naval, and artillery bombardment, which they had not done in previous amphibious invasion attempts.

On the same day, 18 January, Zhang's 10,000-strong invasion force, plus 3,700 sailors, embarked for Yijiangshan at 1215 on 140 landing craft, escorted by 4 warships, 2 gunboats, 12 torpedo boats, and 6 rocket gunboats. Even though the ROC posi-

---

[49] Dong Fanghe, *Zhang Aiping zhuan* [Biography of Zhang Aiping], vol. 2, 674–75; Han, *Dangdai Zhongguo jundui de junshi gongzuo* [Military Affairs of Contemporary China's Armed Forces], vol. 1, 216–17; and Li, "PLA Attacks and Amphibious Operations during the Taiwan Straits Crises of 1954-55 and 1958," 152.
[50] Ma Guansan, "Remember the Combat Years in the East China Sea," in *Sunjun huige zhan donghai* [Combined Forces Wield Spears and Fight in the East China Sea], ed., Nie Fengzhi (Beijing: PLA Press, 1985), 29. Ma was deputy commander of the ZFC naval force.
[51] Han, *Dangdai Zhongguo jundui de junshi gongzuo* [Military Affair of Contemporary China's Armed Forces], vol. 1, 215–16.
[52] Maj Lu Hui, *Sanjun zhan yijiang* [Combined Forces Battle Yijiang] (Beijing: China United Literature Publishing House, 2014), 126.

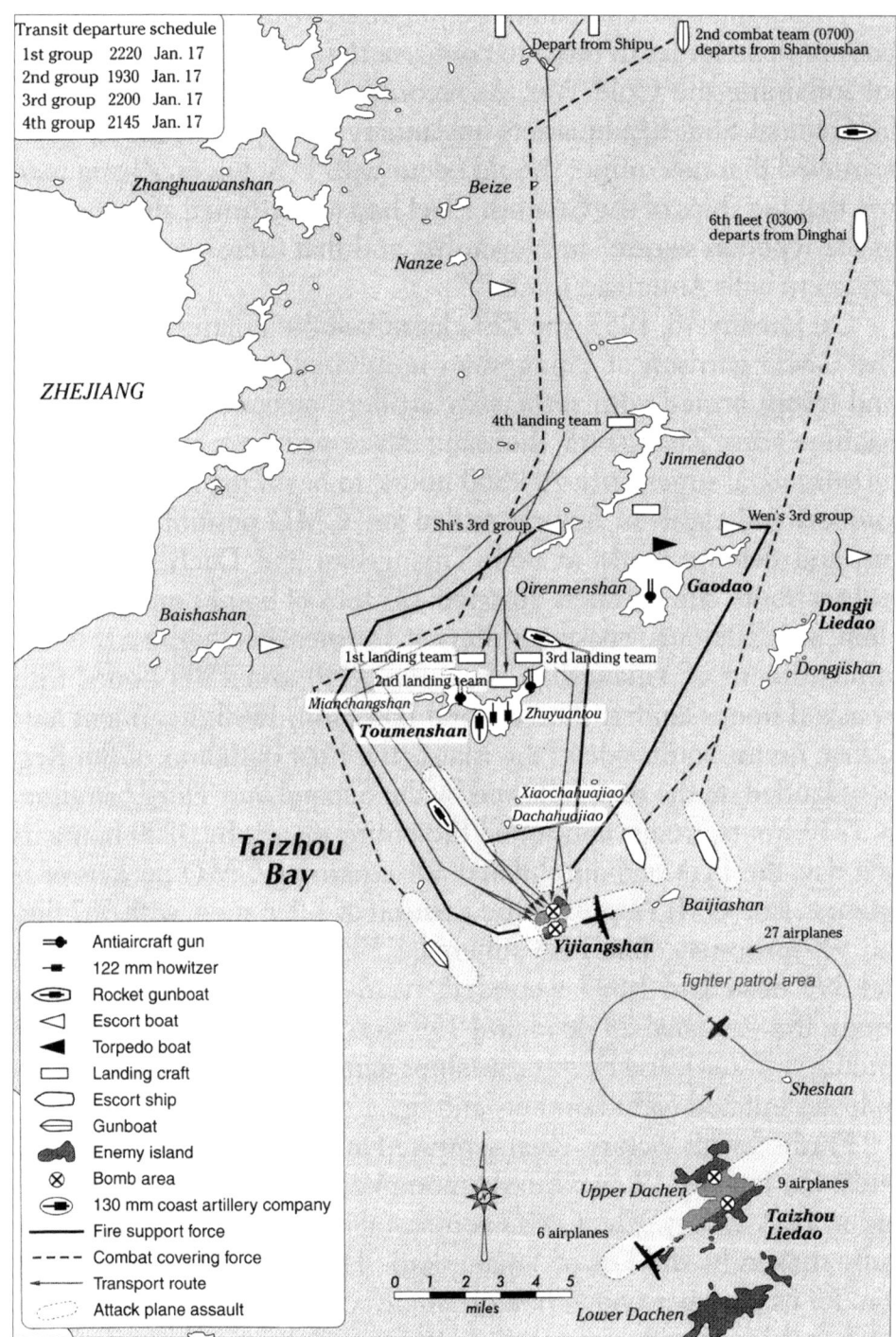

MAP 4
The Battle of Yijiangshan.
Source: Xiaobing Li, *A History of the Modern Chinese Army*
(Lexington: University Press of Kentucky, 2007), 143

tions had been bombed, ROC 60mm rockets hit two of the PLA transports, resulting in more than 30 casualties. Army commanders requested air support, and at 1420, PLAAF MiG-15s conducted low-altitude strikes on the ROC rocket positions in coordination with the landing troops. At 1430, the PLA's first group landed on the west side of the island and overtook defensive positions along the beaches. With support from the second wave of troops, the PLA quickly occupied Hills 180 and 190. At 1730, the entire island fell to the PLA.[53]

All remaining pockets of resistance were eliminated by the morning of 19 January. The entire ROC garrison of 1,086 troops was lost with 567 dead and 519 prisoners. Nevertheless, the PLA suffered 2,092 total casualties. The landing troops had 893 killed and 1,037 wounded, and they had lost nearly 50 percent of their first landing group. Moreover, the navy had 23 dead and 139 wounded.[54] The PLAN only lost 1 landing craft, though 21 ships were damaged. Finally, although the air force suffered no losses, eight bombers and fighters were damaged.[55]

The 1955 landings offered many lessons to the PLA about amphibious landings. First, the Battle of Yijiangshan exemplified for the PLA leadership the effectiveness of joint operations. The PLA landing campaigns illustrated the swift transformation from an army-based attack to a joint operation with an emphasis on air raids, naval support, cross-strait transportation, and landing troops' communication. Lu Xiaoping of the PLA Air Force Command College emphasizes the success of the air support at the Yijiangshan landing: "During the combat implementation, the Air Force units and Army landing force operated in close coordination, attacking the defending enemy forces with flexibility, protecting the frontal charge of the landing unit."[56] Coordination of the different services was not as necessary during the civil war but was critical during later operations.

Second, the PLA succeed because they actively avoided engagement with U.S. forces during the invasion of Taiwan-held islands. PLAAF commander Nie Fengzhi personally ordered his pilots to avoid engagements with American aircraft to ensure that the United States did not enter the Dachens area. Nie recounted, "Throughout the whole campaign we had an excellent result with no involvement with foreign air forces."[57] Major General Xu Yan, PLA's National Defense University, contended that Beijing was convinced by its field generals that the United States would not intervene

---

[53] Commo Yang Zhongyi, *Sulian zhuanjia yu zhongguo haijun hangkongbing* [Soviet Advisors and PLAN Air Force] (Beijing: PLA Press, 2013), 220.
[54] Di Jiu and Ke Feng, *Chaozhang chaoluo: guogong jiaozhu Taiwan haixia jishi* [Records of the CCP-GMD Confrontation in the Taiwan Straits] (Beijing: China Industrial and Commercial Publishing, 1996), 210–12.
[55] Han, *Dangdai Zhongguo jundui de junshi gongzuo* [Military Affairs of Contemporary China's Armed Forces], vol. 1, 220–21.
[56] Lu Xiaoping et al., *The PLA Air Force* (Beijing: China Intercontinental Press, 2012), 52.
[57] LtGen Nie Fengzhi, "Soaring Eagles Strike from the Clouds and Shake the Sea and Sky," in *Sunjun huige zhan donghai* [Combined Forces Wield Spears and Fight in the East China Sea], 16.

in the Yijiangshan landing if they were not directly provoked.⁵⁸ It was critical for the PLA to not provide the United States with any excuse to enter the conflict.

Finally, PLA leadership consistently reassessed their performance and adjusted their amphibious warfare doctrine based on the changing conditions on the battlefield. This rapid evolution characterized the PLA as a modern combat force. The changes in doctrine illustrated an ability to apply lessons on political morale, combat effectiveness, and adaption to amphibious warfare despite an ever-present learning curve. The PLA's rapid adoption of joint operations that involved complex issues related to communication, naval support, and air raids illustrate the PLA's swift transformation from an army-centered force to an effective joint force with modern amphibious warfare capabilities. The PLA quickly recognized the disparity between their weapons and American military technology. By seizing Hainan and all the offshore islands in the East China Sea, the Chinese landing campaigns of the early 1950s, in retrospect, achieved their initial campaign goals. However, the PLA's operational objectives to invade Taiwan were rendered nearly unattainable because of the naval and aviation technological gaps between the PLA and the ROC and United States. The failure to accomplish this strategic goal warranted frustration from Chinese leaders. Beginning in 1954, the PLA engaged in a limited war in the Taiwan Strait by avoiding full-scale war against the technologically superior forces of the United States. The Chinese relied on Russian weapons systems throughout the 1950s; and later in the 1960s, the Chinese attempted improvement and development of their indigenous weapon systems and strategic implements. Systems that would hopefully grant them the advantages they would need to one day take Taiwan.

CONCLUSION

By the late 1960s and early 1970s, Beijing and U.S. relations had changed. U.S. restraint in Vietnam and the continued stalemate on the Korean Peninsula illustrated that America did not want to have the war spread in Asia. The PLA believed that they could keep the United States away from the Taiwan Strait with nuclear deterrence, diplomatic efforts, and aggressive air and naval actions. After his second return to power, Deng Xiaoping and Defense Minister Ye Jianying instructed the PLA to attack South Vietnam-held Paracel Islands in the South China Sea in January 1974. Beijing believed that the South Vietnamese, who were embroiled in their war with the north, and the United States, who were doing everything in their power to leave Southeast Asia, would not counter the advance. On 19 January 1974, the Chinese troops landed on four Paracel islands and defeated the 160-man Army of the Republic of Vietnam garrison, killing or wounding 110 and capturing 49, including 1 American

---

⁵⁸ MajGen Xu Yan, "Did the War to Resist the U.S. and Aid Korea Alter the Solution of the Taiwan Issue?," in *Xu yan jianggao Zixunji* [Self-selected Lecture Notes of Xu Yan], ed., Xu (Beijing: Guofang daxue chubanshe [National Defense University Press], 2014), 118, 120–21. Xu is a faculty member at PLA National Defense University and deputy secretary general of the Chinese Military History Society.

advisor. Furthermore, the PLA Navy also sank a Republic of Vietnam Navy frigate and damaged three destroyers. In March 1988, Deng again ordered an attack on the Vietnamese troops at the Spratly Islands in the South China Sea. In 1988, the PLA seized the islands while Vietnam suffered 400 casualties, including 41 prisons of war and 2 two warships.

These two successful landing campaigns illustrated that the PLA could project its new naval power far from the coast. It also showed that China was capable of deep-water deployment beyond the Taiwan Strait. ROC President Ma Ying-jeou told the author during a meeting that by now a landing would be the end of Taiwan's defense.[59] It seemed that any effective defense or a decisive victory should happen in the Taiwan Strait and be determined by naval battles.

After Deng's retirement in 1989, Jiang Zemin rose to be the next leader and he quickly tested the PLA's combat readiness and U.S. responses through the Taiwan Strait missile crisis in 1995–96. Moreover, the show of military force also was designed to influence the 1996 elections in Taiwan. The crisis began when PLA conducted its first round of missile tests from 21 to 28 July 1995, in an area 58 kilometers north of Taiwan.[60] Simultaneously, the PLA concentrated large naval and landing forces and launched military exercises, including a joint amphibious landing exercise in the Taiwan Strait.[61] From 15 to 25 August, the East Sea Fleet deployed 59 warships and naval vessels for a large-scale naval attack and amphibious landing exercise that was viewed by many as a possible invasion. The PLAN launched 192 aviation sorties and scrambled its fighters and bombers during the naval attack exercise. From 31 October to 23 November, the PLA launched another joint amphibious landing campaign, including 63 warships, landing crafts, and support vessels, on Dongshan island off the Fujian coast. The Army's 91st Infantry Division conducted landing and beachhead defense exercises while the PLAAF sent 50 fighters, bombers, and other planes to the joint amphibious landing campaign.[62] Between January and February 1996, the PLA concentrated 100,000 troops along the coast across the strait from Taiwan and launched another large-scale landing exercise to send a stronger signal to both Taipei and Washington. Tensions remained heightened in the strait through the winter of 1995–96.[63] On 8 March 1996, the PLA conducted even more missile tests by firing three DF-15 surface-to-surface missiles just 19 kilometer off Kaohsiung and about 29

---

[59] President Ma Ying-jeou, interview with the author and several other Chinese historians, Taipei, Taiwan, 8 June 2017. Ma was the ROC president from 2008 to 2016.
[60] Zhang Yutao, *Xin zhongguo junshi dashi jiyao* [Chronicle of Major Military Events of China] (Beijing: Military Science Press, 1998), 608.
[61] For a detailed overview of the 1995-96 Taiwan Strait crisis, see Qimao Chen, "The Taiwan Strait Crisis: Causes, Scenarios, and Solutions," in *Across the Taiwan Strait: Mainland China, Taiwan, and the 1995-1996 Crisis*, ed., Suisheng Zhao (London: Routledge, 1999), 127–62.
[62] Zhang, *Xin zhongguo junshi dashi jiyao* [Chronicle of Major Military Events of China], 610.
[63] Ashton B. Carter and William J. Perry, *Preventive Defense: A New Security for America* (Washington, DC: Brookings Institution Press, 2000), 92–93.

nautical miles off Keelung.⁶⁴ This crisis, referred to as the Third Taiwan Strait Crisis, was similar to the Maoist crisis of the 1950s. In this one, China again tested the level of the Western response to its encroaching on Taiwan. The PLA realized they did not have the necessary power to invade Taiwan, but the crisis allowed them to judge potential responses from the United States and its allies.

From Mao Zedong to Xi Jinping, the Taiwan issue has dominated the military's attention. Xi has followed Mao's ideal and redoubled his grand strategy of reunification at the CCP's 20th National Congress with a consistent definition of national goals. Historians should not miss its historical roots and scope. For Xi Jinping, the Cold War was never over. As Xi moves into his third term, the PLA has continued to strengthen Chinese capabilities for rapid assault, air-ground integrated attacks, and long-distance maneuvers, all of which are critical for any future invasion of Taiwan. The PLA has increased its types of fighter jets, its naval strike force, army aviation troops, mechanized marine units, special operations forces, and cyber warfare units, all designed for offensive operations. Yet, a major U.S. intervention would still threaten the PLA's potential victory over Taiwan, an important lesson learned from previous amphibious campaigns that remain at the forefront of Chinese military thought. Xi Jinping's concept of limited deterrence has replaced Mao's nuclear doctrine of minimum deterrence, something of which Mao left no historical lesson for and of which Xi will be forced to learn on his own. Moreover, as Xi also shifts China's national security strategy from a defensive to a defensive offensive, he removes one of Mao's defensive principles of "never open fire first" in the Taiwan Strait. Xi now justifies any of China's future island attacks as preventive or retaliatory. The years 2024 to 2049 will be the most important and dangerous period for "the great rejuvenation of the Chinese nation" as well as when the PLA will reach the milestone for its modernization and becoming a "world-class force."

---

⁶⁴ Patrick Tyler, *A Great Wall: Six Presidents and China—An Investigative History* (New York: Public Affairs, 1999), 33, 195.

CHAPTER THIRTEEN

U.S. Geostrategic Deterrence and A2/AD at Work in the American Civil War, 1861–1865

Howard J. Fuller

INTRODUCTION

This chapter will examine how hastily mobilized U.S. coastal and naval defenses nevertheless accomplished a primary strategic objective of President Abraham Lincoln's administration during the crisis of the American Civil War: deterring potential (naval) intervention by the maritime powers, especially Great Britain.

The *Trent* affair of late 1861 underscored the Union's vulnerability to British naval power in the form of overseas force projection—the deployment of imperial troops to Canada—as well as sweeping the seas of American commerce, challenging the U.S. Navy's blockade of Confederate ports, and threatening Northern coastal cities with naval bombardment. America's impressive Third System of coastal forts, initiated because of British offensives during the War of 1812, were the largest geostrategic defenses of their kind throughout the nineteenth century. But they were not complete by the outbreak of the Civil War in 1861 (triggered by the immediate issue of Fort Sumter commanding the approaches to Charleston, South Carolina). Yet, in the heated crisis of the Civil War, Congress facilitated hasty upgrades to the existing and newer forts, and all were armed with monster 15-inch guns. The Union Navy meanwhile rapidly mobilized an unparalleled flotilla of brown water (riverine) ironclads—especially the monitors—similarly armed with ironclad-killing weapons.

As this chapter will document, these developments indeed preyed on European political and military decision-makers at the time. As a result, the United States during the Civil War could boast the largest force of coastal defense vessels and forts in the world, when America needed the assurance of geostrategic isolation (and the pretensions of the Monroe Doctrine) the most.

*　*　*

The theory that the British Royal Navy largely sustained the Monroe Doctrine (and that Americans should be grateful for even the indirect protection they received from the British Empire while they were still floundering in the nineteenth century) was enshrined by Winston S. Churchill. His four-volume *A History of the English-Speaking Peoples* won the Nobel Prize in Literature in 1953, and here the history said that the "acceptance by the rest of the world" of Monroe's "resounding claims" in 1823 "depended on the friendly vigilance of the 'British man-of-war,' but this was a fact seldom openly acknowledged." Further, Churchill believed that for "the best part of a century the Royal Navy remained the stoutest guarantee of freedom in the Americas. Thus shielded by the British bulwark, the American continent was able to work out its own unhindered destiny."[1] The beauty of this passage is that it assumes British naval supremacy (even in American waters), calls that power benign, and then infers American crassness. Given the Cold-War context of Churchill's later years, and his complicated friendship with Franklin D. Roosevelt before and during the Second World War, perhaps he felt the $30 billion the British Empire received from the United States as part of the 1941 Lend-Lease Act was fitting justice. America morally owed at least that much to the "British bulwark," the old "wooden walls," which even in 1940 were desperately upholding civilization against brutal European despotism bent on conquering the whole world. This was of course a line that fit perfectly well with the likes of earlier prime ministers: William Pitt, George Canning, or Henry John Temple, 3d Viscount Palmerston. By the 1950s, it was now America's turn to quietly

---

[1] Winston S. Churchill, *A History of English-Speaking Peoples*, vol. 4 (New York: Dodd, Mead, 1958) 29-30. Churchill was awarded the Nobel Prize just as he was completing his final volumes. He was 79. Churchill had already asserted the notion in the House of Commons on 8 March 1948: "From Trafalgar onwards, for more than 100 years Britannia ruled the waves. There was a great measure of peace, the freedom of the seas was maintained, the slave trade was extirpated, the Monroe Doctrine of the United States found its sanction in British naval power—and that has been pretty well recognised on the other side of the Atlantic—and in those happy days the cost was about £10 million a year." "House of Commons Debate, 8 March 1948, vol. 448," Hansard, UK Parliament, cols. 803-981. Ten years later, Sir Joseph Percival William Mallalieu echoed: "It is absolute folly for us to have it [the hydrogen bomb] and I am not impressed with the argument that by giving it up we should increase our dependence on the United States. The United States has been dependent on us for the best part of a century—dependent on the Royal Navy. It was an American President who propounded the Monroe Doctrine, but it was the Royal Navy which maintained it and the fact that America was dependent upon us did not notably stunt her growth in the 19th century." "House of Commons Debate, 4 March 1958, vol. 583," Hansard, UK Parliament, cols. 978-1127.

FIGURE 1
Iron-clad monitors afloat during the Civil War era.
Source: Harpers Weekly

FIGURE 2
Lord Palmerston.
Source: W. & D. Downey

deploy men-of-war around the globe, policing the seas and promoting—one might say "projecting"—freedom.²

It is all very Whiggish, but the historical truth is much more problematical. The fact was America's greatest enemy in the nineteenth (and eighteenth) century was the British Empire. The Royal Navy may have been friendly, but often it was not. Anglo-American relations until the era of so-called rapprochement (ironically, a French expression) were very

---

² A conference held at the Joint Services Command and Staff College (Shrivenham) on 7–8 December 2006 titled "First Force for Good: The British System of Imperial Defence, 1856–1956" was later published in Greg Kennedy, ed., *Imperial Defence: The Old World Order, 1856–1956* (London: Routledge, 2008).

much like a cold war, and rarely did the strategic interests of both powers see eye to eye. For all the mutually enriching trade, shared customs, and language, a viable tension existed, almost Oedipal (ironically Churchill's mother, Jennie, was American). The Monroe Doctrine never found its sea-legs until the United States was finally able to thwart—to counterdeter—the Royal Navy. During the Civil War, Palmerston was worried that his high-handedness during the *Trent* affair in 1861 was about to come back and slap him in 1865.[3]

The circumstances of the Monroe Doctrine are well known enough. The Holy Alliance powers of Russia, Austria, and Prussia (that is, the Romanovs, Habsburgs, and Hohenzollerns) had declared in 1815, after the final defeat of Napoléon Bonaparte, that it was thanks to the French Revolution that so much blood and fire had spread across Europe. With a Bourbon king back on the throne of France, these powers (along with the Hanoverian King George IV of the UK) pledged to prevent by force any future revolutions and republics. Democracy was explicitly rejected as mob rule and antithetical to the divine right of kings—all that was decent in god-fearing societies that respected the natural order of clergy, nobility, and compassionate paternalism. When popular revolution broke out in Spain against the tyrannical King Ferdinand VII, France invaded the country in 1823 to restore the *ancien régime*. Protestant Britain had already come to suspect the absolutist streak in the Holy Alliance, being a constitutional monarchy effectively controlled by Parliament, which had fought and won its own civil war against King Charles I nearly 200 years before, with a "Glorious Revolution" against Catholic King James II in 1688. As such, the acting British representative in the Alliance congress, the Duke of Wellington, abstained from offering support to France or Spain. Additionally, the British foreign secretary, George Canning, was determined to prevent an extension of French power, via Ferdinand, into the New World colonies that had been a largely successful revolt against Spanish imperial rule during the Napoleonic Wars. Rumors abounded that French support in Spain might be paid for with Cuba, for example. What if the Spanish monarchy also desired the reconquest of Latin America? Because of Britain's close relationship with Spain and especially Portugal, British trade with South America had become very

---

[3] On 8 November 1861, two Confederate emissaries traveling to Europe were intercepted by the USS *San Jacinto* (1850) aboard the Royal mail packet steamer RMS *Trent* and forcibly removed. Britain was outraged and demanded their release with an apology. British troops were dispatched to Canada and naval units routed to Bermuda in preparation for war if the United States refused. President Abraham Lincoln, noting "one war at time," agreed with Secretary of State William Seward to return the Confederates to British custody; see, for example, Norman B. Ferris, *The Trent Affair: A Diplomatic Crisis* (Knoxville: University of Tennessee Press, 1977); and Gordon H. Warren, *Fountain of Discontent: The Trent Affair and Freedom of the Seas* (Boston, MA: Northeastern University Press, 1981). As described by Kenneth Bourne, the *Trent* affair was "the most dangerous single incident of the Civil War and perhaps in the whole course of Anglo-American relations since 1815," in *Britain and the Balance of Power in North America, 1815–1908* (London: Longmans, Green, 1967), 251.

lucrative indeed.[4] And neither Canning, Wellington, nor King George wanted to see a revival of French colonialism across the Atlantic.

The role here of the United States, as Canning saw it, was pivotal. President James Monroe and his cabinet had meanwhile watched events in Europe and Latin America very closely and nervously. Summoning Richard Rush, the U.S. ambassador to Britain, on 16 August 1823, Canning mentioned that both powers should declare their firm opposition to any expedition by France or Spain against Central or South America. This was an historic opportunity, the first Anglo-American collaboration since the American Revolution and then the equally bitter War of 1812, which Rush had urged on Madison's government. Canning's belief, Rush later relayed to Secretary of State John Quincy Adams, "was founded upon the large share of the maritime power of the world which Great Britain and the United States shared between them, and the consequent influence which the knowledge that they held a common opinion upon a 'question' on which such large maritime interests, present and future, could not fail to produce upon the rest of the world." Four days later, the British foreign secretary followed up this cunning strategic flattery with a letter asserting, "For ourselves, we have no disguise. We conceive the recovery of the Colonies by Spain to be hopeless." The United States had already recognized their independence, though Canning could only admit that British recognition as well would "be one of time and circumstances." There were too many European complications for Great Britain, as opposed to the United States, to consider first. But here he also affirmed "We aim not at the possession of any portion of them ourselves."[5] Well before the accession of Queen Victoria, the British Empire was feeling colonially bloated worldwide.

Both Rush and Adams, however, were suspicious of Canning's hesitance to acknowledge the new republics. Sure enough, as even Canning's most famous biographer and champion Harold Temperley noted, the hope that "a monarchy might arise in Mexico and perhaps elsewhere" helps explain Canning's deliberations here—and the political and social wedge that eventually undermined Anglo-American cooperation in the early nineteenth century and led to the formulation of the Monroe Doctrine.[6] Rush was prepared to sign off on a formal declaration with Britain, he responded to Canning, but only after Britain joined the United States in recognizing the colonies (currently as republics). Canning again met with Rush on 16 September, this time pressing that "the United States . . . were the first power established on the Continent, and now confessedly the leading Power." What sort of message would American

---

[4] See Leonard Axel Lawson, "The Relation of British Policy to the Declaration of the Monroe Doctrine" (PhD diss., Columbia University, 1922), 76–86, 101–3.
[5] See Harold Temperley, *The Foreign Policy of Canning, 1822–1827: England, the Neo-Holy Alliance, and the New World* (London: G. Bell and Sons, 1925; Routledge, 2006 reprint), 110–13, 122; and Harold Temperley, "Documents illustrating the Reception and Interpretation of the Monroe Doctrine in Europe, 1823-2," *English Historical Review* 39, no. 156 (October 1924): 590–93.
[6] Temperley, *The Foreign Policy of Canning, 1822–1827*, 113.

FIGURE 3
Portrait of John Quincy Adams, ca. 1844.
Source: William Hudson Jr., oil on canvas, National Portrait Gallery

"indifference" therefore send to Europe?[7] But by then, Adams had already come to his decision. The strategic picture for the United States was also complicated. Imperial Russia had announced two years earlier that fishing and trading rights from Russian Alaska all the way to the 51st parallel north were subject to Tsar Alexander I alone, the most rabid of the absolutist monarchs. And what about U.S. ambitions? If Texas

---

[7] Temperley, *The Foreign Policy of Canning, 1822–1827*, 122.

or Cuba eventually applied to join the Union, would a standing Anglo-American pledge prevent further expansion of republican democracy?[8] While both Thomas Jefferson and Madison urged Monroe to accept Canning's offer, inasmuch as British seapower could actually be exploited by the American need to ward off European intervention in Latin America, Adams (rightly) calculated that Britain's cooperation ultimately "rested not upon her principles, but her interest."[9] In that respect, Britain would risk war with France alone to uphold the status quo of its trading interests in South America and a fateful reextension of French influence globally. And indeed, Canning had warned the new French minister to Britain, Jules de Polignac, earlier that March and now again in early October with a memorandum threatening to instantly recognize the Latin American republics if France interfered "by force or by menace . . . in the dispute between Spain and the Colonies."[10] So Adams argued to Monroe's cabinet on 7 November that Russia's peremptory attitude about the North Pacific "afforded a very suitable and convenient opportunity for us to take our stand against the 'Holy Alliance,' and at the same time to decline the overture of Great Britain. It would be more candid as well as dignified," he assured his listeners, "to avow our principles explicitly to Russia and to France than to come in as a cock-boat in the wake of the British man-of-war."[11]

That message was then underscored in the president's annual address to Congress on 2 December 1823 that, while the United States did not seek to interfere in the internal affairs of European governments and societies, the "political system of the Allied Powers [was] essentially different in this respect from that of America." Therefore the United States would consider "any attempt on their part to extend their system to any portion of this hemisphere as dangerous to our peace and safety." Existing colonies and dependencies of monarchical powers were exempt from this view; the U.S. government had not interfered "and shall not interfere." But any attempt to oppress or control those former colonies, now republics, which the American government in Washington had already seen fit to recognize, could not be seen "in any other light than as the manifestation of an unfriendly disposition toward the United States."[12]

Interestingly, Monroe's speech (largely written by John Quincy Adams), highlighted that he could never believe "that our southern brethren, if left to themselves,

---

[8] John Quincy Adams, *Memoirs of John Quincy Adams: Comprising Portions of His Diary from 1795 to 1848*, vol. 6 (Philadelphia, PA: J. B. Lippincott, 1875), 177–78.
[9] Temperley, *The Foreign Policy of Canning, 1822–1827*, 123.
[10] Temperley, *The Foreign Policy of Canning, 1822–1827*, 115–16. In March, Canning had specified that blocking Spanish or French reinforcements to Latin America would be Britain's unique leverage: "There our naval superiority would tell. There a maritime war would be to a purpose." Lawson, "The Relation of British Policy to the Declaration of the Monroe Doctrine," 69.
[11] Charles Francis Adams, ed., *Memoirs of John Quincy Adams*, vol. 6, *Comprising Portions of His Diary from 1795 to 1848* (Philadelphia, PA: J. B. Lippincott, 1875), 177–79.
[12] President James Monroe, "Annual Message, 2 December 1823, Annals of Congress," 18th Cong., 1st Sess., Senate Journal, National Archives, 12–19, hereafter Monroe message.

would adopt [monarchism] of their own accord."[13] Temperley took some satisfaction in noting two points to this doctrine: first, that influential leaders in Latin America, namely General Simón Bolívar, were ready to install a monarchy in the greater Colombian state that stretched from modern day Panama across the northern portion of South America; Bolívar going so far as to assure Canning that "we must look to England for relief" and that he would wholly accept a British-mandated sovereign, backed by a constitution. This was because, in his own frustrating experience with "the people" in power, South America was "perhaps the least fitted for Republican Governments," with all its "Indians and Negros who are more ignorant than the vile race of Spaniards we are just emancipated from." In derogatory language typical of the age, he predicted that a "country represented and governed by such people must go to ruin."[14] A few months later, in 1825, another Spanish aristocrat-turned-revolutionary general, José de la Riva-Agüero of Peru, had also presented to Prime Minister Robert Banks Jenkinson, Earl of Liverpool, some "Political reflections on the future destiny of Spanish America as regards Great Britain" calling for the British establishment of "two great monarchies in Mexico and Peru, which countries are formed for that Government by their education and the character of their inhabitants." At least two generations would have to pass away in those lands, he was certain, until republicanism could take root or the indigenous populations to become noble enough for American-born princes to take up American thrones. Hoping to play on British pride as well as paranoia, Agüero then warned that the archenemy of British interests and society in the meantime were in fact the new republics of Latin America, led by the democratic United States, whose navy would increase "so that she will ultimately dispute the dominion of the sea with Great Britain." Europe would then join America in destroying the British Empire, starting with Ireland, until finally "the Continental powers will make Great Britain change her constitution, so hostile to their anti-liberal views."[15] It is impossible to know how seriously Liverpool and Canning regarded these entreaties, but the fact they were carefully preserved is significant.

Second, Temperley stressed that Monroe and subsequent American presidents—including Adams—immediately prevaricated; the United States did not necessarily pledge to go to war against any enemy of any American republic any time. For one, war was technically in the hands of Congress according to the U.S. Constitution. Noninterference abroad was also a stated core element of U.S. national policy. And Britain's interests were likewise opposed to European aggrandisement in the New World (at the expense of its own). "So it is really true," observed Temperley in 1925, "that [Adams] was proclaiming the Monroe Doctrine beneath the shelter of the Brit-

---

[13] Monroe message.
[14] Temperley, *The Foreign Policy of Canning, 1822–1827*, 555–58.
[15] "Political Reflections on the Future Destiny of Spanish America as Regards Great Britain," presentation to Liverpool, "Enclosure in Senor Riva Aguero's of 27 May 1825," Add MS 38300, Liverpool Papers, vol. 111, Official Correspondence, March–13 November 1825, fols. 98–103, British Library.

ish fleet."[16] How far back in the literature does this belief go? One pamphlet from 1921 denied it, citing numerous incidents where Great Britain "consented to the effort of some other nation to test the doctrine, e.g., Great Britain's attempts on Cuba 1825, on Venezuela from 1840 to 1895, her joint attempts with Spain and France resulting in the French invasion of Mexico in 1862, and her joint effort with Italy and Germany against Venezuela in 1902."[17] Another work the following year, however, maintained that "upon the firmness of British opposition to intervention depended the success of the policy of the United States," though Britain was operating from ultimately "commercial interests" while Adams had cited "political liberalism."[18] During the First World War, with American intervention on the side of the Triple Entente a pressing issue, British naval historian Julian S. Corbett stressed that "there was little force to support the new doctrine except the naval power of England. But that power was behind it heart and soul till it was strong enough to stand alone." Perhaps taking this cue, a study of *American Diplomacy* from 1916 likewise concluded that the confidence of the Monroe Doctrine "rested more on the efficiency of the British navy than on our own strength." This was the gambit Adams had calculated: "Thus to use one's own resources of a rival power, while yielding nothing to her rivalry, is daring; but, if justified, it is the highest manifestation of the diplomatic art."[19] Yet, the overall assertion in question was awkwardly flipped on its head in 1907 when a British member of Parliament suggested that "Canada was defended not only by the British Navy but also by the American Navy, owing to the Monroe doctrine. Therefore, Canada relied upon two navies and paid for neither."[20] When Robert Gascoyne-Cecil, Lord Salisbury, in 1895 defended his government's actions over the British Guiana/Venezuela border dispute, he assured the British public there was no conflict between the Monroe Doctrine and British policy, and he had gladly submitted to the idea of international arbitration since U.S. interests in that hemisphere were understood.[21] A British study from 1898 did not advance the notion that the Royal Navy protected the Monroe Doctrine; quite the opposite: "Despite the outcry of the Argentine," for example, Great Britain "had occupied and retained the Falkland Islands."[22] President

---

[16] Temperley, *The Foreign Policy of Canning, 1822-1827*, 124; and Adams, *Memoirs of John Quincy Adams*, vol. 6, 203-4.
[17] Thomas H. Mahony, *The Monroe Doctrine: The Vital Necessity of Its Continued Maintenance* (New Haven, CT: Knights of Columbus Historical Association, 1921), 73.
[18] Lawson, "The Relation of British Policy to the Declaration of the Monroe Doctrine," 5, 111.
[19] Julian Corbett, *The Spectre of Navalism* (London: Darling & Son, 1915), 6; and Carl Russell Fish, *American Diplomacy* (New York: Henry Holt, 1916), 212-13. Corbett added, without need for any evidence it seems, "In that hour, so fateful for the world, America trusted implicitly British 'Navalism' at its height," 6-7.
[20] "House of Commons Debate, 15 December 1907, vol. 169," Hansard, UK Parliament, cols. 424-92.
[21] "House of Commons Debate, 11 February 1896, vol. 37," Hansard, UK Parliament, cols. 73-164, as paraphrased by William Vernon Harcourt.
[22] W. F. Reddaway, *The Monroe Doctrine* (New York: G. E. Stechert, 1924), 141. This was amended by an article from 1905 that argued when the doctrine was first iterated American natural and human resources were already developed enough to enforce it, including an "ample" navy with capable sailors; Alfred Spring, "The Monroe Doctrine," *American Law Review* 39 (1905): 495-516.

Andrew Jackson, ignoring the various pleas of Latin America at the time, stood by and did nothing.

Before the 1890s, all seems quiet on this argument. Indeed, it seems to have cropped up only in relation to the First and especially the Second World Wars, when the United States *had* finally intervened in European affairs and was a formal military and naval ally with the UK. Only then did the story of the Monroe Doctrine's "debt" to the Royal Navy emerge, as Britain slipped further into debt to America, though, as one particularly angry and suitably obscure work from 1938 was titled, the Anglo-American relationship was now about how *England Expects Every American to Do His Duty*.[23]

History by this point simultaneously buried the story about how the greatest threat to the Monroe Doctrine was never Russia, Spain, or France but Britain. As Jefferson wrote to Monroe on 24 October 1823, the question of Canning's offer of a jointly declared policy was "the most momentous which has ever been offered to my contemplation since that of Independence." Yet, it was "Great Britain . . . which can do us the most harm of any one, or all on earth; and with her on our side we need not fear the whole world." Even the prospect of someday obtaining Cuba, which he confessed he always regarded as "the most interesting addition which could ever be made to our system of States," was not worth the enmity it would create with the British.[24] Later that same week, another former president, James Madison, also wrote to Monroe that it was "not improbable that G. Britain would like best to have the merit of being the sole Champion of her new friends [in Latin America], notwithstanding the greater difficulty to be encountered but for the dilemma in which she would be placed [another European war]. She must in that case, either leave us as neutrals to extend our commerce & navigation at the expence [sic] of hers, or make us enemies, by renewing her paper blockades & other arbitrary proceedings on the Ocean." Such a dilemma he hoped would "not be without a permanent tendency to check her proneness to unnecessary wars." Likewise, the issue of eventual American growth into former Spanish imperial territories had to be considered—Cuba but also Puerto Rico. As Madison understood Canning's proposal, an Anglo-American pact of nonexpansion would still leave Great Britain free "in relation to other Quarters of the Globe."[25] This was the only balance of power that mattered outside of Europe. Even if Russia thought to complain of Britain's policy on the Spanish colonies, Canning wanted the

---

[23] Quincy Howe, *England Expects Every American to Do His Duty* (London: Robert Hale, 1938), which argues among other things that the Monroe Doctrine was the product of British manipulation whereby "the United States underwrote Britain's stake in Latin America," 9, 25-26.

[24] Jefferson to Monroe, 24 October 1823, in Paul Leicester Ford, ed., *The Works of Thomas Jefferson*, vol. 10, *1816-1826* (New York: G. P. Putnam's Sons, 1899), 277-79.

[25] Madison to Monroe, 30 October 1823, in Gaillard Hunt, ed., *The Writings of James Madison*, vol. 9, *1819-1836* (New York: G. P. Putnam's Sons, 1910), 157-60. Madison repeated these considerations to Richard Rush on 13 November 1823.

British ambassador in St. Petersburg to politely question Karl Vasilyevich, Count Nesselrode, the Russian state secretary, in turn about rumours the tsar was bargaining for the *Île a Vache* (Cow Island) off the coast of Haiti, "and it may be amusing at least that you should let [him] see that you know of his intrigue with the Negurs."[26]

At the eighth annual meeting of the American Society of International Law, in 1914, Charles Francis Adams Jr.—the grandson of John Quincy Adams—voiced his concern that the Monroe Doctrine had been warped by what he called "Mommsen's Law" going into the twentieth century. Theodor Mommsen was a famous German classical historian of his generation and was exceedingly prolific (with more than 1,500 publications to his name as well as 16 children). To Charles Adams, it was perfectly natural "that a people which has grown into a state absorbs its neighbors who are in political nonnage, and a civilized people absorbs its neighbors who are in intellectual nonnage." This was a "natural law" of human nature at work in ancient Rome against barbarians, "just as England with equal right has in Asia reduced to subjection a civilization of rival standing but politically impotent, and in America and Australia has marked and ennobled . . . extensive barbarian countries with the impress of its nationality." John Quincy Adams, conversely, was opposed to the annexation of Texas and the Mexican-American War, for example, just as he was vehemently opposed to slavery (i.e., the enforced subjugation of "inferior" races—something the Romans were also very good at). Charles Adams was likewise concerned that the spirit of the "Ostend Manifesto" was also alive and well.[27] In 1854, this circular sought to rationalize the forcible seizure of Cuba in the name of the Monroe Doctrine if Spain did not sell it to the United States. If the Spanish colony succumbed to slave revolts, the whole island would be quickly "Africanized and become a second [Haiti], with all its attendant horrors to the white race." It would then surely spread to the Southern states of the Union.[28] Since then-president Theodore Roosevelt in his State of the Union address of 1904 had declared that the Monroe Doctrine gave the United States the right act as "an international police force" when faced with "chronic wrongdoing or an impotence which results in a general loosening of the ties of civilized society" in any

---

[26] Canning to Charles Bagot, 20 August 1823, in Josceline Fitzroy Bagot, ed., *George Canning and His Friends: Containing Hitherto Unpublished Letters, Jeux D'Esprit, Etc.* (London: John Murray, 1909), 195-96.
[27] Charles Francis Adams, *The Monroe Doctrine and Mommsen's Law* (Boston, MA: Houghton Mifflin, 1914), 28-31, 34-35.
[28] See, for example, Robert E. May, *Slavery, Race and Conquest in the Tropics: Lincoln, Douglas, and the Future of Latin America* (Cambridge, UK: Cambridge University Press, 2013), 116-18, describes the document penned by several prominent American Democrats such as James Buchanan, Pierre Soulé, and John Y. Mason as "disgracefully expansionist"; and Robert E. May, *The Southern Dream of a Caribbean Empire, 1854-1861* (Baton Rouge: Louisiana State University Press, 1973; Gainesville: University Press of Florida, 2002 reprint). Well into the American Civil War, an article by British professor Goldwin Smith on "England and America," *Atlantic Monthly*, December 1864, warned Northerners how both the Monroe Doctrine and the Ostend Manifesto "are still ringing in our ears" as expressions of territorial ambition and possible "violence," 765.

state in the Western Hemisphere. This closely coincided with U.S. strategic interests in the building of the Panama Canal. The younger Adams therefore challenged his audience to recall that the dream of America in 1823 was to be entirely distinct from European affairs and practices, from imperialism and balance-of-power politics to the suppression of liberty and self-determination, however "messy" it may seem to autocratic rulers.[29] Somewhere along the line, the ideology had been flattened by the realpolitik of strategic interests. America in the twentieth century had not come in the wake of the British man-of-war, it had become one.

Hence, William Dunning maintained in his 1914 study of the British Empire and the United States that the Monroe Doctrine had pitted Adams against Canning, America against Britain, inasmuch as these two powers—and these two alone—had become "serious rivals for the controlling interest in American affairs."[30] Indeed, after learning of the U.S. president's message to Congress in December 1823, Canning wrote to Sir Charles Bagot in St. Petersburg that while he agreed with America about the need to protest against "the forcible or authoritative interference of any Foreign power in the dispute between Spain and Spanish America," he did not agree with Monroe (or Adams) in "objecting to an attempt to recover her dominions on the part of Spain herself."[31] Two years later, his views had hardened. For while the "general maxim" of Anglo-American interests was that they were both the same, he informed the British ambassador at Washington that "we must not be the dupes of this conventional language of courtesy." In other words, there was diplomacy and statesmanship, and there was the pursuit of power. "The avowed pretension of the United States to put themselves at the head of the confederacy of all the Americas," he continued, "and to sway that confederacy against Europe (Great Britain included), is *not* a pretension identified with our interests, or one that we can countenance as tolerable."[32] While the United States might be admittedly the "leading power" in its own hemisphere, that portion of the globe was connected to others by the sea, whose rule was the pretension of the Royal Navy. Certainly, tensions already existed over numerous territorial disputes. "Why, do you not *know* that we have a claim to the mouth of the Columbia River?" exclaimed Lord Stratford Canning (the foreign secretary's cousin) to Adams in 1821. "I do not *know*," replied the American secretary of state, "what you claim nor what you do not claim. You claim India; you claim Africa; you claim—" "Perhaps," cut in the British ambassador sardonically, "a piece of the moon." "No," said Adams coolly, "I have not heard that you claim exclusively any part of the moon; but there is not a spot on *this* habitable globe that I could affirm you do not claim; and

---

[29] Adams, *The Monroe Doctrine and Mommsen's Law*, 38-41.
[30] William Archibald Dunning, *The British Empire and the United States: A Review of Their Relations during the Century of Peace Following the Treaty of Ghent* (New York: Charles Scriber's Sons, 1914), 54-55.
[31] Entry for 9 January 1824, in Bagot, *George Canning and His Friends*, 208.
[32] Canning to Charles Richard Vaughan, 8 February 1826, in Dunning, *The British Empire and the United States*, 54-56.

there is none which you may not claim with as much color of right as you can have to Columbia River or its mouth."[33]

Despite their fierce mutual (and personal) suspicions, neither Adams nor Canning continued to dominate Anglo-American relations for very long. The former lost his presidential reelection bid against Andrew Jackson in 1828; and while Canning had risen to prime minister the year before, he died after only four months in office. According to the Columbia University doctoral dissertation of Leonard Axel Lawson in 1922, "The Relation of British Policy to the Declaration of the Monroe Doctrine" was instrumental and really came down to seapower. "Not the United States, but England, was the real barrier to allied intervention in Spanish America," he concluded, because "her possession of the largest navy in the world gave practical effectiveness to her own opposition." Conversely, had the British Empire fully supported European intervention in Latin America, those new states "would have been destroyed, and probably the southern continent would have been parcelled out amongst its conquerors, against which an independent protest by the United States would have been conspicuously ineffective."[34] The specter of British seapower certainly weighed on the minds of American statesmen like Jefferson. Unfortunately, Lawson did not back up his theory with reference to actual evidence. Therefore, the practical limits of seapower in changing the map of the world remains counterfactual and hypothetical not historical. For there is still the defiance of Adams, while Canning never dictated policy to Rush the American ambassador, he supplicated, even with so much apparently unstoppable power at his disposal.[35] Even against France, the prospect of yet another war was never something to be taken lightly or brushed away with the magic wand of British naval supremacy. At the height of the Maine boundary dispute with Britain, culminating in the controversial Webster-Ashburton Treaty of 1842, King Louis-Philippe of France warned the American government "though you could do each other much harm, the results of War in the present improved State of the art of destruction, are more uncertain than ever." This included steam-powered warships armed with shell-firing guns. "Great Britain was omnipotent on the Ocean & could destroy [U.S.] towns," he added, though he admitted the financial repercussions alone of an Anglo-American war in the Victorian Age would be devastating in return. When the U.S. ambassador to France, Lewis Cass, discussed

---

[33] Adams, *Memoirs of John Quincy Adams*, vol. 5, 251–52; and Harlow Giles Unger, *John Quincy Adams* (Boston, MA: De Capo, 2012), 202–3.

[34] Lawson, "The Relation of British Policy to the Declaration of the Monroe Doctrine," 142–43.

[35] British naval historian William James thought this respect was misplaced, presenting to Canning on 9 January 1827 a copy of his controversial *The Naval History of Great Britain: From the Declaration of War by France in February 1793, to the Accession of George IV, a New Edition, with Additions and Notes Bringing the Work down to 1827* (London: William Clowes, 1902), and noting British superiority at sea during the War of 1812. In his estimation, "the Americans will never be a naval power of any magnitude" because westward expansion would eventually fragment the country "long before the republic becomes formidable from density of population." Edward J. Stapleton, ed., *Some Official Correspondence of George Canning*, vol. 2 (London: Longmans, Green, 1887), 144.

the possibility of war with Great Britain to the French ambassador, François Guizot, he was told that French interests apprehended "the maritime pretensions which England has asserted during war, and which she will no doubt repeat when she finds herself engaged in hostilities"; namely, "her fictitious blockades, and her contraband of war." In such event, Cass was told, "France would be driven to war to defend those principles which she has always maintained and never will abandon."[36] Two months later, the British ambassador to the United States, Alexander Baring, Lord Ashburton, cautioned the foreign secretary (George Hamilton-Gordon, Earl of Aberdeen) "not to mistake or undervalue the *power* of this country":

> *You will be told that it is a mass of ungovernable & unmanageable anarchy, and so it is in many respects. To a common observer it might be a matter of doubt how this confederative can hold together another year. Bankrupt finances. Bad administration & jobbing in every department. A loose, ill-connected mass of conflicting interests, in short apparently nothing for the eye of confidence to rest upon. Yet with all these disadvantage . . . the energies & power of the country would be found to be immense in the case of war and that the jarring elements would unite for that purpose. This is [Sir Henry] Fox's opinion, who as you know, is no admirer of any thing here.*[37]

Aberdeen in turn desired a peaceful settlement with America as soon as possible as it would "greatly improve our relations with France, which are now rather uncomfortable, from the weakness of the Govt. and the general hostility of the Chamber & the country."[38] Though Palmerston later attacked the treaty pronounced by Sir Robert Peel's government as "placing the United States in a better military position in regard to us than they occupied before, and by inducing them to think that we shall yield whenever they hold out," Aberdeen saw fit to congratulate Ashburton. "The good temper in which you have left them all, and the prospect of a continued peace, with I trust improved relations," he noted, "far outweigh in my mind the value of any additional extent of Pine Swamp."[39]

---

[36] Entry on 15 February 1842, "copy of private letter from Mr. Wheaton to Mr. Webster, reporting conversation on Sunday evening (the 13th) with the King of France," Add MS 43123, Aberdeen Papers, vol. 85, Correspondence of Lord Ashburton, 1841-43, British Library; and Cass to Webster, 12 March 1842. See also TNA/ADM 7-712, 12 January 1841, enclosed clipping from a New York newspaper forwarded by the British consul there, boasting of the capabilities of the newly launched paddlewheel steam-frigate USS *Missouri* (1841) as the forerunner of "a steam navy that would be superior to any in the world" and a strategic check to "the rapid strides which Great Britain is making for empire, not only in China, but all over the world."

[37] Ashburton to Aberdeen, 29 May 1842, Add MS 43123, Aberdeen Papers, vol. 85.

[38] Aberdeen to Ashburton, 18 June 1842.

[39] See "Treaty of Washington, 21 March 1843, vol. 67," Hansard, UK Parliament, cols. 1162–285; and Aberdeen to Ashburton, 26 September 1842. John Quincy Adams noted in his diary the "long and searching" House of Commons debate that evening culminated in "pitched battle" between Palmerston and Peel; entry dated 22 April 1843, in Adams, *Memoirs of John Quincy Adams*, vol. 11, 368–69.

Thus, scarcely one month *after* Monroe declared the firmly worded doctrine of the United States in relation to foreign powers, he also addressed Congress on a "plan of the peace establishment of the Navy." Here, the "great object in the event of war is stop the enemy at the coast." And for this, he stressed that

> *our fortifications must be principally relied on. By placing strong works near the mouths of our great inlets in such positions as to command the entrances into them . . . it will be difficult, if not impossible, for ships to pass them, especially if other precautions, and particularly that of steam batteries, are resorted to in their aid.*
> *. . . nor can it be doubted that the knowledge that such works existed would form a strong motive with any power not to invade our [neutral] rights, and thereby constitute essentially to prevent war.*[40]

Monroe's eighth annual message to Congress on 7 December 1824 likewise noted that the last war with Great Britain "admonished us to make our maritime frontier impregnable by a well-digested chain of fortifications, and to give protection to our commerce by augmenting our Navy to a certain extent." This included new roads interconnecting the various defensive zones north and south and with the coast to the interior. Seven years later, Monroe wrote to John Quincy Adams that affairs in Europe seemed shaky as always. Reform and the Corn Laws, Ireland, and a large, poor industrial class all tended to distract the British from American interests as well,

> *provided we sustain the attitude, on land and sea, which we have done since the late war. If we complete our fortifications, & have a force to occupy & keep them in order, and sustain our navy at the point contemplated, exhibiting squadrons in the several seas which they have hitherto visited, I have no doubt we shall command their respect, especially if our govt. pursues a pacific policy.*[41]

Monroe passed away later that summer. But America's expansive Third Tier system of coastal fortifications, stretching from Maine to Louisiana, were indeed underway. Massive harbor structures like Fort Sumter in Charleston, South Carolina, concentrated rows of heavy seacoast artillery behind brick and granite walls five-feet thick and often several stories high. By the early 1840s, new guns like the 10-inch calibre Columbiad could fire exploding shells up to 4.8 kilometers, or ricochet 125-pound

---

[40] Monroe to Congress—Naval Establishment, 30 January 1824, in Stanislaus Murray Hamilton, ed., *The Writings of James Monroe*, vol. 7, *1824–1831* (New York: G. P. Putnam's Sons, 1903), 7.
[41] Message to Congress, 7 December 1824, 48–49; and Monroe to Adams, 14 February 1831, 222–24 both in Hamilton, ed., *The Writings of James Monroe*, vol. 7.

solid shot along the surface of the water.[42] It was the most ambitious coastal fortifications scheme in the world, and work proceeded slowly and intermittently during succeeding presidential administrations up to the end of the American Civil War in 1865.[43] Very few of these stone sentinels were ever fully armed with the guns or garrisons they were planned for. Professional military advisors meanwhile warned the country was never quite safe enough. But the message Monroe and Adams had in mind did get through: America was determined to defend itself and not be reliant on the forbearance of successive British prime ministers and the Royal Navy in the nineteenth century.

Occasionally, British naval officers would carefully scrutinize American defenses and declare they might be bypassed or overwhelmed. On 2 October 1855, with Sevastopol fallen and the British mobilized for war greater than they had been in 40 years, the First Lord of the Admiralty, Rear Admiral Sir Maurice Berkeley, wrote that screw blockships (old converted wooden ships of the line, cut down for more guns and less speed) would "have New York in flames almost as soon as I get sight of the Land."[44] However, these were isolated sentiments. The proof was that, fearing a steam-powered French invasion of the British Isles in 1859–60, Palmerston wholeheartedly advocated a series of fortifications along the south coast of England, starting with the dockyards. Noteworthy too is Foreign Secretary Lord Clarendon's letter to First Lord of the Admiralty Sir Charles Wood a month after Berkeley's boast. While public meetings in Britain against upsetting Anglo-American relations regarding the John Crampton affair and various Central American issues would likely "lead to war with

---

[42] See, for example, Emanuel Raymond Lewis, *Seacoast Fortifications of the United States: An Introductory History* (Annapolis, MD: Naval Institute Press, 1979), 58–59. Jefferson Davis's 1 December 1856 Report of the Secretary of War, 34th Cong., 4th Sess., estimated some "23,000 pieces of ordnance and 3,000 gun-carriages in addition to those at the forts and arsenals" would be required to fully arm every fort under construction and newly ordered in the United States—an impossible order. By 1851, however, an 11-inch shell-firing gun had been developed by Lt John A. Dahlgren of the U.S. Navy and figured prominently in American men-of-war throughout the Civil War era. See, for example, "Records of the Bureau of Ordnance," 20 October 1851, Record Group 74, Entry 39, Box 1, vol. 1, National Archives, Washington, DC.

[43] To reinvigorate efforts at coastal fortification (predominantly against possible British intervention during the Civil War), Congress in March 1862 ordered that 1,550 copies be made of Executive Documents No. 243 (April 1836), no. 206 (May 1840) and no. 5 (December 1851) as part of its comprehensive study of American military and naval defenses. These massive reports compiled by successive U.S. secretaries of war and U.S. Army fortifications experts like BGen Joseph G. Totten, and signed off by presidents Andrew Jackson, Martin Van Buren, and Millard Fillmore, all argued for fixed coastal defences of important places as "preferable to vessels-of-war," 206; 37th Cong., 2d Sess., House of Representatives, Ex. Doc. no. 92. See also Robert S. Browning III, *Two If by Sea: The Development of American Costal Defense Policy* (Westport, CT: Greenwood Press, 1983); and Samuel J. Watson, "Knowledge, Interest and the Limits of Military Professionalism: The Discourse on American Coastal Defence, 1815–1860," *War in History* 5, no. 3 (1998), 280–307, https://doi.org/10.1177/0968344598005003.

[44] Berkeley to Sir Charles Wood (First Lord of the Admiralty), 2 October 1855, Halifax Papers (Hickleton), Borthwick Institute of Archives, York University, A4-74. See also "A Tour of 2,000 Miles of the United States of America, commencing October 16, and ending November 23, 1826," which downplayed American ordnance, for example, but warned of the strategic danger posed by the Erie Canal in TNA/ADM7-712.

the U.S.," Clarendon felt there was "nothing so likely to prevent war as shewing the Yankees that we are not afraid of them."[45]

Afraid of them? Two days before he had assured the U.S. ambassador, James Buchanan, that "nothing could be further from their intention than any, even the most remote idea of a menace" in the recent sending of naval reinforcements to the West Indies and North American stations. Clearly this was a bald-faced lie, but it was also statesmanship. The foreign secretary agreed that if Anglo-American relations were allowed to spiral out of control that a war would not only be catastrophic to both sides, ruining decades of painstaking diplomacy for decades to come, but "the Despotisms on the continent would be highly gratified with such an unnatural war."[46] This echoed Buchanan's missive to Clarendon early the year before regarding Britain's claims of a protectorate of the Mosquito Coast in Central America and that the Monroe Doctrine had not been enforced to avoid a "collision" with the British Empire. "We can do each other the most good, and the most harm," he added, "of any 2 nations in the world."[47]

* * *

Yet, how to preserve peace and avoid war? The role of mere seapower—brute strength—in modern international relations was not just strategically but politically problematic in the extreme. Buchanan for his part wrote to the U.S. secretary of state, William L. Marcy, that a British fleet sent out to bully America would actually play into their hands by alarming the British public against the provocations of Palmerston's government while a war against Russia was still in progress.[48] This was a prescient analysis, as confirmed by Kenneth Bourne in his classic study of *Britain and the Balance of Power in North America* (1967).[49] For the year before, Palmerston had urged the British cabinet under Aberdeen that in "dealing with Vulgar minded Bullies, and such unfortunately the people of the United States are" only superior force mattered. Once Sevastopol

---

[45] Clarendon to Wood, 10 November 1855, Halifax Papers, A4-57. Sir John Crampton was Britain's minister to the United States from 1852 but was replaced in 1856 by the demand of the U.S. Government which charged him with recruiting American citizens to fight against Russia during the Crimean War (despite America's neutrality in the conflict).

[46] Buchanan to Marcy, 9 November 1855, in William R. Manning, ed., *Diplomatic Correspondence of the United States: Inter-American Affairs, 1831-1860*, vol. 7 (Washington, DC: Carnegie Endowment for International Peace, 1936), 621-22. Buchanan went on to state that "were the decision to depend upon [Clarendon], I am persuaded the fleet would be withdrawn; but this would not be in character with Lord Palmerston." Obviously, Clarendon held a more convincing smile.

[47] Buchanan to Clarendon, 6 January 1854, "Correspondence between Great Britain and the United States, Respecting Central America—1854-1856," in *British and Foreign State Papers, 1855-1856*, vol. 46 (London: William Ridgway, 1865), 244-55.

[48] Buchanan to Marcy, 2 November 1855, in *British and Foreign State Papers, 1855-1856*, vol. 46, 620. Despite the fall of Sevastopol in August, the Siege of Kars (in Asia Minor) ended in a major Russian victory on 28 November 1855.

[49] Bourne, *Britain and the Balance of Power in North America, 1815-1908*, 179-83.

U.S. Geostrategic Deterrence and A2/AD

had fallen much of Britain's newly mobilized naval power would be "let free," especially with the French currently as wartime allies and unlikely to interfere. "The U.S. have no navy which we need be afraid," he continued, "and they might be told that if they were to resort to privateering we should, however reluctantly, be obliged to retaliate by burning all their Sea Coast Towns."[50] Others in the cabinet, however, had their doubts if victory would be so easy or indeed accomplish much of anything. Even with the Crimean War safely concluded in 1856, Anglo-American tensions remained along with the need for a firm policy decision. Charles Wood agreed with Palmerston in June that "we have no real *interest* in Nicaragua, Greytown or Mosquito." Nor did he believe the British government in London could prevent the ultimate fall of various Central American states to American filibuster adventurers any more than the U.S. government in Washington could. "The main question" to him was how to get out of their predicament gracefully. The longer American and British warships hovered around these confined waters the greater the risk of an incident. Thus, "We ought to *prevent* collision, not have to resent and avenge it." Half-measures would not do the trick; either British naval units must quietly withdrawal altogether or be openly reinforced by a squadron "larger than any force the U.S. can muster."[51] America might be fortified at home, piling up rocks as such and pointing all its guns seaward, but only Britain had a navy prepared at that moment to impose its will in Latin America.

At any rate, such deployments would not escape the scrutiny of Parliament, Wood reminded the prime minister. When asked, he could "put the answer so as to make the sending little more than reliefs." But this sort of cover-up was a dangerous and short-term ploy at best. The First Lord of the Admiralty therefore thought "it may be well to *tone down* as painters say the high colouring which our gobe-mouches are disposed to give to everything." It did not help matters later that September when Clarendon told him there was little chance of bringing some of these far-flung units home for, in addition to naval forces needed to put order "in that chaos of Yankee villainy," there was a "good deal of South American work on hand, both Atlantic and Pacific, for those rascally Govts. must be taught not to believe in the *Times* assertions of our decrepitude & that we are not to be robbed & insulted with impunity."[52] An increasingly chauvinist, gunboat diplomacy, seeking to humiliate lesser peoples—even great ones—back into line on a point of justice was finally too much for other mid-Victorian elites. The war with Russia was considered not so much a triumph for Britain as it was for France; it had also cost Aberdeen his ministry and Palmerston was careful not to share his predecessor's fate when a coalition of antiwar wolves began to circle in the House of Commons. "In fumbling for a tougher line," notes

---

[50] "Memorandum on a Draft from Ld Clarendon to Mr. Crampton about American Destruction of Grey Town," 10 September 1854, MS 62 Palmerston Papers (Broadlands), University of Southampton, Hartley Library, MM/US/7-11.
[51] Wood to Palmerston, 4 June 1856, Palmerston Papers, Southampton, GC/WO/66-83, Sir Charles Wood, March–July 1856.
[52] Clarendon to Wood, 29 September 1856, Halifax Papers, A4-57.

Kenneth Bourne, Palmerston and Clarendon "had conjured up the most decisive intervention yet of merchant and radical opinion on American affairs . . . in the press and parliament alike, not just a passing phase of Anglo-American relations but the whole concept of an anti-American balance of power policy had been challenged and publicly defeated."[53]

Well, this was only the beginning of the end. Palmerston certainly had his "revenge" as such with the outbreak of civil war in America, five years later in 1861 and especially during the *Trent* crisis by the end of that year. His attitude to the Monroe Doctrine had been one of complete contempt his entire professional career. In 1837, he dismissed as "nonsense" the so-called "rights" of Maine to land, which was still in question on the boundary with Canada. Such "Land Jobbers," he wrote to the British ambassador to the United States, "must learn to be more reasonable, and they will become so when they find that we do not care for their swagger; that we are resolved to keep the whole, till an amicable arrangement is made; and that we are quite strong enough to do so."[54] Even so, as he explained to Lord John Russell in 1839, there were only two ways of settling the matter (including the *Caroline* affair) before the next election might see a less amenable president in the White House than Martin Van Buren: "going to war and forcing the Americans to give us what we ask; the other by negotiating."[55] The first mode he considered "out of the question, and the latter requires time." However, when the Webster-Ashburton Treaty was finally concluded in 1842, Palmerston (now out of office as foreign secretary) assured the Earl of Minto, the former First Lord of the Admiralty, that Peel's government had made "unnecessary Sacrifices of Things which are not only losses to us, but in the Hands of the Americans will prove Instruments of future aggression against us." However much peace-lovers may celebrate, peace to him was rendered insecure "by even multiplying possible Points of Difference and by giving the Americans additional Means of annoying us, and therefore fresh Temptation to do so."[56] There could never be a true equilibrium between the two powers, only shifting advantages in a neverending contest of rival state interests.

This seemed to be the case for Aberdeen about the Oregon boundary dispute with the comparatively aggressive administration of President James K. Polk. But here too, the British government was determined "to cede nothing to force or menace, and are fully prepared to maintain our Rights." The Royal Navy maintained presence in British Columbia throughout 1845, while Parliament voted to recruit 40,000 sailors and marines without opposition.[57] The Americans, on the other hand, were more worried about accepting the Republic of Texas into the Union, and a likely war with Mex-

---

[53] Bourne, *Britain and the Balance of Power in North America, 1815–1908*, 200.
[54] Palmerston to Henry Stephen Fox, 19 November 1837, Palmerston Papers, Southampton, GC/FO/162-170, From/to Henry Stephen Fox.
[55] Palmerston to Russell, 25 October 1839, Russell Papers, British National Archives (Kew), hereafter TNA, PRO 30-22, 3D.
[56] Palmerston to Minto, 10 October 1842, Palmerston Papers, GC/MI/575-592.
[57] Aberdeen to Pakenham, 2 April 1845, British Library, Add MS 43123, Aberdeen Papers, vol. 85.

ico, reported the British ambassador, Sir Richard Pakenham. Aberdeen then noted another political factor that obviated the need for naval power-posturing to secure a favorable agreement on Oregon; the imminent repeal of the Corn Laws, "the access of Indian corn to our Markets," he was certain "would go far to pacify the warriors of the Western States" and help Polk obtain the sanction of two-thirds of the U.S. Senate to ratify the treaty.[58] Mutual market interests and growing socioeconomic pressures had to be weighed in the balance of long-term strategic considerations. They might also tip the scales of war toward peace, while the Royal Navy swung the other way.

Thus, it was only when Pakenham confirmed to Buchanan, as U.S. secretary of state, that "a fleet of thirty sail of the line and a large force of Steam Vessels were about to be fitted out as a preparation for any thing that might happen" that Buchanan became angry as well as fearful. "The appearance of such a force on our Coast," he proceeded to say, "or any other menacing demonstration would play the very Devil." The British ambassador had to agree; it would only feed American paranoia, wreck the treaty negotiations, and galvanize Congress into rampant war spending. Then again, "an attitude of dignified & imposing preparation which shall prove to the American people that England is determined to 'stand no nonsense'," he wrote to Aberdeen, "will be attended with the best results, and be sufficient, with a little patience and forbearance, to bring matters to a favourable conclusion."[59] British naval power had to be present, that was all. It could not be pushed. It had to be allowed to insinuate itself into American decision-making, already complicated by domestic factors. Was a potential war with Great Britain a *politically* good choice or not? What were the clear issues at stake?

Within months, the Oregon question was also resolved to everyone's satisfaction and the British fleet once again disappeared beyond the horizon. "The positive impatience shown by Mr. Buchanan to sign and conclude," reported Pakenham, "convinces me that the fear, lest any complication should arise out of the Mexican War, has done a great deal in inducing the American Govt. to accept Your Lordship's proposal without alteration. The bare suggestion of a reference to England was sufficient to overcome every difficulty that was talked of." The foreign secretary replied that it was "not the apprehension of any embarrassment in consequence of the Mexican War, which led to this decision; but that it was entirely owing to the impending change of the administration in this country, & a desire to settle the whole affair with us before our departure." The repeal of the Corn Laws had split the Tory party and wrecked Peel's government; everyone on both sides of the Atlantic knew that Palmerston would

---

[58] Aberdeen to Pakenham, 3 December 1845. See also J. L. Worley, "The Diplomatic Relations of England and the Republic of Texas," *Quarterly of the Texas State Historical Association* 9, no. 1 (July 1905): 1-40; Robert S. Hicks, "Diplomatic Relations with Mexico during the Administration of James K. Polk," *Annual Publication of the Historical Society of Southern California* 12, no. 2 (1922): 5-17; and Ephraim Douglass Adams, ed., *British Diplomatic Correspondence Concerning the Republic of Texas, 1838-1846* (Austin: Texas State Historical Association, 1918).

[59] Pakenham to Aberdeen, 20 February 1846.

soon be back in the Foreign Office and what that would mean.⁶⁰ The Americans had to be contained as much as the Russians did in world affairs. The Royal Navy was there to hold the line on existing British possessions in North America, from Oregon to Maine to Central America. The twentieth-century notion that the "friendly vigilance of the British man-of-war" was about "shielding" American growth was a preposterous humbug from Palmerston's own perspective. Who would ever believe it? Sure enough, three years later Palmerston assured the Nicaraguan minister that U.S. intervention in the matter of San Juan (or Greytown, as it was forcibly renamed by the British the year before) "was of no importance." Nicaragua should distinctly not look to its "Big Brother" to the north for help. " 'We have been disposed,' [Palmerston] added with a contemptuous laugh, 'to treat the United States with some degree of consideration, but in reference to this question, it is a matter of total indifference to Her Majesty's Government, what she may say or do'."⁶¹

Again, this was not entirely true. Palmerston was also very good at bluff and bluster. He knew exactly how big the U.S. Navy was at any given moment, and kept himself informed about the latest American advances in steam power and ordnance. Nevertheless, he could be reasonably sure that not everyone had access to the same information he did. An experienced statesmen, he also knew how far the Americans might be pushed—and not pushed—on what they considered vital interests closer to home. Cuba stood as a prime example. John Quincy Adams suspected Britain of wishing to buy the island from Spain in 1822, and indeed Canning considered occupying Havana in the event of an Anglo-Spanish war in 1826.⁶² When rumors circulated in 1840 that Britain might annex or seize Cuba, the secretary of the state, John Forsyth, wrote that Spain "may securely depend upon the military and naval resources of the United States to aid her in preserving or recovering it." His successor in office, Daniel Webster, again warned in 1843 that the United States "never would permit the occupation of that Island" by British forces. Buchanan was more explicit in declaring in 1848 that the United States "should be compelled to resist the acquisition of Cuba by any powerful maritime state, with all means which Providence has placed at our command."⁶³ After successful combined operations in the Mexican-American War, foreign powers had to take these kinds of warnings more seriously. The U.S. Navy might be reduced to harrying British commerce again, but defending Canada against

---

⁶⁰ Pakenham to Aberdeen, 13 June 1846; and Aberdeen to Pakenham, 30 June 1846.
⁶¹ As relayed by Ephraim George Squier, U.S. chargé d'affaires in Guatemala, to John M. Clayton, Secretary of State, 10 September 1849, in Manning, *Diplomatic Correspondence of the United States*, vol. 3, 360–70.
⁶² See, for example, Mahony, *The Monroe Doctrine*, 47fn2; Canning to Liverpool, 6 October 1826, British Library, Add MS 38568, Liverpool Papers; and Stapleton, Some Official Correspondence of George Canning, vol. 2, 144.
⁶³ Mahony, *The Monroe Doctrine*, 54; and Elihu Root, "The Real Monroe Doctrine," *American Journal of International Law* 8, no. 3 (1914): 427–42.

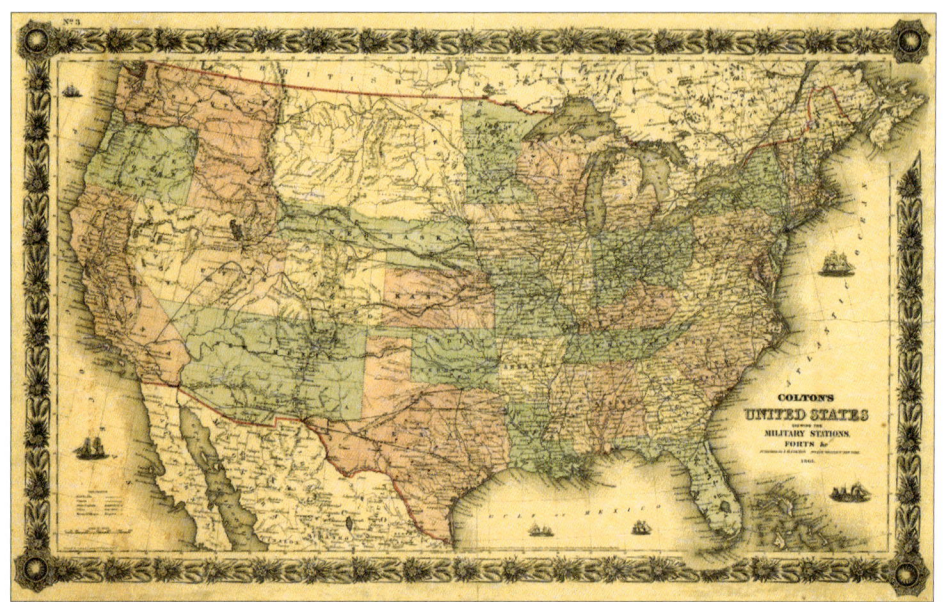

MAP 1
Strategic map of the United States, 1855.
Source: J. H. Colton, Geographicus Antique Maps

a Yankee invasion was more problematic than ever for British imperial forces.[64] Even before the American invasion of Mexico by sea and land, a local British naval officer at Quebec warned the Admiralty in March 1845 that the United States now had "a powerful Steam Marine, which is increasing and might do great mischief in a very short." London could no longer rely on ice on the St. Lawrence River from the sea to Montreal "to prevent a Hostile Force from Boston appearing before that place in 10 days."[65] American canal and rail networks in the Great Lakes region were giving them an immense advantage for both defense and attack. The Erie Canal feeding into the Hudson, pouring out into New York Harbor, for example, made the defenses of the city that much more important—and the Americans were building strong forts commanding both entrances, one defensive line after another offering multiple cross-fires

---

[64] See, for example, J. M. Gregor, *Siege and Bombardment of Vera Cruz and Surrender of that City and the Castle of San Juan De Ulloa to the American Forces, 29th March, 1847* (Norfolk, VA: W. C. Shields, 1847), which noted "this was the first enterprize [sic] of the United States for the invasion by sea of an enemy's territory. In point of numbers, it was on an imposing scale; and its requisite appurtenances in every department were commensurate with its magnitude," 6; and Cdr S. F. Du Pont, USN, *Report on the National Defenses* (Washington, DC: Gideon, 1852) in Manuscript Collections, U.S. Naval Historical Center, Washington Navy Yard, Washington, DC. Du Pont attacked the notion of a coastal defense role for American naval forces as against "the spirit of this nation"; "Steam, this new element in the affairs of the world, has very materially changed our position with reference to other nations," 28.

[65] Capt Edward Boxer to Admiralty, 20 March 1845, TNA/WO (War Office) 1-553; and the enclosed report by Hamilton, 25 July 1845.

against attacking warships that might attempt to run the gauntlets. No one would safely get through the treacherous shallows of Hell Gate (the East River and Long Island Sound) in the first place without a good local pilot.[66]

*   *   *

The outbreak of the Civil War was the greatest test of the Monroe Doctrine throughout American history. And here, the British Royal Navy perched at Halifax and Bermuda and watched. Great Britain declared its neutrality in the conflict, as is well known, and by then mid-Victorian annoyance with America was at its peak; not just with the quarrelsome, ambitious and impertinent Yankees of the North but with the equally aggressive and expansionist South, even more vile because of the way it clung to slavery. Suffice it to say that Palmerston's reaction to French intervention against Mexico—installing an Austrian Hapsburg over a "Mexican Empire" and therefore constituting a fundamental and direct challenge to the Monroe Doctrine—was very positive indeed. As he wrote to Earl Russell the foreign secretary in June 1862, the establishment of a European monarchy would be "a great blessing for Mexico and a godsend for all Countries having anything to do with Mexico." More to the point, it would also "stop the North Americans whether the Federal or Confederate States in their projected absorption of Mexico. If the North and South are definitely disunited and if at the same Time Mexico could be turned into a prosperous Monarchy I do not know any arrangement that would be more advantageous for us."[67] The events of the Civil War, including the Emancipation Proclamation, hardly changed the old prime minister's views. "The establishment of good and orderly Government in that Country under the rule of the New Emperor will not only be a real blessing to Mexico, but a great advantage to Europe," he wrote to King Leopold II of Belgium, Britain's closest ally on the continent (and Queen Victoria's uncle), in August 1864. Thanks to the war, with its rising costs in blood and treasure—and apparent stalemate on both the Virginia and Georgia fronts—"The United States seem to have come to the conviction that they are unable to prevent it and had better therefore say as little as possible on the subject."[68]

Interestingly enough, and perhaps not so well known, is that this French-appointed emperor of Mexico, the Archduke Ferdinand Maximilian, was prior to this the royal commander of the Imperial Austrian Navy. In early August 1861, he traveled in-

---

[66] Lt Dalrymple Fanshawe's report, 30 November 1841, TNA/ADM (Admiralty) 7-626. The 1860 *Report of the Secretary of War* specified that the magnificent pentagonal-sided, multitiered Fort Schuyler, guarding the outer approaches to the East River on Throggs Neck, was completed and "essentially ready for its entire armament"; LtCol R. E. DeRussy to John B. Floyd, 14 November 1860, in *Senate Executive Documents*, 36th Cong., 2d Sess. (Washington: George W. Bowman, 1861), xvii, 258.
[67] Palmerston Russell, 19 June 1862, Russell Papers, TNA/PRO 30-22, 22.
[68] Palmerston to the King of the Belgians, 28 August 1864, Palmerston Papers, Southampton, Private Letterbox, 1862 (from 30 April 1862).

cognito to Great Britain, requesting permission to visit the new (and already famous) ocean-going ironclad HMS *Warrior* (1860), just as it was preparing for sea trials, "and to the private Yards where iron ships are building."[69] He had already been approached by Mexican nobles to establish a European crown in their country, and within a few months, Spain, France, and Britain jointly despatched naval and military forces to collect debts against the faltering republican government there (the French did not leave). One wonders how his impression of European naval strength and technological advances finally helped convince him to take up Napoleon III's offer to become a client-emperor so close to the United States and in the face of the Monroe Doctrine.

But by the 1860s, there was remarkable upheaval both at Britain's doorstep (on the continent) and abroad; a time when the mid-Victorians were concerned not only about another great war not only with France but other powers—namely Russia, Prussia, and the United States. These were very different *kinds* of powers, requiring varying applications of naval and maritime force to cope with. What was good against a strong maritime state and colonial, imperial power like France might prove useless against a continental power, especially remote ones like Russia and America. Thus, Britain's iron-hulled yet stately HMS *Warrior* (1860) was a ship borne of necessary design compromises, with maximum speed and strategic range the declared aims, but at the expense of being partially armored—the stern and rudder were especially exposed to enemy gunfire—as well as being very expensive, and too large in that only Portsmouth naval base at high tide could really accommodate such a super-sized men-of-war. As Palmerston wrote to the Duke of Somerset—First Lord of the Admiralty—the gigantic, 9,000-ton ironclad frigate was "a fine yacht, but not an efficient Ship of War." Toe to toe against French ironclads like the barque-rigged yet fully armored FS *Gloire* (1859), he could only imagine "the *Warrior* and [HMS] *Black Prince* [1861], with their two Pasteboard ends knocked to shivers: the underwater compartments filled with water, everything above waterline smashed to Fragments. . . . Naval Men may see this Matter in a different Light, but to a Simple Landsman this seems to be the inevitable Course of Things."[70]

In this respect, the British Royal Navy could *not* boast the most powerful warships afloat, especially because they were able to go anywhere. Rival powers like France then the United States claimed a new form of privileged seapower status; a more localized naval supremacy based on superior firepower and armor protection. Monitors especially but also mines and other forms of torpedoes similarly threatened to turn a large seagoing ironclad's strengths into acute weaknesses, attacking exposed points both above and below the waterline. The Crimean War (1853–56) already saw Russia turn to what might be called a modern antiaccess, area-denial (A2/AD) strate-

---

[69] Special Minutes from the Board, 27 July 1861, TNA/ADM 3/269.
[70] Palmerston to Somerset, 11 June 1862 and 27 March 1861, Somerset Papers Collection, Aylesbury, Buckinghamshire Record Office, D/RA/A/2A/38 and D/RA/A/2A/37.

gy, for nullifying the ability of British or French fleets to simply steam up to and burn rich strategic targets like Sevastopol or St. Petersburg.[71]

In any case, the Civil War saw the United States mobilize its natural and human resources far beyond anything in its history; the Army, the Navy, and improved coastal fortifications, all heavily armed.[72] Even before the outbreak of hostilities, *Scientific American* reported the casting of an experimental new "Mammoth Cannon," weighing 35-tons—the first of the famous 15-inch Rodman smoothbores.[73] The British reaction was predominantly defensive, yet with attempts to constrain the conduct of the belligerent powers if possible. Without a trace of hypocrisy, Palmerston wrote to Russell in January 1862 that some moral argument might be applied in an effort to prevent the Union from going through with its stated intention of blocking Charleston Harbor and other Southern ports in rebellion with ships filled with stones, thereby permanently destroying their value as commercial centers and so be "an Injury to all the commercial Nations of the World having Intercourse with the North American continent." Palmerston insisted that neither England nor France would ever contemplate such barbarity against one another. Although during the Crimean War, the Russian capital of St. Petersburg was threatened with naval bombardment—and Washington, DC, was indeed burned by British amphibious forces in 1814—these were more political targets just as military arsenals and naval bases were fair game. But "the French would never think of blocking up Liverpool or Bristol," Palmerston was sure, "and we should never dream of blocking up Bordeaux or Marseilles." Attacking civilian warehouses and private property would be a "proceeding which would revolt the Feelings of all Mankind in this Age."[74]

Unopposed, the Union blockade had meanwhile become the largest, most effective of its kind in modern history, as officially reported in early December 1864 by Secretary of the Navy Gideon Welles to President Lincoln.[75] Europe now watched with slowly dawning horror as the grand, slaveholding "Old South" in America—a

---

[71] See, for example, D. Bonner-Smith and A. C. Dewar, eds., *Publications of the Navy Records Society*, vol. 83, *Russian War, 1854—Baltic and Black Sea, Official Correspondence* (London: Navy Records Society, 1943); "Operations in the Baltic, 13 March 1856, vol. 141," Hansard, UK Parliament, 48-119; " Sir Charles Napier at Acre, 4 April 1856, vol. 141," Hansard, UK Parliament, 480-522; and "The Return from the Baltic!," *Punch* 27 (1854): 117.

[72] See, for example, Records of the Office of the Chief of Ordnance, "Summary of Statement of Ordnance & Ordnance Stores on Hand in the Forts of the United States, September 30th, 1862," Record Group 156, Entry 102, National Archives.

[73] *Scientific American*, 7 January 1860, 26; and *Scientific American* 2, no. 25, 16 June 1860. Within a week, this same journal noted the unparalleled danger posed by the new ironclad frigates built and building in France and Britain, respectively.

[74] Palmerston to Russell, 10 January 1862, Russell Papers, TNA/PRO 30-22, 22.

[75] *Report of the Secretary of the Navy*, 38th Cong., 2d Sess. (5 December 1864); Welles declared the Union blockade of the Confederacy, stretching more than 5,700 kilometers from Virginia to Texas, was "greater in extent than the whole coast of Europe" from Cape Trafalgar (the southern tip of Spain) to the North Cape of Norway.

proud, aristocratic civilization of landed gentry and sharp class divides—rapidly crumbled into utter ruin on a scale no one had witnessed before. "The end has come," one member of South Carolina's ruling elite wrote gloomily in her diary that September, "We are going to be wiped off the earth." Before the end of February 1865, both Charleston, South Carolina, and Wilmington, North Carolina, had fallen, and she never wanted to see a newspaper again: "Shame, disgrace, beggary, all have come at once, and are hard to bear—the grand smash!" The only ray of hope she still entertained was for sheer self-interest on the part of the Old World powers, "England must know if the United States of America are triumphant they will tackle her next, and France must wonder if she will not have to give up Mexico."[76] The recent report by Welles to Congress seemed to agree, casting Great Britain in particular as America's archenemy. "Almost every vessel employed in violating the blockade has been constructed in England with great skill, regardless of cost, and with sole reference to engaging in this illicit trade," he stressed, "the profits of which are almost as remunerative as those attending the slave trade."[77]

Thus, in the 1860s, it was the threat of war with the Northern states in particular that seemed to confirm a sea change in the balance of power dynamic. "If it can be done, no Time should be lost in preparing for a Storm which as far as political Forecasting goes has been foretold as likely to follow the Conclusion of Peace between the Federals and Confederates," Palmerston confided to Somerset on 6 September 1864. And while he had "no doubt that we shall find Means to send across the Atlantic and into the St. Lawrence Guns strong enough to send their floating batteries to the Bottom," he grumbled to the secretary of state for war less than a week later that the Admiralty could only point to *"what is to be"* in terms of heavy guns "which would smash and sink the Monitors."[78] By 9 March 1865, opposition members of parliament in the House of Commons declared the United States was now "the most formidable naval power in existence." If Anglo-American relations unfortunately burst into war, then none of the ships found on Britain's North America and West Indies stations were either iron-plated or mounted a gun that could penetrate iron armor. Hence, if its commander, Admiral Sir James Hope, "were [to be] called on to blockade one of the North American ports," admonished Royal Navy Captain Sir John Dalrymple-Hay, "he could not do so for one single day with such [wooden] ships—a single Monitor would come out and set fire to those under his command."[79]

---

[76] Entries for 21 September 1864, 22 and 26 February 1865, Mary Boykin Chestnut, *A Diary from Dixie*, eds., Isabella D. Martin and Myrta Lockett Avary (New York: D. Appelton, 1906), 284-85, 305-6, 308.
[77] *Report of the Secretary of the Navy*.
[78] Palmerston to Somerset, 6 September 1864, Somerset Papers, Aylesbury, D/RA/A/2A/40; and Palmerston to Ripon, 11 September 1864, Palmerston Papers, Broadlands (University of Southampton), MS 62, Private Letterbook, April 1862 to March 1865.
[79] "Supply—Navy Estimates, 9 March 1865, vol. 177," Hansard, UK Parliament, 1373-1456; and Capt Sir John Dalrymple-Hay, RN (3d Baronet), conservative member for Wakefield (1862-65), and the chairman of the Iron Plate Committee.

MAP 2
Map of monitors staged along the East Coast, ca. 1863.
Source: L. Prang and Company, Boston, MA

Increasingly, America's newfound power led to direct accusations and threats against Great Britain for violations of neutrality and at least "moral support" of the Confederate cause, enough for Lord Richard B. Lyons, the British ambassador in Washington, to warn Russell that the North was fighting against the South with one hand and quietly building up for a major maritime war against European powers on the other. A new imperial commission on Canadian defenses acknowledged the peril but nevertheless recommended an equally vast and expensive series of armaments by land and sea that, if they did not deter the Yankees, would at least buy the empire time while it mustered reinforcements. This led to another round of serious debate in the House of Commons; William Ewart Gladstone as chancellor of the Exchequer, and an unruly member in Palmerston's cabinet, questioned the need for futile colonial investments that might provoke an arms race and then war. Was this not an opportunity, he wrote in a cabinet memorandum, "to show whether all that has recently been said about our calling on the Colonies to bear their full share of military burdens has a meaning or not, and whether we do or do not mean to alter our system of Colonial defence with reference to altered circumstances of capability, power, and privilege." Nor was it simply about "money," about selfish Liberal penny-pinching versus "duty." The real peril here, Gladstone suggested, was in "committing the honour of this country to the assumption of an attitude, which it may be unable or unwilling permanently to maintain."[80]

This perhaps comes closer to the root of the various problems at work in the Anglo-American balance of power from at least the announcement of the Monroe Doctrine in 1823 to the end of the Civil War in 1865. And not without irony, Gladstone's analysis was in many respects preempted by Benjamin Disraeli many years before when discussing the vexatious quandary over Greytown and the strategic "neck" of Central America. Here, "the Monroe Doctrine," he told Parliament, was not especially "suited to the age in which we live." The process of what we now call "globalisation" made, in his view, "one great family of the countries of the world."[81] When John Quincy Adams pinned American foreign policy to American political and social ideals, he only carried on a tradition of American isolationism and exceptionalism that tended to exacerbate the differences—not the fundamental similarities—between peoples. This was something Canning had also hinted at; that as the two great maritime (trading) nations on Earth, might they not be the closest of partners rather than the greatest of rivals in the new epoch of human history?[82]

Since then there could always be found British and American statesmen who worried more about power, worried over peace, and in their way helped push the two transatlantic countries closer to war instead—seapower and the British Royal Navy

---

[80] Defence of Canada, printed memo, confidential, 12 July 1864, Palmerston Papers, Southampton, CAB/183-193.
[81] "House of Commons Debate, 16 June 1865, vol. 142," Hansard, UK Parliament, cols. 1499-513.
[82] See H. W. V. Temperley, *Life of Canning* (London: James Finch, 1905), 179, 181-82.

acting as fulcrum, the "hinge of fate," to paraphrase Churchill here, tilting either way. Disraeli therefore also objected to any British policy that was "founded on the idea that we should regard with extreme jealousy the so-called 'aggressive spirit' of the United States" or that was innately "hostile to the legitimate development of their power." The annexation of California to him stood as a prime example, asking the House of Commons "whether the balance of power [had] been injured" by its absorption into the Union, or "whether there is any event since the discovery of America which has contributed more to the wealth, and through the wealth, to the power of *this* country, than the development of the rich resources of California by means of the United States?" Good statesmanship in his view was, on the one hand, recognizing "the necessity of an increase in their power," while on the other, America needed to learn that international law, such as it was—and such as it was becoming—would do more to ensure an inherent need to grow (Manifest Destiny) than winding itself up in the Monroe Doctrine—mighty yet alone, democratically "free" but imprisoned by the sea.[83]

---

[83] "House of Commons Debate, 16 June 1865, vol. 142," Hansard, UK Parliament, cols. 1499-513.

CHAPTER FOURTEEN
A Groundswell of Support in the Pacific

Deploying Small Wars Doctrine
amid the Rise of Amphibious Warfare

Evan Zachary Ota

At the outset of the United States' entry into World War II, the U.S. Navy faced a vexing problem in the Pacific. With the Pacific Fleet crippled at Pearl Harbor and facing an immediate material disadvantage compared to the Imperial Japanese Navy, the United States desperately fought to maintain air and sea connections with Allies in the Pacific and regain the initiative from Japan. While the United States mobilized for war, local and indigenous security forces provided an immediate and critical Allied advantage on key terrain in the Pacific.

This chapter explores how indigenous populations in American Samoa and the Solomon Islands supported Allied naval forces to regain the initiative from Imperial Japan during 1941–42. This chapter argues that Allied prewar investments in security structures, the U.S. Marine Corps' institutional adeptness at incorporating local security forces, and, most importantly, the cooperation of local security forces on key terrain yielded a decisive advantage for the Allies in the early and uncertain days of the war. By exploring these civil-military interactions during amphibious operations, this research seeks to illuminate how naval forces can develop advantageous civil environments on contested shores.

Modest but farsighted Allied investments in local security apparatuses enabled this local support in the early and uncertain days following America's entrance in the Pacific War. In American Samoa, the Department of the Navy rapidly expanded a prewar initiative to co-opt local support and incorporate indigenous forces into the defense of the archipelago. Through a Royal Australian Navy initiative begun in the 1920s, indigenous security forces in the Southwest Pacific provided critical intelligence and support to Allied amphibious forces during their first counteroffensive in Solomon Islands.

Furthermore, the Marine Corps' adeptness in working alongside foreign security forces facilitated cooperation between Allied forces and foreign populations. Through years of service in Haiti, Nicaragua, the Dominican Republic, American Samoa, and the Philippines in the early twentieth century, a generation of Marines gained insights and experiences on cooperating with foreign populations in contested environments. These experiences formed the basis for academic rigor and discourse, which resulted in a formalized manual and a professional military education program that guided the Service's approach to cooperating with foreign security forces.

The first part of this chapter examines the role of the Samoan people in securing their islands and contributing to sea control along the vital routes between Australia and the United States. First, the author will discuss how the U.S. Navy's prewar administration of American Samoa affected wartime cooperation with the population. This section will then detail the U.S. Navy's creation of Samoan naval forces, including the establishment of the Fita-Fita Guard from the naval militia and later the 1st Samoan Battalion of the Marine Corps Reserve, and how the Marine Corps' experience in small wars influenced the development of these local forces. Finally, this section will explore the Samoan population's wider support for the Allied war effort as Naval Station Tutuila in Pago Pago Harbor expanded into an advanced naval base in 1942.

The second section of this chapter will explore how and why Solomon islanders supported the Allies during amphibious operations on Guadalcanal. Research will examine preinvasion structures and organizations in the Solomon Islands, to include the British Solomon Islands Protectorate Defence Force and the Royal Australian Navy's coastwatcher program. This section will then detail how and why Solomon islanders informed Allied operations, resupplied Allied forces, rescued Allied aviators, protected local populations, and disrupted Japanese forces, even on neighboring islands, after the Allies concluded operations on Guadalcanal.

AMERICAN SAMOA AND THE FITA-FITA GUARD

As the United States acquired overseas territories after the Spanish-American War, the U.S. Navy assumed greater responsibility to govern these outposts. Correspondingly, the Marine Corps assumed a greater role in securing these overseas naval bases. As personnel and funding fluctuated during the following years, the Department of the Navy applied novel approaches to secure these bases and, in locations such as Guam and American Samoa, govern the local population.

On 6 July 1900, the U.S. Navy authorized Commander Benjamin F. Tilley, gover-

FIGURE 1
1stSgt Nelson Huron of the U.S. Marine Corps and Fita-Fita Guards, Tutuila, Samoa, ca. 1925.
Beginning in 1918, Marine first sergeants commanded the Fita-Fita Guard, and Samoans
referred to this lone Marine as "Chief of the Fita-Fitas."
Source: Bain News Service, Library of Congress Prints and Photographs Division

nor and commandant of Naval Station Tutuila, to enlist 58 Samoans and establish the Samoan Naval Militia.[1] Two years later, the U.S. Navy authorized the formation of a local band, comprised of a U.S. Navy bandmaster, a U.S. Navy musician, and 14 Samoans.[2] This collective formation would soon become known as the Fita-Fita Guard and Band, named after the Samoan word for a soldier.

Marines assigned to Naval Station Tutuila soon took on an additional role in securing the island, including training and leading Samoans to defend their island. Although the Fita-Fitas enlisted as landsmen in the U.S. Navy, a Marine first sergeant led and drilled the organization.[3] Together, Marines and the Fita-Fita Guard

---

[1] Capt T. F. Darden, USN (Ret), *Historical Sketch of the Naval Administration of the Government of American Samoa, April 17, 1900–July 1, 1951* (Washington, DC: Department of the Navy, 1952), 1.
[2] Darden, *Historical Sketch of the Naval Administration of the Government of American Samoa, April 17, 1900–July 1, 1951*, 2.
[3] 1stSgt Cecil R. Bates, "The Fita-Fita Guard," *Leatherneck*, October 1940, 6–9. The term *landsman* refers the lowest rank of the U.S. Navy in the nineteenth and early twentieth centuries. It also referred to new recruits with little or no experience at sea.

and Band carried out a wide range of duties on behalf of the Navy administration. In addition to their primary duty as guards of the naval station, Fita-Fitas served as radio operators, corpsmen, firefighters, drivers, cooks, yeomen, and augmented the crews of naval vessels anchored off Tutuila.[4]

The Fita-Fita Guard and Band became a means for social mobility in the U.S. Naval administration, and Fita-Fitas accordingly gained prominence in Samoan society. Fita-Fitas featured prominently in evening parades and traveled throughout the Pacific for organized sports. Four Fita-Fitas were local chiefs, and one went on to be a district governor.[5] Material benefits also accompanied social benefits. Fita-Fitas shopped in the Navy exchange, lived in well-made barracks, dined with their American counterparts in the mess, and generated remittances for families dispersed throughout the islands.[6]

The Fita-Fitas gained prestige in the community and established a reputation for loyalty and devotion among the U.S. Navy on which future generations of Samoans would build. Samoans deemed the Fita-Fitas *mamalu o le malo o le malo*, the "prestige of the government," due to the members of Samoan royalty in their ranks and the position Fita-Fitas held within the Navy administration.[7] U.S. Marine First Sergeant Cecil R. Bates, commander of the Fita-Fita Guard, noted that the Samoans in his charge were "an excellent supplement to the U.S. Navy in Samoa." More importantly, the Fita-Fitas were important interlocuters between the Navy and the community. As First Sergeant Bates noted, Fita-Fitas were "acutely attuned to every happening and event on the entire island," and that "news travels faster between them than can be despatched [sic] over our up-to-date communication systems."[8] Samoans, Fita-Fitas, and Marines would increasingly cooperate during the course of the U.S. Navy's administration of the islands.

SMALL WARS

As the U.S. Navy governed American Samoa, the Marine Corps undertook numerous campaigns in Central and South America to stabilize friendly governments and counter insurgent forces. Collectively known as *small wars*, these campaigns indoctrinated a generation of Marines with the lessons of operating with and amongst foreign populations.[9] In this capacity, Marines developed proficiency in recruiting, training,

---

[4] Darden, *Historical Sketch of the Naval Administration of the Government of American Samoa, April 17, 1900–July 1, 1951*, 2–3.
[5] Darden, *Historical Sketch of the Naval Administration of the Government of American Samoa, April 17, 1900–July 1, 1951*, 1.
[6] Darden, *Historical Sketch of the Naval Administration of the Government of American Samoa, April 17, 1900–July 1, 1951*, 3.
[7] Toeutu Faaleava, "Fitafita: Samoan Landsmen in the United States Navy, 1900–1951" (PhD diss., University of California, Berkeley, 2003), 163.
[8] Bates, "The Fita-Fita Guard," 6–9.
[9] For more on small wars concepts at the time, see U.S. Marine Corps, *Small Wars Manual* (Washington, DC: Government Printing Office, 1940).

and integrating local security forces in combined operations. The Marine Corps' experience in Samoa, Haiti, the Dominican Republic, Nicaragua, and numerous locations around the world generated invaluable lessons for the greatest power competition of the modern era.

Marines recruited, trained and fought alongside constabularies during successive campaigns in Nicaragua, Haiti, and the Dominican Republic between 1915 and 1933. In an organizational maneuver employed in all three countries, Marine officers and noncommissioned officers accepted commissions in these foreign constabularies while maintaining their duties—and pay—in the U.S. Marine Corps.

The Marine Corps' first major foray into small wars occurred in Nicaragua. After decades of smaller interventions, the Navy and Marines landed in force in August 1912. Major Smedley D. Butler landed his Marine battalion of 13 officers and 341 enlisted on 14 August 1912.[10] Lieutenant Alexander A. Vandegrift was one of the company grade officers in Major Butler's battalion.[11] After defeating the revolutionary forces, the Marine battalion departed Nicaragua and a detachment of Marines remained in the embassy at Managua.

Marines also landed in Haiti soon after their first major intervention in Nicaragua. On 28 July 1915, 340 Marines and sailors from the USS *Washington* (ACR 11) landed at Port-au-Prince to secure the city.[12] Soon after, in 1916, the Marine Corps organized 250 officers and 2,500 enlistees into the Gendarmerie d'Haiti.[13] Lieutenant Colonel Smedley Butler again led a Marine battalion and simultaneously served as the major general commandant of the gendarmerie.[14] Lieutenant Vandegrift again served in Butler's battalion and also served in the gendarmerie. In his latter role, Vandegrift recruited and trained two companies of constabulary with the aid of a U.S. Marine fluent in French.[15]

On the eastern half of Hispaniola, the island containing the modern-day countries of Haiti and the Dominican Republic, Marines undertook a mission similar to that in Haiti. On 7 April 1917, the U.S. military governor of Santo Domingo, Rear Admiral Harry S. Knapp, authorized the formation of a constabulary to consist of no more than 88 officers and 1,200 enlistees.[16] The constabulary, renamed the Policía Na-

---

[10] Bernard C. Nalty, *The United States Marines in Nicaragua* (Washington, DC: Historical Branch, G-3 Division, Headquarters Marine Corps, 1958), 7.
[11] Nalty, *The United States Marines in Nicaragua*, 7.
[12] "US Occupation of Haiti, 1915–1934," Naval History and Heritage Command, 29 July 2020. The USS *Washington* would be renamed *Seattle* in November 1916 so that *Washington* could be used on a new *Colorado*-class battleship.
[13] "US Occupation of Haiti, 1915–1934," Naval History and Heritage Command.
[14] Travis Prendergast, "Assessment of the American-led Constabulary during the American Occupation of Haiti from 1915–1934 in Comparison to Later Occupations," *Small Wars Journal*, 16 September 2019.
[15] Robert B. Asprey, *Once a Marine: The Memoirs of General A. A. Vandegrift, U.S.M.C.* (New York: W. W. Norton, 1964), 49.
[16] Capt Stephen M. Fuller, USMCR, and Graham A. Cosmas, *Marines in the Dominican Republic, 1916–1924* (Washington, DC: History and Museums Division, Headquarters Marine Corps, 1974), 46.

FIGURE 2
A platoon of the Guardia Nacional Dominicana conducts an equipment inspection.
Source: Dominican Republic Papers, Reference Branch, Marine Corps History Division

cional Dominicana (PND) in 1921, grew to a maximum strength of 800 enlistees under the leadership of Marine lieutenant colonel Presley M. Rixey Jr.[17] Lieutenant Colonel Rixey raised the stature and strength of the policía by establishing two district training centers and a school for PND officers. Lieutenant Colonel Rixey charged First Lieutenant Edward A. Fellowes to lead officers' school, but Fellowes knew no Spanish and was unfamiliar with the culture. To accomplish his mission, First Lieutenant Fellowes solicited cultural and language expertise. An American-born PND major raised in Santo Domingo aided Fellowes by communicating with the junior officers in their native Spanish language.[18] The combined team of Marines and Dominicans proved effective. Between 1917 and 1921, the Guardia Nacional Dominicana conducted 5,500 patrols and killed 320 enemy fighters at the cost of 27 guardia.[19]

Marines shared their expertise in fighting small wars, and soon the Marine Corps captured these lessons in manuals, textbooks, and professional military education. Major Earl H. Ellis was one of the first to write about these small wars in his article,

---

[17] Fuller and Cosmas, *Marines in the Dominican Republic, 1916–1924*, 46.
[18] Fuller and Cosmas, *Marines in the Dominican Republic, 1916–1924*, 49.
[19] Fuller and Cosmas, *Marines in the Dominican Republic, 1916–1924*, 48.

A Groundswell of Support in the Pacific

"Bush Brigades," published in the *Marine Corps Gazette* in 1921.[20] Even in these early writings, Ellis recognized the importance of the local population to generate intelligence for military operations. "A flying column should never be sent into the bush unless amply provided with CASH," Ellis stated, because "with it can be purchased knowledge of the terrain and movements of the enemy, and food."[21] Furthermore, Ellis recognized the importance of good relations with the population in order to achieve military objectives. "When Uncle Sam occupies the territory of a small nation," Ellis observed, "he wants to interfere as little as possible with the lives of the people–in fact, he wants to be considered the good angel."[22] Marine officers soon grappled with the issue of encountering foreign populations in the conduct of military operations.

The academic discourse in Marine Corps professional publications soon reached the operating forces. As the campaign against César Augusto Sandino and his army in Nicaragua persisted, the Marines returned to develop a more enduring presence. On 8 May 1927, President Diaz of Nicaragua requested Marines to assist in the development of a local constabulary, the Guardia Nacional de Nicaragua.[23] The Marines readied the first company of the guardia for duty by 1 July, and this company served in combat by the end of the month.[24] The guardia dispersed throughout the country to maintain security, with a company assigned to each of the country's political departments and platoons and squads assigned to significant towns and villages.[25] By the end of September 1929, the Guardia Nacional consisted of 1,846 members organized into three battalions.[26]

Harnessing the increasing civil-military cooperation between U.S. forces and the local population, Marine captain Merritt A. Edson led one of the Corps' most famous engagements of the Nicaragua campaign. The "Coco Patrol," as it would become known, enacted many of Lieutenant Colonel Ellis' recommendations for "flying columns" of inland patrols by "Bush Brigades." While the Coco Patrol would gain Edson fame, his actions following the initial patrol demonstrated his savviness operating with local forces and cooperating with the local population. After establishing a defense at the critical river junction of Poteca, Edson integrated local scouts into the head of his patrols, leveraged the little existing Spanish language capability in his Marines, hired local laborers and informants, and even enlisted the aid of mission-

---

[20] Richard C. McMonagle, "The *Small Wars Manual* and Military Operations Other Than War" (master's thesis, U.S. Army and General Staff College, 1996), 47.

[21] E. H. Ellis, "Bush Brigade," *Marine Corps Gazette* 6, no. 1 (March 1921): 1–15; and quoted in Brett Friedman, "The Bush Wars: Ellis on Population-Centric Counterinsurgency," *War on the Rocks*, 30 March 2015. Emphasis in original.

[22] Friedman, "The Bush Wars."

[23] Nalty, *The United States Marines in Nicaragua*, 15.

[24] Nalty, *The United States Marines in Nicaragua*, 15.

[25] Nalty, *The United States Marines in Nicaragua*, 15.

[26] Nalty, *The United States Marines in Nicaragua*, 28.

FIGURE 3
LtCol Earl H. Ellis and compatriots in the Dominican Republic, ca. 1919,
during his efforts to form the Guardia Nacional.
Source: Earl H. "Pete" Ellis Collection (COLL/3249) at the Archives Branch, Marine Corps History Division

aries in his fight against the Sandinistas.[27] Edson's approach was so successful that 20 local families requested to move to the secure areas around the Marine encampment at Poteca.[28] After years of overseas duty, Captain Edson returned to the United States to interject the lessons gained from the Coco Patrol into the growing discourse about small wars.

The Marine Corps' professional military education also reflected the growing importance of small wars to the Service's core identity. The 1934–35 Company Officers' School curriculum dedicated 172 hours of instruction to landing operations, 94 hours to small wars, and 71 hours to Spanish language out of a total of 1,056 hours.[29] The incorporation of Spanish language training for all students undoubtedly reflected the importance of cooperating with populations in Nicaragua and Santo Domingo and the lack of organic language expertise within the Marine Corps. Although landing operations increasingly dominated the curriculum of Marine Corps schools as the international security situation deteriorated in the late 1930s, this instruction on small wars impacted the cadre of officers who would later serve as the basis for an enlarged

---

[27] Jon T. Hoffman, *Once a Legend: "Red Mike" Edson of the Marine Raiders* (Novato, CA: Presidio Press, 1994), 88.
[28] Hoffman, *Once a Legend*, 88–89.
[29] McMonagle, "The *Small Wars Manual* and Military Operations Other Than War," 56.

FIGURE 4

1stLt Lewis "Chesty" Puller, quartermaster of the 1st Mobile Battalion of the Guardia Nacional de Nicaragua, with William A. Lee and two Nicaraguan Guardia soldiers. Puller would command the 1st Battalion, 7th Marines, during the Guadalcanal campaign.
Source: official U.S. Marine Corps photo

Marine Corps. Only 1,400 officers and 18,000 enlisted Marines comprised the Corps of 1939; but by July 1942, the Service nearly doubled in size.[30]

Although the study of amphibious landings increasingly dominated the curriculum of Marine Corps schools as the war in Europe approached, the Service did not discard the lessons gained through small wars. Instead, Commandant Thomas Holcomb assigned a team to select team to revise *Small Wars Manual*. Colonel William H. Rupertus led this doctrine review team that comprised of Major Merritt Edson, Major Vernon M. Guymon, and Major Ernest E. Linsert.[31] Two Marines of the committee, in particular, distinguished themselves in small wars and would later feature prominently in Marine Corps history. Colonel Rupertus served in Haiti during 1919 with the 1st Provisional Marine Brigade, as an inspector of constabulary for the Gendarmerie d'Haiti, and later as the chief of police in Port-au-Prince.[32] Major Edson dis-

---

[30] Aaron B. O'Connell, *Underdogs: The Making of the Modern Marine Corps* (Cambridge, MA: Harvard University Press, 2012), 29.
[31] McMonagle, "The *Small Wars Manual* and Military Operations Other Than War," 59.
[32] "Major General William H. Rupertus, USMC (Deceased)," History Division, Marine Corps University, accessed 29 September 2023.

FIGURE 5
BGen William H. Rupertus (left) and Col Robert C. Kilmartin (right) pose at Tulagi, Solomon Islands, after Marines captured the islands from the Japanese.
Source: Thayer Soule Collection (COLL/2266) at the Archives Branch, Marine Corps History Division

tinguished himself in the 1928–29 campaign against the Sandino rebels in Nicaragua, where he received the Navy Cross for an action against the rebels in which "superior forces of bandits were driven from their prepared positions and severe losses inflicted upon them."[33]

The Marines consolidated the best practices in the *Small Wars Manual*, released in 1940. The *Small Wars Manual* provided Marines a broad framework and approach for interacting with civil populations, raising local security forces, and incorporating local security forces into military operations. The manual included considerations such as how to employ "mobile columns and flying columns," the "establishment of advanced bases inland," the development of "native troops," and cooperation with "prominent native civilians."[34] The *Small Wars Manual* encapsulated the academic theories of Ellis and best practices of Edson.

These expeditions in small wars generated lessons that the Marines carried with them into World War II. The lessons of co-opting local leaders, cooperating with local

---

[33] "Major General Merritt Austin Edson, USMC (Deceased)," History Division, Marine Corps University, accessed 29 September 2023.
[34] *Small Wars Manual*, 3–5.

populations, and building collaborative frameworks carried over into high-intensity conflict. When Marines needed to fight alongside allies and gain support on contested terrain, they leveraged their doctrine and years of experience within their ranks.

ESTABLISHMENT OF THE 1ST SAMOAN BATTALION

The cadre of officers that retained small wars expertise would be critical as the Marine Corps rapidly expanded to fight a major war. Alexander Vandegrift, then assistant to the Commandant, observed that "expansion ruled our days beginning in the autumn of 1939."[35] In 1940, the Marine Corps mobilized its reserve to meet the growing threat.[36]

The Marine Corps' mobilization efforts focused on rapidly generating combat power in the Pacific. The Marine Corps formed the 7th Defense Battalion on 16 December 1940 specifically for the defense of Tutuila.[37] Unlike the previously formed defense battalions already employed in the Pacific, the 7th Defense Battalion specifically included a detachment to train locally recruited Samoans who may be needed in the island's defense.[38]

Subsequently, the Commandant established the 1st Samoan Battalion, Marine Corps Reserve, on 1 July 1941 in Pago Pago.[39] On 16 August 1941, the first Samoan, Sianava Robert Sena'Aetasi, enlisted in the battalion.[40] The 1st Samoan Battalion's organization reflected years of accumulated practice to rapidly establish security forces in Nicaragua, Haiti, and the Dominican Republic.[41] Active duty Marine officers commanded the 1st Samoan Battalion and each of its three rifle companies, while active duty Marine sergeants served as platoon commanders. Each platoon contained three rifle squads led by a Samoan Marine corporal. The battalion also contained a .50-caliber machine gun group of 59 Samoan Marines and an artillery battery of 6-inch guns manned by 45 Samoan Marines. The 7th Defense Battalion developed the Samoan artillery battery, and 45 Samoan Marines were reciprocally assigned to the 7th Defense Battalion.[42] While the 7th Defense Battalion was the first and most influential unit to serve alongside the 1st Samoan Battalion, it would be the first of many Marine units assigned to Samoa as war approached.

The attack on Pearl Harbor laid bare Samoa's vulnerability to Japanese attack.

---

[35] Asprey, *Once a Marine*, 92.

[36] Asprey, *Once a Marine*, 92.

[37] LtCol Frank O. Hough, Maj Verle E. Ludwig, Henry I. Shaw Jr., *Pearl Harbor to Guadalcanal*, vol. 1, *History of U.S. Marine Corps Operations in World War II* (Washington, DC: Historical Branch, G-3 Division, Headquarters Marine Corps, 1970), 67.

[38] Hough, Ludwig, and Shaw, *Pearl Harbor to Guadalcanal*, 67.

[39] Gordon L. Rottman, *U.S. Marine Corps World War II Order of Battle; Ground and Air Units in the Pacific War, 1939-1945* (Westport, CT: Greenwood Press, 2002), 73.

[40] Hough, Ludwig, and Shaw, *Pearl Harbor to Guadalcanal*, 68.

[41] Jack A. Lewis, "The Forgotten Battalion," *Leatherneck*, March 2005, 28–32.

[42] Hough, Ludwig, and Shaw, *Pearl Harbor to Guadalcanal*, 68.

FIGURE 6
A private first class of the 1st Samoan Battalion in dress uniform, ca. January 1944. The Marine Corps emblem and his chevron are visible at the lower edge of his lava lava.
Source: *official U.S. Navy photo now in the collections of the National Archives*

After the post-attack change of leadership in the Pacific Fleet, Admiral Ernest J. King tasked Admiral Chester W. Nimitz with two primary missions: "covering and holding the Hawaii-Midway line" in the Central Pacific and "maintaining communications between the west coast and Australia, chiefly by covering, securing and holding the Hawaii-Samoa line."[43] Admiral King's concerns were valid; the Imperial Japanese

---

[43] Hough, Ludwig, and Shaw, *Pearl Harbor to Guadalcanal*, 84.

Army planned amphibious assaults of "strategic points around the New Caledonia, Fiji, and Samoa Islands" to isolate Australia.[44]

The Imperial Japanese Navy tested the Hawaii-Samoa line soon after the attack on Pearl Harbor. On the night of 11 January 1942, a Japanese submarine fired on Samoa with its deck gun, which wounded one Samoan supporting Tutuila's defenses, Tauiatu Fo'isia.[45] The U.S. military accordingly rushed reinforcements to this key terrain. By the end of January 1942, the 2d Marine Brigade, created specifically to defend Samoa, arrived in Tutuila. The brigade was the first American expeditionary force to depart the United States since the declaration of war.

Due to the severity of the threat to Samoa, Commandant Holcomb established the 3d Marine Brigade for the defense of Samoa and directed the 1st Marine Division to send the "best men, weapons, and equipment" to the new formation. The Commandant intended the 3d Marine Brigade to be "as combat ready as possible" for the mission ahead.[46] This prioritization of Samoa's defenses came at no small cost to the 1st Marine Division. Major General Vandegrift, recently appointed commanding general, recalled that "this brigade, built around the Seventh Marines (reinforced), withdrew from the division a disproportionate number of officers, noncommissioned officers and men trained in amphibious warfare."[47]

The growing number of Samoan enlistees, however, helped offset the personnel tax required to defend Samoa. Both the Fita-Fita Guard and the 1st Samoan Battalion rapidly expanded, and by January 1943, 515 Samoans enlisted in the battalion.[48]

With growing organic capacity for Samoa's defense, the 7th Marine Regiment transferred to Guadalcanal to rejoin the 1st Marine Division at the epicenter of the Pacific War and the 3d Marine Brigade was subsequently disbanded on 8 November 1943.[49]

The Department of the Navy's investment in Samoa, and especially the Marine Corps, yielded and outsized the return during the uncertain early months after America's entry into World War II. For the persistent but relatively modest investment of one Marine first sergeant who cultivated indigenous Samoan forces in the Fita-Fita Guard, the U.S. Marine Corps was able to grow a force of more than 500 Marines when critically needed. More importantly, this investment resulted in the Marines being able to favorably shape environments conducive to advanced naval bases in

---

[44] John Miller Jr., *Guadalcanal: The First Offensive*, U.S. Army in World War II: The War in the Pacific (Washington, DC: U.S. Army Center of Military History, 1995), 5.
[45] Faaleava, "Fitafita," 200.
[46] Asprey, *Once a Marine*, 100.
[47] *Final Report on Guadalcanal Operation, Phase I* (San Francisco, CA: Headquarters, 1st Marine Division, 1943), 1.
[48] Lewis, "The Forgotten Battalion," 28–32.
[49] "3D MEB Lineage," 3rdmeb.marines.mil, accessed 15 December 2023.

FIGURE 7
The Fita-Fita Guard of American Samoa, 27 April 1943. The guard had been part of the U.S. Navy since 1900 and rapidly expanded to meet wartime demands.
Source: official U.S. Navy photo, now in the collections of the National Archives

the Pacific. This modest investment during the course of decades yielded significant results when the Marine Corps needed every servicemember it could muster to a counteroffensive against the Imperial Japanese Army and Navy.

COASTWATCHERS

As the United States reinforced its position in the southwest Pacific, so did Australia. While U.S. Marines captured best practices from their small wars and developed new methods for amphibious assaults, the Royal Australian Navy launched a new initiative to secure the northern approaches to their country.

Through an initiative begun in 1921, the Royal Australian Navy established a network of observers across northern Australia to report on maritime activities. This network of observers expanded into the Southwest Pacific after the outbreak of war in Europe. Officially codenamed Operation Ferdinand after the mythical bull who watched rather than fought, the members of this program came to be known as the coastwatchers. More than 1,000 coastwatchers screened a 4,023-kilometer arc across Papua New Guinea, the Solomon Islands, and Vanuatu by December 1941.[50] Contrary to their fairy-tale namesake, many Coastwatchers would fight in addition to observing.

District officer Martin Clemens landed at Guadalcanal on 12 February 1942, and soon became the senior—and only—British government representative on the island.[51] Clemens and nine police officers formed the cadre of the British Solomon Islands Protectorate Defence Force (BSIPDF) resisting the Japanese invasion on Guadalcanal.[52] Sergeant Major Jacob C. Vouza, a retired chief of the armed constabulary on the neighboring island of Malaita, joined the cadre in May 1942.[53] Vouza, originally from the village of Tadhimboko on Guadalcanal, would be a key leader in the team of coastwatchers and scouts. By August 1942, 23 coastwatcher stations observed and reported enemy actions from the Solomon Islands.[54]

GUADALCANAL

The Marines who landed on Guadalcanal were uniquely prepared to incorporate the coastwatchers and members of the BSIPDF into their operations. When the 1st Marine Division landed in the Solomon Islands on 7 August 1942, it carried a wealth experience operating within foreign civilian populations and fighting alongside foreign security forces. The division commanding general, division assistant commanding general, division chief of staff, division operations officer, division logistics officer, one of two infantry regiment commanding officers, and five of eight infantry battalion commanders had served in either Haiti, the Dominican Republic, Nicaragua, or multiple other small wars. These leaders likely did not arrive on Guadalcanal by chance. Vandegrift noted that senior Marines "not only knew every officer in the Marine Corps but knew a great deal about them. From time to time we erred in their placement; generally we did not."[55]

Major General Vandegrift, commanding general of the 1st Marine Division, had years of experience cooperating with indigenous forces. In fact, as a Marine lieutenant in Haiti, Vandegrift accepted a commission as a major in the Haitian Gendarmerie.[56] Vandegrift raised and trained two companies of gendarmerie by using his French language skills, along with those of a fluent Marine sergeant and the interpretation of a savvy recruit.[57] Vandegrift observed that his companies "behaved splendidly," and

---

[50] Anna Annie Kwai, *Solomon Islanders in World War II: An Indigenous Perspective* (Acton: Australian National University Press, 2017), 28, http://doi.org/10.22459/SIWWII.12.2017.
[51] Donald Richter, *Where the Sun Stood Still: The Untold Story of Sir Jacob Vouza and the Guadalcanal Campaign* (Calabasas, CA: Toucan Publishing, 1992), 75.
[52] Martin Clemens, *Alone on Guadalcanal: A Coastwatcher's Story* (Annapolis, MD: Naval Institute Press, 1998), 46; and Richter, *Where the Sun Stood Still*, 75.
[53] Richter, *Where the Sun Stood Still*, 50, 76.
[54] Kwai, *Solomon Islanders in World War II*, 29.
[55] Asprey, *Once a Marine*, 93.
[56] Asprey, *Once a Marine*, 49.
[57] Asprey, *Once a Marine*, 49.

FIGURE 8
Solomon Islanders guide a Marine Raider patrol across Guadalcanal, ca. 1942, which was critical to the 2d Raider Battalion's patrol, led by LtCol Evans F. Carlson, that decisively defeated Japanese resistance.
Source: Thayer Soule Collection (COLL/2266) at the Archives Branch, Marine Corps History Division

even recalled one gendarme who saved his company commander's life at the expense of a severed arm.[58] Vandegrift later observed "most of us profited from our long years in Haiti. Whether in the Gendarmerie or the brigade, Marines learned valuable lessons in jungle and guerilla [sic] warfare. We learned many cunning and wily tricks the hard way, but we also invented many ourselves."[59] The Marine leaders' accumulated knowledge in small wars would soon pay dividends in one of the most crucial amphibious landings of the Pacific War.

Although the 1st Marine Division contained veterans skilled in leveraging human networks, these leaders critically lacked an understanding of the operating environment and their enemy. Prior to the landings, division intelligence officer Lieutenant Colonel Frank B. Goettge solicited information of the landing beaches and coastal

---

[58] Asprey, *Once a Marine*, 50.
[59] Asprey, *Once a Marine*, 58.

terrain from prewar inhabitants of Guadalcanal. Although this human intelligence filled a critical knowledge shortfall by providing rough bathymetric data for the initial landings, it was not perfect. The Marines' sources mistook the Ilu River for the Tenaru River, for example, and the mistake remained in the Marines' terminology and maps for the remainder of the operation. Aerial surveillance produced photographic images of Guadalcanal, but weather inhibited the images so greatly that, when converted to topographic maps, many areas were listed simply as "cloud."[60] Prisoners of war captured early in the landings proved few and unhelpful, and a raid to capture more Japanese soldiers resulted in the loss of the division intelligence officer, one of the critical Marine Japanese linguists, the 5th Marine Division's intelligence officer, and most of the division intelligence section.[61]

The difficulty of patrolling compounded the Marines' lack of intelligence of the environment and enemy activity. Patrolling, always a difficult proposition, was even more difficult on Guadalcanal due to the dense vegetation that masked terrain features and even sunlight. As one Marine officer later stated, "If I were training my unit again, I would really have some high-class patrol training. I would do everything with these patrols I could possibly think of to include losing them and making them go across country without maps or compasses."[62]

On 10 August, however, Allied fortunes changed when the coastwatchers linked up with the newly arrived Marines. While returning a downed airman to the Allies, Sergeant Major Vouza first contacted Company G, 2d Battalion, 1st Marine Regiment, who were defending along the Ilu River to the east.[63] Captain Martin Clemens soon linked up with the Marines, and Major General Vandegrift assigned Clemens with the responsibility for "all matters of native administration and of intelligence outside the perimeter."[64] Clemens quickly integrated into the division intelligence section (D-2) to manage collections from the coastwatchers and local population.

Lieutenant Colonel Edmond J. Buckley, division intelligence chief, and Captain Clemens devised a plan to maximize the contributions of the local scouts. The D-2 plan divided Guadalcanal into four sectors and assigned armed constabulary members to each section. Captain Clemens charged Sergeant Major Vouza and 18 armed constabulary members with scouting the Volanavua area east of the Marine defenses. Sergeant Langabea and 17 armed constabulary took charge of Aola, farther east.[65] Both Vouza and Langabea would regularly meet with village leaders to gather information and stay abreast of recent developments. The D-2 plan adhered to the key tenets of the *Small Wars Manual*, which advised that "native troops are especially valuable for

---

[60] Merrill B. Twining, *No Bended Knee: The Battle for Guadalcanal*, ed. Neil Carey (New York: Presidio Press, 1996).
[61] Clemens, *Alone on Guadalcanal*, 17–18.
[62] *Fighting on Guadalcanal*, FMFRP 12-110 (Washington, DC: Department of the Navy, 1991), 38.
[63] Clemens, *Alone on Guadalcanal*, 19.
[64] Clemens, *Alone on Guadalcanal*, 199–200.
[65] Richter, *Where the Sun Stood Still*, 214.

reconnaissance and security missions," and "prominent and well-informed civilians will furnish valuable information."[66]

By 12 August, Sergeant Major Vouza guided his first Marine patrol, led by Second Lieutenant John J. Jachym of Company A, 1st Battalion, 1st Marines.[67] The scouts guided the Marines through unfamiliar and unforgiving territory. The scouts, with their intimate knowledge of the local terrain and population, played a key role in Marine patrols. Major General Vandegrift described the reports derived from these patrols as "first rate intelligence."[68]

The Marines on Guadalcanal increasingly turned to the coastwatchers and armed constabulary to develop the enemy situation. On 20 August, Captain Charles H. Brush led a patrol of 50 Marines from Company A, 1st Battalion, 1st Marines, east of the battalion defense to gain intelligence on potential threats.[69] Danial Pule, the senior indigenous coastwatcher and Captain Clemens' second in command, guided the Brush patrol with three other scouts.[70] The patrol killed 16 Japanese officers and soldiers in a sharp skirmish.[71] More importantly, the Brush patrol gained valuable intelligence that indicated an imminent Japanese attack on the Marine defenses near the Ilu River.

That same day, Sergeant Major Vouza was scouting near Koli Point, also east of the Marine defenses. Japanese forces, however, discovered Vouza and captured him. Although tied to a tree, laid on a red ant nest, beaten with rifle butts, interrogated, and bayonetted in the abdomen, chest, and throat, Sergeant Major Vouza did not divulge any information about the Marines. Left for dead, Vouza dragged himself toward friendly lines.[72]

Private William Bewley of Company G recognized the wounded Vouza and immediately moved him to the battalion command post.[73] There, Sergeant Major Vouza relayed his information about the Japanese force to Lieutenant Colonel Edwin A. Pollock, commanding officer, 2d Battalion, 1st Marine Division. Lieutenant Colonel Pollock, another veteran of the Coco River patrol led by then-Captain Edson in Nicaragua, immediately recognized the importance of Sergeant Major Vouza's message and warned his defenses of an imminent Japanese attack.[74] The subsequent engagement, the Battle of the Tenaru (21 August 1942), would be one of the key battles in the Guadalcanal campaign.

The Marines' nascent friendship with local scouts likely saved Sergeant Major

---

[66] *Small Wars Manual*, 12.
[67] Richter, *Where the Sun Stood Still*, 165.
[68] Asprey, *Once a Marine*, 114.
[69] Richter, *Where the Sun Stood Still*, 174.
[70] Clemens, *Alone on Guadalcanal*, 206.
[71] Richter, *Where the Sun Stood Still*, 175.
[72] Richter, *Where the Sun Stood Still*, 180.
[73] Richter, *Where the Sun Stood Still*, 180.
[74] "In Memoriam," *Fortitudine* 12, no. 2, Fall 1982, 15.

FIGURE 9
Cpl Wembling sports a captured Japanese sword and canteen as he poses with two local police.
Source: Thayer Soule Collection (COLL/2266) at the Archives Branch, Marine Corps History Division

Vouza's life on the night of 20 August. Soon, the local population would save many Marines through their timely and accurate reporting. As Vandegrift later commented, "Thanks to Sergeant Major Vouza we now held a pretty good idea of the enemy force."[75] With an information advantage gained through the cooperation and support of the local population, the Marines outmaneuvered the Imperial Japanese Army.

The scouts' reporting in August and September 1942 drove one of the most decisive actions of the Guadalcanal campaign. After the scouts reported enemy activity near Tadhimboko, east of the Marine perimeter, Major General Vandegrift decided to conduct a reconnaissance in force.[76] Vandegrift chose a fellow small wars veteran to lead the mission.

Lieutenant Colonel Merritt Edson, who assisted in writing the *Small Wars Manual* in 1939, led the 1st Raider Battalion at the time. Having recently rejoined the division after the successful amphibious assault on Tulagi, Edson undertook the task of reconnoitering the reported enemy activity. On 8 September, with Sergeant Major Vouza

---

[75] Asprey, *Once a Marine*, 141.
[76] Asprey, *Once a Marine*, 150.

guiding the U.S. Navy ships, the 1st Raider Battalion conducted an amphibious landing at Taivu Point.[77] Edson reported the force "landed in the rear echelon of a sizeable [Japanese] force."[78] With two Solomon Islands scouts, Selea and Olorere, in the lead echelon, 1st Raider Battalion drove off the rear guard of the Japanese Army's force led by General Kiyotake Kawaguchi.[79] Coastwatcher Dick C. Horton, who formerly served in the Taivu Point area, accompanied Edson's officers to the location of the Japanese headquarters.[80] The documents captured by the raid force confirmed earlier Solomon Islander scout reports of a 3,000-strong Japanese force intent on dislodging the Marines and approaching from a southwesterly route.[81]

Armed with this new intelligence, Edson and the division operations officer conferred the next day and determined the most likely axis of advance for the Japanese force. Edson retained 1st Raider Battalion, assumed command of the remnants from 1st Parachute Battalion, and established a defense along a prominent ridge in the center of the expected enemy axis of advance. Solomon Islands scouts picketed the forward areas along the likely axis of advance with orders to "report any enemy infiltrators, their number, and location right away."[82] In the ensuing fight from 12 to 14 September, known as the Battle of Edson's Ridge, this ad hoc force defeated the main effort attack of the Kawaguchi force, estimated at 1,500 in strength, and preserved the Marines' tenuous hold on Guadalcanal. Edson recalled that the Tadhimboko raid had "much to do" with the victory at Edson's Ridge.[83] Just as at the first Battle of the Tenaru, Solomon Islands scouts and cooperative local populations generated the intelligence that drove successful operations.

In the following months, reinforcements enabled the 1st Marine Division to take the offensive. On 17 September the 7th Marine Regiment, recently relieved of their duties in American Samoa by a growing force of Samoan Marines and sailors, landed on Guadalcanal. The 7th Marine Regiment also brought two legendary small wars veterans, Lieutenant Colonel Lewis B. Puller and Lieutenant Colonel Herman H. Hanneken, who commanded 1st Battalion and 2d Battalion, respectively. As an enlisted Marine, Hanneken simultaneously served as an officer in the Haitian gendarmerie and famously captured a rebel leader, for which he was awarded the Medal of Honor.[84] Puller also served in the Gendarmerie d'Haiti and was later awarded two Medals of Honor during two successive tours in Nicaragua.[85]

---

[77] Richter, *Where the Sun Stood Still*, 226.
[78] Hoffman, *Once a Legend*, 189.
[79] Hoffman, *Once a Legend*, 189; and Richter, *Where the Sun Stood Still*, 224.
[80] Clemens, *Alone on Guadalcanal*, 225.
[81] Hoffman, *Once a Legend*, 191.
[82] Richter, *Where the Sun Stood Still*, 244.
[83] Hoffman, *Once a Legend*, 191.
[84] Asprey, *Once a Marine*, 56.
[85] "Lieutenant General Lewis B. Puller," History Division, Marine Corps University, accessed 29 September 2023.

U.S. Marines were not the only reinforcements on Guadalcanal, however. With the Imperial Japanese Navy limiting U.S. amphibious shipping to Guadalcanal, the Marines turned to the local population to generate combat power. Just as Marines had done in Nicaragua, Haiti, Santo Domingo, and Samoa, the 1st Marine Division raised and incorporated local forces into their operations.

The first of these nascent organizations to be raised was the Whaling group, named after the prominent outdoorsman, Marine Lieutenant Colonel William J. Whaling. The Whaling group combined Solomon Islands scouts, recently trained Marine scout snipers, and the 3d Battalion, 7th Marine Regiment, into a "Bush Brigade" force tailored to take offensive action against the Imperial Japanese Army.[86]

A growing logistics force of Solomon islanders was equally as important to the campaign as were these new combat formations. In November 1942, Australian Royal Air Force commander Charles V. Widdy, former inhabitant of Guadalcanal, began recruiting a local labor force to formalize the assistance the population provided to U.S. forces since the initial landing.[87] The Solomon Islands Labour Corps assisted in the construction of roads, airfields, and the unloading of cargo from ships, which was often done manually.[88] In addition to their work in the Marine perimeter, Solomon Islanders also supported Marine patrols and enabled the type of extended activity envisioned by Lieutenant Colonel Ellis.

On 4 November, the 2d Raider Battalion joined the division on Guadalcanal. Two days after their arrival, Lieutenant Colonel Evans F. Carlson led the battalion on a series of extended-duration raids to clear the remaining Japanese forces to the east and south of the Marine defenses on Guadalcanal.[89] Australian Army major John V. Mather, fluent in the local pijin language, served as chief liaison officer to the Solomon scouts and coordinated local support.[90] Solomon scout Tabasui guided the Carlson patrol through his hometown of Tadhimboko, and Sergeant Major Vouza guided the patrol the remainder of the way.[91] By 24 November, Sergeant Major Vouza led 150 Solomon islanders who supported the 2d Raider Battalion as scouts, guides, and couriers.[92] In addition to scouting, Vouza's local force transported supplies inland from the coast and evacuated casualties.[93] With two attached Korean-American linguists

---

[86] Hoffman, *Once a Legend*, 227.
[87] Maj John L. Zimmerman, *The Guadalcanal Campaign* (Washington, DC: Historical Division, Headquarters Marine Corps, 1949), 174.
[88] David Welchman Gegeo, "World War II in the Solomon Islands: Its Impact on Society, Politics, and World View," in *Remembering the Pacific War*, ed. Geoffrey M. White, Occasional Paper no. 36 (Honolulu: University of Hawai'i at Manoa, 1991), 28.
[89] Richter, *Where the Sun Stood Still*, 327.
[90] Richter, *Where the Sun Stood Still*, 327. *Pijin* refers to Solomon Islands Pijin, which is a fusion of the indigenous language with English and French.
[91] Richter, *Where the Sun Stood Still*, 327.
[92] Richter, *Where the Sun Stood Still*, 342.
[93] Twining, *No Bended Knee*, 179.

FIGURE 10
In January 1944, Sir Jacob Vouza, honorary sergeant major in the U.S. Marine Corps,
presents a plaque to the commander of U.S. forces on Guadalcanal.
Source: University of Hawaii Manoa Library Digital Image Collections

fluent in Japanese, Lieutenants Park and Lee, Carlson wielded a combined-arms team of regional, cultural, and language experts to great effect.[94]

Lieutenant Colonel Merrill Twining, the 1st Marine Division operations officer, noted that "Carlson won victory after victory over the stubbornly resisting Japanese."[95] The 2d Raider Battalion inflicted more than 400 casualties on the Japanese force, estimated to be 500 in strength, at the cost of 17 Marines lost.[96] Carlson later observed of the Solomon scouts that "their service was invaluable. The information they provided almost invariably proved correct."[97] As Major General Vandegrift recalled, "Carlson's patrol . . . accomplished everything I hoped for by the time it returned to the perim-

---

[94] Richter, *Where the Sun Stood Still*, 337.
[95] Twining, *No Bended Knee*, 180.
[96] Twining, *No Bended Knee*, 180.
[97] Richter, *Where the Sun Stood Still*, 327.

FIGURE 11

Following the campaign on Guadalcanal, Solomon Islanders continue to support the Allied offensive in Melanesia. In September 1944, these Melanesian guides directed a raid behind Japanese lines that was launched and recovered from the submarine on which they were aboard.

Source: University of Hawaii Manoa Library Digital Image Collections

eter in early December."[98] "Of the entire coast watcher organization," Vandegrift continued, "I can say nothing too lavish in praise."[99] Of one Solomon scout, in particular, Vandegrift remarked, "The redoubtable Vouza. There was no one like him."[100]

---

[98] Asprey, *Once a Marine*, 202.
[99] Asprey, *Once a Marine*, 137.
[100] Richter, *Where the Sun Stood Still*, 279.

CONCLUSION

The Marines' ability to cooperate with local populations grew as the campaign in the Pacific progressed. By December 1942, approximately 400 Solomon scouts supported Allied operations on Guadalcanal.[101] The Solomon Islands Labour Corps, initiated in 1942, grew to a peak strength of 3,700 Solomon islanders by 1944.[102] Alongside highly educated naval officers who specialized in military government, the Marines formalized the process of securing terrain and passing responsibility for local populations from the assault troops to military government units. The Marines applied the lessons learned from Samoa and Guadalcanal to their later roles as island commanders of Guam, Tinian, and Saipan.

The defense of Samoa and the Battle of Guadalcanal demonstrated the importance of specialists trained in operating amongst foreign populations, of local security forces incorporated into U.S. forces, and of frontline troops imbued with an appreciation for the contributions of their allies and partners. Furthermore, relatively modest investments in local and Allied security networks paid outsized dividends in the critical, early stages of conflict in the Pacific. A Marine first sergeant in American Samoa and a British district officer in Solomon Islands enabled the generation of combat power that was comparable to a defense battalion and an infantry battalion, respectively. In manpower alone, the Solomon Islands Labour Corps alone exceeded that of a Marine logistics regiment. While these formations rapidly grew at the outset of war, these products required sustained, persistent cultivation of human networks prior to conflict.

The security arrangements described in this chapter were far from perfect, however. Imperialism and ethnic discrimination colored the relationships between U.S., British, Australian, and indigenous forces in Samoa and Solomon Islands. As Fita-Fita Guard Jonathan Fifi'i recalled, "We did the same kind of work as the Americans and the British, but we weren't allowed to wear the same uniforms." Furthermore, Fifi'i remembered only receiving hand-made rank insignia instead of the professionally manufactured chevrons of Caucasian servicemembers. "The white officers all wore their stripes sewn onto their shirts, but all we got were those pieces of khaki," Fifi'i recalled. "I was ashamed to wear it like that, so I would just carry it around in my hand."[103] A member of the Solomon Island Labour Corps more directly describes his relationship to the British Solomon Islands Protectorate: "We were an oppressed people in the Solomon Islands. We had been oppressed for some time up to that point."[104] Contemporary security frameworks must facilitate more equal cooperation between U.S. and foreign forces and in ways that mutually benefit each nations' sovereign interests.

---

[101] Clemens, *Alone on Guadalcanal*, 46.
[102] Gegeo, "World War II in the Solomon Islands," 29.
[103] Jonathan Fifi'i, "Remembering the War in the Solomon Islands," in *Remembering the Pacific War*, 41.
[104] Fifi'i, "Remembering the War in the Solomon Islands," 43.

Recent experience gives Marines reasons for optimism, however. Like their forebearers who landed on Guadalcanal in 1942, Marines today also carry lessons from two decades of combat in and among local populations. The U.S. Marine Corps of the 2020s, much like the Corps of 1920s, is also grappling with how to capture the experiences of recent combat while simultaneously preparing for a potential new, more horrifying conflict with emerging concepts and capabilities. Similarly, contemporary Marines are wrestling with the core identity of the Marine Corps. The debates about *Force Design 2030* echo the discourse between small wars and landing operations at Marine Corps schools in the 1930s. As then-Major Leo P. Spaeder asked Commandant David H. Berger, "Sir, who am I?"[105]

While concepts and capabilities may change, however, the human nature of war remains the same. As the tides of naval combat change from amphibious operations to expeditionary advanced base operations, a steady stream of support springs forth from security cooperation with allies and partners on key terrain. This groundswell of support was the United States' greatest advantage in the early months of the Pacific War and may be so again in the early days of another potential conflict in the Indo-Pacific.

Allies and partners embody the deep, personal affection gained through cooperation, and they are the best measure of success. Twenty-six years after fighting alongside Marines to defend Guadalcanal, Sergeant Major Sir Jacob Vouza and Sir Martin Clemens traveled to the United States as guests of honor of the 1st Marine Division Association. When he passed, Vouza was buried in the uniform provided by U.S. Marines so many years ago. His family also inherited his commitment to his friends, when 80 years after Marines first landed on Guadalcanal, Vouza's granddaughter, Gina, joined allies and former enemies to commemorate the battle. Her daughter, Sir Vouza's great granddaughter, sat by her mother's side throughout the sweltering day's ceremonies and events.

If war remains a "human enterprise" and "a clash of opposing human wills," U.S. forces must actively engender the loyalty and support of future generations.[106] The geopolitical landscape of Solomon Islands in 2024 is not what it was in 1924, and forces from the United States, Australia, and New Zealand can no longer assume access and support from Pacific Island countries. A battle for influence and access is ongoing across the Pacific, and allied efforts to shape advantageous operating environments in the future must include more than simply memorializing the past.

---

[105] Leo Spaeder, "Sir, Who Am I?: An Open Letter to the Incoming Commandant of the Marine Corps," *War on the Rocks*, 28 March 2019.
[106] *Warfighting*, Marine Corps Doctrinal Publication 1 (Washington, DC: Department of the Navy, 1997), 13, 67.

CHAPTER FIFTEEN
Prelude to Stalin's Third Crushing Blow
The Kerch-Eltigen Landing, 1943

Timothy Heck

Soviet photographer Yevgeny Khaldei, who captured the famous image of the Soviet banner over Reichstag, participated in the Kerch-Eltigen landing operation. He wrote of Red Army attempts to take Kerch on the Crimean peninsula: *There was a landing at Kerch in November 1943, but the fighting went on through December, January, February and March. Only in April 1944 did we take the city. For six months we were in a "meat grinder." An offensive was prepared to take a particular hill, and I spent the night before it in the trenches with the soldiers. In the morning the cook arrived with a large bowl of porridge but nobody wanted to eat. Everyone was thinking: "What's going to happen in half an hour, during the offensive? Am I going to live, will I see my wife, my children, my parents?" I didn't take any pictures, I just couldn't. Then the offensive took place. They didn't take the hill, and the dead were left on the ground. In the trench where I was staying, less than half the men returned.*[1]

The Nazi-Soviet front of the Second World War is dominated by narratives of sweeping land battles, frozen steppes, and armored thrusts culminating in the bloody battle for Berlin in 1945. Naval engagements, even the German *Kriegsmarine* and Soviet Red Navy, are largely overlooked with the exception of the sinking of the MV

---

[1] Michael Jones, *Total War: From Stalingrad to Berlin* (London: John Murray, 2011), 298.

Wilhelm Gustloff (1937) and *Goya* (1939) in the Baltic in 1945. This interpretation of what most Americans refer to as the eastern front overlooks the amphibious operations that occurred during the war. Overall, the Soviet Union conducted more than 100 amphibious landings during the war. Of these, two operational-level landings occurred in Crimea. While the failed Soviet Kerch-Feodosia landing operation and the subsequent Battle of the Kerch Peninsula (26 December 1941–19 May 1942) are moderately well-known thanks to the participation of General Erich von Manstein's *11th Army* and his subsequent widely read if problematic memoir, this chapter focuses on the lesser-known Kerch-Eltigen landings in late 1943 and the subsequent successful campaign to liberate Crimea.[2]

SOVIET THOUGHT AND PREPARATION

Despite the important role Russian Navy sailors played in the Bolshevik Revolution and its antecedents, the Soviet Navy was decidedly an afterthought in Soviet military prioritization and doctrine. Only in the last days of 1937 was the navy separated into its own commissariat, separating from the People's Commissariat for the Defense of the Soviet Union. Prewar procurement largely focused on creating an ocean-going navy, with cruisers and destroyers being identified in December 1940 as the best general-purpose ships for naval operations.[3] Limited, if any, efforts were put into developing an amphibious capability despite the stated role of the navy in supporting the maritime flank of army operations. As a result, Admiral Sergey G. Gorshkov later remarked, "Fleets entered the war without a single ship of special construction and only one brigade of naval infantry. The fleets had no special gunfire support ships for amphibious landings."[4]

The 1940 Soviet *Field Regulations* stated that "a landing can have as its aim the encirclement and defeat of elements on the hostile littoral flank, and also the fulfillment of independent operational missions for the creation of a new front."[5] Soviet naval objectives for amphibious landings during the second period of the Great Patriotic War was largely to "help the Soviet ground forces breach the heavily fortified coastal areas on the enemy's maritime flanks, to seize beachheads for offensive op-

---

[2] For more on Kerch-Feodosia, see David M. Glantz, "Forgotten Battles of the German-Soviet War (1941–45), Part 6: The Winter Campaign (5 December 1941–April 1942): The Crimean Counteroffensive and Reflections," *Journal of Slavic Military Studies* 14, no. 1 (2001): 121–70, https://doi.org/10.1080/13518040108430472. See also Erich von Manstein, *Lost Victories: The War Memoirs of Hitler's Most Brilliant General* (Minneapolis, MN: Zenith, 2004).

[3] V. I. Achkasov and N. B. Pavlovich, *Soviet Naval Operations in the Great Patriotic War, 1941–1945*, trans. U.S. Naval Intelligence Command Translation Project (Annapolis, MD: Naval Institute Press, 1981), 15.

[4] John G. Hibbits, "Admiral Gorshkov's Writings: Twenty Years of Naval Thought," in Paul J. Murphy, ed., *Naval Power in Soviet Policy*, Studies in Communist Affairs vol. 2 (Washington, DC: Government Printing Office, 1978), 20. Originally cited in Sergey Gorshkov, "The Soviet Navy in the Great Patriotic War," *Voyennaya Mysl'*, May 1965.

[5] Quoted in Raymond L. Garthoff, *Soviet Military Doctrine* (Santa Monica, CA: Rand, 1953), 369.

erations, and to capture ports, enemy bases, and strongpoints."⁶ Landings, overall, were divided into four categories: strategic, operational, tactical, and sabotage, as determined by the size of participating forces and their objectives. The landing at Kerch-Eltigen was operational in nature as it was intended "to strike behind enemy lines and envelop enemy positions in depth . . . [or for] seizing a certain coastal sector of enemy territory."⁷

THE STRATEGIC AND OPERATIONAL PICTURE IN LATE 1943

Following defeat at Kursk in July–August 1943, Axis forces in Ukraine and the Caucasus were on a consistent retreat as ever-strengthening Soviet forces pushed against them. The strategic consequences weighed heavily in Berlin and Bucharest, where Adolf Hitler and Romanian *Conducător* Marshal Ion Antonescu had differing views of the importance of Crimea. Crimea, in Hitler's eyes, had long been viewed as part of *Reichsland*, an integral part of Germany and not a subjugated colonial area like much of the rest of eastern Europe.⁸

Coloring Hitler's interpretation of Crimea was the recent capitulation of Italy on 8 September 1943. The loss of Crimea, he feared, "could have determined Romania's exit from the alliance, which would have meant an unimaginable blow to the Third Reich."⁹ He was also looking east to Turkey, where the German-Turkish Treaty of Friendship (1941) and the Clodius agreement, which saw Turkish chromite, vital to German weapons production, were at risk if the German foothold in the Black Sea was lost.¹⁰ As such, Hitler saw a vested interest in maintaining German control of the peninsula even as, by 26 October 1943, Field Marshal Paul Ludwig Ewald von Kleist, commander of *Army Group A* under which the *17th Army* in Crimea served, was advo-

---

⁶ Achkasov and Pavlovich, *Soviet Naval Operations in the Great Patriotic War, 1941-1945*, 97. Soviet historians divide the Great Patriotic War into three phases. The first phase (22 June 1941-18 November 1942) covers the German invasion and subsequent near-destruction of the Red Army prior to the launching of Operation Uranus to defeat Axis forces around Stalingrad. The second phase (19 November 1942- 31 December 1943) starts with Operation Uranus, includes the Battle of Kursk, and ends with strategic initiative firmly in Soviet hands. The third phase ends with Germany's defeat in 1945 and is characterized by operationally mature and well-equipped Soviet forces.

⁷ Achkasov and Pavlovich, *Soviet Naval Operations in the Great Patriotic War, 1941-1945*, 97.

⁸ Alexander Dallin, *German Rule in Russia, 1941-1945: A Study in Occupation Policies* (London, UK: MacMillan, 1957), 280.

⁹ Benone Andronic, "Warfare Actions of the Large Romanian Military Units for Defense and Evacuation of Crimea in World War II," *Annals-Series on Military Sciences* 13, no. 1 (2021): 131.

¹⁰ For more on the German-Turkish Friendship Pact and the Clodius agreement, see Edward Weisband, "A Brief Analysis of the Economic Picture," in *Turkish Foreign Policy, 1943-1945: Small State Diplomacy and Great Power Politics* (Princeton, NJ: Princeton University Press, 1973), 88-116; and Gül İnanç, "The Politics of 'Active Neutrality' on the Eve of a New World Order: The Case of Turkish Chrome Sales during the Second World War," *Middle Eastern Studies* 42, no. 6 (2006): 907-15, https://doi.org/10.1080/00263200600923005.

cating for its evacuation, citing the circumstances the *6th Army* found itself in.[11] The *17th Army*'s commander, Generaloberst Erwin Jaenecke, also pushed for evacuation, possibly because he "did not want to preside over another Stalingrad."[12] Three days later, despite the operational concerns of his commanders, Hitler telegraphed Antonescu stating that Crimea "will be defended 'at all costs'."[13] As late as December 1943, with Soviet forces trapped, Hitler was still calculating the strategic value of Crimea, stating that if Germany was unable to hold Crimea, "the consequences will be catastrophic. They'll be catastrophic in Rumania."[14]

ROMANIAN STRATEGIC POSITION

In November 1942, Romanian military strength was all but gutted in the Soviet Union when Operation Uranus cut through the *3d* and *4th Romanian Armies* on Stalingrad's northern and southern flanks. On 10 January 1943, after "a three-hour tirade" from Hitler about the poor performance of Romanian troops, Antonescu countered that 200,000 Romanians were dead, 18 divisions had been destroyed on the Don and Volga rivers, and that 4 Romanian generals had been killed in action, 3 of whom "met their death in hand to hand combat."[15] The meeting was not portentous of Romanian-German cooperation. Near simultaneously, other German allies began to falter in support too.

After the subsequent Soviet victory at Stalingrad in February 1943, the Romanian Army only had eight divisions in the Soviet Union, all of whom were to be withdrawn west of the Bug River and, perhaps more damningly, were not to be supplemented with additional formations.[16] Six of these divisions were part of the *17th Army*, located at the Kuban bridgehead on the Taman Peninsula. The remaining two were defending the Crimean Peninsula. Ongoing Soviet operations pressured the Axis defenders with the major push starting on 10 September 1943. By 2 October, all six Romanian divisions (approximately 50,139 troops) withdrew to Crimea, followed shortly thereafter by the evacuation of the remaining German units.[17] By 9 October, more than 177,000 German troops, 25,000 Russian auxiliaries, and 27,000 Soviet civilians, along with

---

[11] Andronic, "Warfare Actions of the Large Romanian Military Units for Defense and Evacuation of Crimea in World War II," 129.

[12] Samuel W. Mitcham Jr. and Gene Mueller, *Hitler's Commanders* (Lanham, MD: Scarborough House, 1992), 101.

[13] Quoted in Andronic, "Warfare Actions of the Large Romanian Military Units for Defense and Evacuation of Crimea in World War II," 131.

[14] Gen Walter Warlimont, *Inside Hitler's Headquarters, 1939–1945* (Novato, CA: Presidio Press, 1991), 390.

[15] Albert Seaton, *The Russo-German War, 1941–45* (New York: Praeger Publishers, 1971), 392.

[16] Seaton, *The Russo-German War, 1941–45*, 392.

[17] Dennis Deletant, *Hitler's Forgotten Ally: Ion Antonescu and His Regime, Romania, 1940–1944* (New York: Palgrave Macmillan, 2006), 99. Note: Seaton, *The Russo-German War, 1941–45*, reports seven divisions, 393.

vitally needed heavy equipment, were evacuated.[18] Included in this evacuation were 72,899 horses.[19] Simultaneously, following the evacuation of the Kuban bridgehead, the mixed *Fliegerkorps (Air Corps) I* was transferred into mainland Ukraine, "leaving the Crimea's defense to *FlFü Krim* . . . and *SeeFlFü Schwarzesmeer* . . . supported by *9. Flak Division* (55 batteries)."[20]

Despite strong showings in recapturing Novorossiysk, which included landing nearly 9,000 troops using 129 landing craft, Soviet efforts to interfere with the withdrawal were lackluster at best.[21] While the Soviet Air Force was "very active," the "Black Sea Fleet was inert."[22] General Wolfgang Pickert, commander of the *9th Flak Division*, remarked that "if the Soviet Navy had shown any determination to interrupt the passage over the Kerch Straits the situation might have been otherwise. Not one surface attack was made against the ferrying operations."[23] As a result, the Germans and Romanians escaped largely intact but not without their limitations.

Overall, the Axis forces in Crimea were in a state of disrepair. Crimea had been a rear area until Soviet advances in the early fall, and its occupiers were mostly rear-area and support troops. By late October, more than 200,000 Axis military personnel were in Crimea, but only approximately 40,000 were combat troops. Summarizing the support troops, one author remarked that "over 27,000 personnel were assigned to quartermaster and logistics units, *Fliegerkorps I* had over 5,000 Luftwaffe personnel and the *Kriegsmarine* had over 4,000 in Crimea. In addition, the [*Schutzstaffel*] SS, [*Sicherheitsdienst*] SD, and *Abwehr* [military intelligence] still had a very strong presence in Crimea, with over 6,000 assigned personnel, but their military effectiveness was negligible."[24] Similarly, the combat value of the Romanian formations was mixed at best. The German *17th Army* only thought that "the 1st Mountain, 2nd Mountain, and 9th Cavalry Divisions [were] suitable for rear security duties only."[25] Compounding the difficulties were a variety of chains of command that hampered Axis response, something seen in other theaters of the war.

---

[18] Grant T. Harward, *Romania's Holy War: Soldiers, Motivation, and the Holocaust* (Ithaca, NY: Cornell University Press, 2021), 186.
[19] R. L. DiNardo, *Mechanized Juggernaut or Military Anachronism?: Horses and the German Army of WWII* (Mechanicsburg, PA: Stackpole Books, 1991), 66.
[20] E. R. Hooton, *Eagle in Flames: The Fall of the Luftwaffe* (London: Brockhampton Press, 1999), 198.
[21] Lawrence Paterson, *Schnellboote: A Complete Operational History* (Barnsley, UK: Seaforth Publishing, 2015), 256.
[22] Seaton, *The Russo-German War, 1941–45*, 379–80.
[23] Seaton, *The Russo-German War, 1941–45*, 380, citing Wolfgang Pickert, *Vom Kuban-Brückenkopf bis Sewastopol Flakartillerie im Verband der 17. Armee* (Heidelberg: Scharnhorst-Buchkameradschaft, 1955), 57.
[24] Robert Forczyk, *Where the Iron Crosses Grow: The Crimea, 1941–44* (Oxford, UK: Osprey Publishing, 2014), 242–43.
[25] Harward, *Romania's Holy War*, 186–87. Other sources said the seven Romanian divisions "were considered to be of the highest quality, given both their experience on the battlefield and their equipment." See Andronic, "Warfare Actions of the Large Romanian Military Units for Defense and Evacuation of Crimea in World War II," 134.

In the eastern Kerch peninsula, General der Infanterie Karl Allmendinger commanded *V Armeekorps*, which was significantly smaller than its name would indicate. His formation was anemic, consisting solely of Generleutnant Martin Gareis' weakened *98th Infanterie Division*, a battalion of tank destroyers, the Romanian *6th Cavalry Division*, and parts of the Romanian *3d Cavalry Division*, whom *17th Army* thought were capable of frontline combat when reinforced by German assets.[26] Air and naval support could also be expected, though Allmendinger would have to navigate the Byzantine command relationships to employ it effectively.

SOVIET PLANS AND LIMITATIONS

Across the Kerch Strait, however, the Soviets were growing ever stronger. Soviet forces of the North Caucasus Front (which, along with the 56th Army, was renamed the Independent Coastal Army on 15 November 1943), supported by the Red Navy's Black Sea Fleet and Azov Flotilla, in addition to the Soviet Air Force's 4th Air Army, planned to seize beachheads in Crimea. The planning for landings at Kerch-Eltigen echoed other Soviet amphibious operations. In a postwar analysis, Admiral of the Fleet Ivan Isakov identified several conditions common to Soviet amphibious operations during the war. Of note for Kerch-Eltigen, the landings were expected to "mostly be carried out in the autumn . . . in stormy weather." Further, there was a lack of specialized landing craft in the Black Sea, "only fishing boats, launches, and seiners, a fact which rendered the landing operations difficult, and especially so when encumbered by the army's heavy equipment."[27]

From a naval perspective, the Black Sea Fleet was limited in its effectiveness. In September 1943, its submarine fleet stopped attacking Axis shipping off Crimea's southern coast due to the types of ships the Axis was using and, instead, prowled the sealanes connecting Odessa with Sevastopol. They did this with minimal results, with seven submarines sinking three transports and four landing barges for the loss of three of their own submarines to surface action.[28] Furthermore, after the sinking of three destroyers by German dive bombers on 6 October 1943, Soviet naval operations in the Black Sea changed dramatically. Of the 900 crew, only 170 were rescued in the "most serious surface ship loss since 1941."[29] Less than a week later, Joseph Stalin "issued a directive condemning the 'unnecessary' loss of three major warships. The Black Sea Fleet was to carry out no operations without the authority of the local army group commander, and 'long-range operations of the fleet's major surface ships are to take place only with permission of the Stavka VGK [Headquarters, Supreme High

---

[26] Harward, *Romania's Holy War*, 186.
[27] Quoted in Garthoff, *Soviet Military Doctrine*, 373.
[28] Rolf Erickson, "Soviet Submarine Operations in World War II," in James J. Sadkovich, ed., *Reevaluating Major Naval Combatants of World War II* (New York: Greenwood Press, 1990), 170.
[29] Evan Mawdsley, *The War for the Seas: A Maritime History of World War II* (New Haven, CT: Yale University Press, 2020), 150.

Command]'."³⁰ As a result, any ability to either influence the deep fight or interfere with Axis operations off the coast of western Crimea were hamstrung.

On 12 October, the Stavka issued the order to the North Caucasus Front to plan and prepare for landing on the Kerch Peninsula.³¹ The Soviets believed that the Germans would evacuate from Crimea shortly after the 4th Ukrainian Front sealed the Perekop Isthmus. Much to their surprise, this was not the case and, instead, against the advice of his commanders and the Romanians, Hitler reinforced German forces in Crimea. Unaware of this, the Soviets began planning to pursue the *17th Army* across Crimea and destroy it once and for all. At least four courses of action were debated.

The first, second, and third were grandly ambitious and wildly impractical given the restrictions on the Soviet Navy and its lack of assets. One course of action had Soviet troops landing at Sevastopol to block any German evacuation and restore the port city to Soviet control. Sevastopol, site of a famed siege during the Crimean War (1854-55), had taken on a renewed place in Soviet mythos after the German invasion of 1941 and subsequent siege of the city in 1941-42.³² As such, this plan would likely have had significant historical and emotional appeal to the Soviet Navy. The second proposed a similar landing at Yalta, on the southern tip of the peninsula, which was almost equally ambitious and unrealistic. The third, which was initially proposed by General Colonel Ivan E. Petrov, landed a large number of troops at Feodosia.³³ The fourth was a landing around Kerch-Eltigen followed by a push to the west toward Feodosia, site of the 1941-42 Soviet landings, which would set conditions for the 4th Ukrainian Front to push south across the Perekop.

Shortly after the Kuban bridgehead was overwhelmed, General Colonel Petrov, commander of the North Caucasus Front, tasked his staff with planning a cross-strait operation. His initial vision was an operation nearly identical to the 1941 amphibious landings, with parts of two armies being transported to beaches north and south of Kerch. An attack on such a broad front, he believed, would "make the Germans disperse their forces to fight Soviet sea-borne troops without a chance to concentrate for a strong counterattack."³⁴ In a rather sympathetic biography published well after events, Petrov is quoted as saying, "We'll pounce on them in a few places at

---

³⁰ Mawdsley, *The War for the Seas*, 150.
³¹ S. Ivanov, "Crimean Landing," *Soviet Military Review* (November 1973): 57.
³² See, among others, Serhii Plokhy, "The City of Glory: Sevastopol in Russian Historical Mythology," *Journal of Contemporary History* 35, no. 3 (2000): 378, https://doi.org/10.1177/002200940003500303. Adm Pavel Nakhimov, who died during the 1854-55 siege, was about to take a renewed place in the Soviet naval pantheon as the Kerch-Eltigen operation occurred. In 1944, the Soviet Union created a naval medal for valor named in his honor. See "Presidium of the Supreme Council of the USSR Decree: About the Establishment of Military Medals: Ushakov Medals and Nakhimov Medals," 3 March 1944, LibUSSR.ru, accessed 15 December 2023.
³³ S. A. Zonin, *Loyalty to the Ocean* (Moscow: Politizdat, 1986), 86.
³⁴ V. Vladimir Karpov, *The Commander*, trans., Yuri S. Shirokov and Nicholas Louis (London: Brassey's Defence Publishers, 1987), 149.

the same time. Each of our attacks should be strong enough to make them think it is our main strike. In the meantime we will rip their belly where they expect it least of all."[35]

His vision, however, exceeded his capabilities. The Black Sea Fleet and Azov Flotilla were far from capable of supporting that kind of landing. There existed no specialized landing craft in the Soviet navy and the DUKW amphibious trucks provided as part of the 1941 Lend-Lease Act went elsewhere.[36] Instead, Petrov's staff planned for a more reasonable landing operation given their materiel shortfalls.

Regardless of capabilities, the Soviets had superiority in key areas over the German and Romanian forces. The 56th and 18th Armies had nearly 130,000 troops, more than 600 howitzers and 90 Guards rocket mortars. Comparative ratios put them at approximately 1.5 times stronger in terms of personnel, twice as strong in armor, and four-fold in heavy artillery.[37]

The plan, according to S. A. Zonin's biography of Black Sea Fleet commander Lev Vladimirsky, was presented to the Stavka on 13 October 1943.[38] The plan had two operational objectives that were to be accomplished simultaneously, with operations commencing on 31 October 1943. The main effort consisted of the 56th Army landing in the vicinity of Yenikale, northeast of Kerch. They were supported by almost 400 of their own howitzers and guns providing fire support from the Chushka Spit, across the narrowest part of the strait, supplemented by spotlights.[39] In the south, the 318th Mountain Rifle Division with support from the 386th Naval Infantry Battalion was to seize a beachhead near Eltigen after crossing from the Taman Peninsula under the cover of its artillery. Allegedly, Marshal Semyon K. Timoshenko, chairman of the Stavka, remarked "that a successful landing by the 318th Division guaranteed the lib-

---

[35] Karpov, *The Commander*, 150. While Karpov's biography might have been almost hagiographic, Petrov was a skilled commander. John Erickson described Petrov as "able and energetic." Erickson, *The Road to Berlin: Continuing the History of Stalin's War with Germany* (Boulder, CO: Westview Press, 1983), 197.

[36] Forcyzk, *Where the Iron Crosses Grow*, 252. Relevant Lend-Lease Act materiel, even if it had been available, might not have been of use to the Soviets. George C. Herring writes in *Aid to Russia, 1941-1946: Strategy, Diplomacy, and the Origins of the Cold War* (New York: Columbia University, 1973), that in January 1944 an American naval officer providing technical assistance to the Soviets found that of 90 valuable diesel engines for patrol craft, only 3 had been installed, the Soviets lacked hulls to mount 45 others, and the remaining 42 were "deteriorating from rust." Nevertheless, the Soviets requested an additional 50 engines in 1944. Herring, *Aid to Russia, 1941-1946*, 128. The Americans were not the only Westerners frustrated by Soviet naval operations. British naval liaison officer Capt Robert Garwood was, during the landings at Kerch-Eltigen, "sequestered at Sukhumi, a long way from the fleet's command staff or its major warships at Poti and Tuapse," let alone its amphibious forces. Martin H. Folly, "From Sevastapol to Sukhumi-and Back Again: British Naval Liaison in Action with the Red Navy in the Black Sea, 1941-1945," *War in History* 28, no. 4 (2021): 882, https://doi.org/10.1177/0968344519871971.

[37] Ivanov, "Crimean Landing," 57.

[38] Zonin, *Loyalty to the Ocean*, 87.

[39] Fredrich Ruge, *The Soviets as Naval Opponents, 1941-1945* (Annapolis, MD: Naval Institute Press, 1979), 115.

eration of the Crimea."⁴⁰ Of note, Timoshenko was not the only high-ranking Soviet officer present. Nikolai Kuzentsov, head of the Soviet Navy during the war, flew to the Black Sea Fleet to supervise preparations, allegedly on Stalin's orders.⁴¹

Overhead, the Soviet Air Force was expected to provide support and interdict Axis counterattacks, aircraft, and shipping. Furthermore, while the need for coordination with aviation had been identified as early as 1929 and incorporated into Soviet doctrine, little combined effort seems to have occurred.⁴² It is possible that no prelanding reconnaissance was performed, which played a tragic role in the fate of many men of the 318th Mountain Rifle Division. General Konstantin Vershinin, commander of 4th Air Army, gave his forces a variety of prelanding roles, including reconnaissance, preparatory attacks on Axis airfields, and the construction of bases on the Taman Peninsula from which to operate.⁴³ Once the landings commenced, 4th Air Army was expected to suppress German and Romanian positions, cover ships in transit across the Kerch Strait, and fly close-air support missions.⁴⁴

Simultaneously, attention in Berlin was also on Crimea. On 26 October 1943, during the evening situation report, Hitler and General Kurt Zeitzler, chief of staff of the *Oberkommando des Heeres* (German Army High Command), analyzed likely Soviet operations in Crimea. Hitler remarked, "The biggest danger on the Crimea, as I see it, is not sea landings but airborne landings–he could drop an airborne brigade on the isthmus of Feodosiia [*sic*]."⁴⁵ This belief was not unreasonable as the Soviets dropped a commando unit of 500 troops near Cape Illy on 21 October, which was a failure.⁴⁶ Zeitzler concurred with Hitler, stating the Soviets were unlikely to attack western Crimea due to a lack of shipping. Later in the conversation, Hitler provided guidance down to the divisional level and encouraged Zeitzler to "keep the (coastal) batteries sufficiently manned," even if they were manned by some of the Russian auxiliaries evacuated from Kuban.⁴⁷

As the plans were being drawn up, the Red Army prepared for the landing. Some of the training was political, other was military. First, the political, where we meet a soon-to-be famous comrade: Leonid Brezhnev. Brezhnev had previously been assigned as a political commissar in the 18th Army where he had performed admirably at Novorossiysk. In preparation for the Kerch-Eltigen landings, he tasked his men with conducting "strenuous Party-political work," which encouraged the troops to "act boldly

---

⁴⁰ Leonid Il'ich Brezhnev, *How It Was: The War and Post-war Reconstruction in the Soviet Union* (Oxford, UK: Pergamon Press, 1979), 42.
⁴¹ Zonin, *Loyalty to the Ocean*, 87.
⁴² John Erickson, *The Soviet High-Command: A Military-Political History, 1918–1941*, 3d ed. (Oxford, UK: Frank Cass, 2001), 317.
⁴³ Konstantin Vershinin, *Fourth Air Force* (Moscow: Voenizdat, 1975), 283–84.
⁴⁴ Vershinin, *Fourth Air Force*, 284.
⁴⁵ Helmut Heiber and David M. Glantz, eds. *Hitler and His Generals: Military Conferences, 1942–1945* (London: Greenhill Books, 2002), 277.
⁴⁶ Heiber and Glantz, *Hitler and His Generals*, 902n772.
⁴⁷ Heiber and Glantz, *Hitler and His Generals*, 281.

and decisively . . . [reminding the troops] of the feats of arms of the Red Army men who stormed the Perekop in 1920, of the heroes of the legendary Sevastopol defence in 1941–42."[48] Furthermore, his troops may have had some role in organizing or leading the 318th Mountain Rifle Division as his memoirs remark on the need "to coordinate all the commandos" and that he sent his instructors to the division, where "they carried out their extremely difficult battle duty assignment."[49]

For military training prior to the landings, Karpov gives Petrov significant credit, stating that

> he resolved to take advantage of this respite to increase the fitness of his troops for the impending sea-borne attack. He ordered all commanding officers to train troops in every detail of embarkation, disembarkation and fighting for a foothold on a beach. Training went on round the clock and nobody complained despite the fatigue of the battles that had just ended. All realized that efficiency would have to offset the disadvantages of an attack against an entrenched enemy relying on a formidable system of fortifications and immense firepower.[50]

THE LANDINGS

Ultimately, even with updated Soviet doctrine, including the *1943 Instructions for Joint Actions by the Ground Forces, the Navy, and Military River Flotillas*, Soviet planners had less than three weeks from the date of the Stavka order to the first wave's embarkation.[51] Whether that rushed process impacted the ability of the Soviets to successfully embark, land, and push inland cannot be known. Regardless, the landings were moderately successful if costly affairs for the troops of the 56th and 18th Armies.

First, the weather on the night of 31 October hampered the embarkation. Of the seven detachments slated to depart from the Novorossiysk naval base, only five appear to have departed. Gorshkov noted that the surf at 56th Army's embarkation points were strong enough to toss boats and ships ashore.[52] Weather in the straits further hampered operations. The initial plan of landing troops simultaneously at several points was quickly thrown off schedule. Vladimirsky told Petrov that with such problems, "we risk the failure of the operation, since the simultaneous landing of all groups of assault forces may not work out. The Germans will beat the landing in parts. Then, in such weather, significant losses are inevitable in small vessels."[53]

By 0300, it was apparent the weather had already significantly impacted the plans. For forces to be landed by the Black Sea Fleet, three elements were at least an

---

[48] N. Larichev, "Liberation of the Crimea," *Soviet Military Review* (April 1979): 40.
[49] Leonid Il'ich Brezhnev, *Leonid I. Brezhnev: His Life and Work* (New York: Sphinx Press, 1982), 38–39.
[50] Karpov, *The Commander*, 149.
[51] Ya. B. Yeshchchenko, "Analyzing the Practice of Conducting Amphibious Assault Operations during the Great Patriotic War (1941–1945)," *Military Thought*, no. 4 (2018): 54–55.
[52] Zonin, *Loyalty to the Ocean*, 89. Forcyzk states the boarding process was bungled by the Azov Flotilla. Forcyzk, *Where the Iron Crosses Grow*, 252.
[53] Zonin, *Loyalty to the Ocean*, 89.

hour behind schedule and only four detachments were ready to deploy. A storm in the Sea of Azov prevented embarkation at the pier of Kordon Ilyich. By 0400, Gorshkov reported "a little more than half of the boats and ships of the [Azov] flotilla will go to the deployment line in the strait. The rest were damaged or thrown ashore by the [waves]."[54] Petrov agreed to delay 56th Army's landings, but elements of the 318th Mountain Rifle Division were already landing. Not until 3 November did 56th Army successfully start landing near Kerch.

Throughout the night of 31 October–1 November, Colonel A. D. Shiryaev's 137th Rifle Regiment of the 318th Mountain Rifle Division embarked at Krotovka and crossed the wider strait in six small flotillas. The embarkaration, however, was almost the only successful event that night for the 137th Rifle Regiment and its commander. First, the flotillas encountered a German minefield off Cape Panagia, which sank two vessels, killing almost 200 troops, none of whom had life jackets.[55] Included in the dead were Shiryaev and most of his regimental staff. Second, due to the ad hoc nature of the Soviet Navy's support, the vessels were separated due to disorientation and differing speeds stretched the flotillas. Indeed, "some troops were even rowing across in longboats."[56] Third, another minefield was encountered at approximately the midway point, resulting in the loss of several more Soviet ships and all onboard. Finally, when approaching the Kamysh Burun beach, the lack of reconnaissance cost more lives. Unknown to or unaccounted for by Soviet planners, a sandbar lay 50 yards offshore but water depths quickly increased thereafter. As they disembarked in the darkness, "troops fell into three yards of deep water, and many heavily laden soldiers drowned."[57]

The Soviets, perhaps not unexpectedly, picked the same landing site used in 1941, so the lack of knowledge of the sandbar is curious. The Soviets had an active reconnaissance and intelligence capability in the form of partisans and naval commandos. Soviet naval scouts (*morskiye razvedchiki*) conducted reconnaissance and coastwatcher-type operations in the southwestern tip of Crimea, observing German shipping and unit movements on the peninsula.[58] Naval commandos were particularly well placed for Soviet purposes and had been for months: "In June 1943 several members of the [Black Sea Fleet ground reconnaissance] detachment parachuted into the southern coast of the Crimea to conduct operations with the partisans. The naval scouts conducted reconnaissance against German airfields, garrisons, and supply centers and radioed the information they obtained back to fleet headquarters. Other patrols from the detachment, which included female radio operators, parachuted into the hills overlooking the south coast near Yalta. Over a period of several months these groups

---

[54] Zonin, *Loyalty to the Ocean*, 89.
[55] Forcyzk, *Where the Iron Crosses Grow*, 252.
[56] Forcyzk, *Where the Iron Crosses Grow*, 252.
[57] Forcyzk, *Where the Iron Crosses Grow*, 253.
[58] See Yuriy Fedorovich Strekhnin, *Commandos from the Sea: Soviet Naval Spetsnaz in World War II*, trans., James F. Gebhardt (Annapolis, MD: Naval Institute Press, 1996), chaps. 3 and 4.

Prelude to Stalin's Third Crushing Blow

directed air strikes against German coastal shipping."[59] While these claims sound impressive, Axis losses, coordination with the Soviet Air Force, and the ineffectiveness of the Black Sea Fleet make such claims somewhat suspicious and contain an element of bluster.

Gorshkov's Azov Flotilla also had commandos who directly supported operations around Kerch. Again, naval scouts arrived ahead of the landing forces, attempting to influence the outcome: "By October 1943, the front was once again on the east shore of the Crimea at Kerch Peninsula, and Azov Flotilla reconnaissance detachment elements, among them a female intelligence agent [Sofia Osetrova], were on the peninsula itself, sending reports back by radio. In November 1943, in support of the Kerch El'tigen operation, the naval scouts went ashore ahead of amphibious landing units to determine the precise location and strength of German defenses."[60] Unfortunately, the language in these sources do not make it clear if the scouts were supporting both beachheads or only the northern one.

Fifteen minutes after the Soviet landing, the German defenders finally noticed the activity on the beach. The Germans called for fire from a battery of four heavy cannons located approximately a kilometer southwest of the beachhead. The Soviet flotilla, improperly armored, suffered heavy losses. Among the losses were the artillery battalion's 12 76mm guns, depriving the Soviets of crucial firepower on the beachhead.[61] Another Soviet battalion landed in the wrong location and had to hike to rejoin their comrades. As the sun rose and the Red Army was forced to suspend landings until nightfall, total Soviet casualties in the crossing alone are estimated at approximately 2,600 of the 5,700 embarked, most of the heavy artillery, and more than one-third of the already scarce naval vessels.[62] Nevertheless, the Soviets were able to gain a foothold and even overran a Romanian battery situated at the north edge of the beachhead.

General Allmendinger believed, incorrectly, that the landing at Eltigen was a battalion-size diversionary force. He ordered *Grenadier Regiment 282* to wipe it out. German counterattacks that morning and afternoon were fierce but ultimately proved ineffective. In the beachhead, Major Dmitri S. Koveshnikov was the senior surviving officer but lacked communications with his higher headquarters on the Taman Peninsula. Furthermore, the surviving Soviet troops were from a variety of units and likely lacked adequate command and control. Nevertheless, by 1130, Koveshnikov established communications across the strait and was able to use artillery fire from the guns on the Taman Peninsula to save the Soviet beachhead. Ranging overhead were fighters and fighter bombers of the Red Air Force's 4th Air Army, who attacked Axis units attempting to mass to attack the landing party. A midday attack led by the

---

[59] Viktor Leonov, *Blood on the Shores: Soviet Naval Commandos in World War II*, trans. James F. Gebhardt (Annapolis, MD: Naval Institute Press, 1993), 154.
[60] Leonov, *Blood on the Shores*, 198n10, 154-55.
[61] Forcyzk, *Where the Iron Crosses Grow*, 253.
[62] Forczyk, *Where the Iron Crosses Grow*, 253.

assault guns of *Sturmgeschütz-Abteilung 191* from the north overran a Soviet penal unit. Major Koveshnikov's front line started to disintegrate. His position was, once again, saved by artillery fire from Taman.[63]

On the night of 1–2 November, an additional 3,200 Soviet troops and nine mortars were successfully landed, including Colonel Vasily F. Gladkov, the 318th Mountain Rifle Division's commander, who took charge of the beachhead. German counterattacks were again fierce and half the Soviet position was lost before cross-strait artillery inflicted significant losses on the German infantry, halting their advance. Fighting on 2 and 3 November resulted in similar outcomes, with Soviet forces being evermore compressed but the Germans unable to finally break them despite support received from *Luftflotte 4*'s (Air Fleet) Junkers Ju 87 Stukas and the presence of the assault guns and 88mm flak guns of the *9th Flak Division*. A stalemate of sorts emerged, with the Germans outnumbered by the surrounded Soviets who lacked the armored vehicles or heavy weapons needed to break out.

THE SECOND LANDING (56TH ARMY)

In the north, the 56th Army conducted its landing operation on 3 November. Here, the precrossing bombardment by artillery located on the Chushka Spit consisted of more than 600 guns and rocket launchers.[64] Elements of the 2d Guards and 55th Guards Rifle Division, as well as the 369th Naval Infantry Battalion, started their crossing, only to be discovered by Axis lookouts and fired on by the guns of *Marine-Artillerie-Abteilung 613*, which inflicted casualties but were unable to stop the landings. Once ashore, the Soviet forces pushed the two companies of *Grenadier Regiment 290* guarding this part of the peninsula back. The 55th Guards attacked and *9th Kompanie* retreated while *11th Kompanie* was routed by the 2d Guards.[65] By the conclusion of the first day, approximately 4,000 Soviet troops were ashore and held a beachhead of seven square kilometers. By 5 November, "troops of the whole corps were concentrated" in that small space.[66]

Petrov kept reinforcing his northern landing site and pushing the 56th Army to expand its foothold. German forces retreated on the ground but the *Luftwaffe* was able to maintain pressure above, limiting the effectiveness of Soviet numerical aviation strength. In terms of fighters, 40 Messerschmitt Bf 109Gs of II./Jagdgeschwader 52 were outnumbered more than 10 to 1 by 4th and 8th Air Armies and the Black Sea Fleet's aviation units. Nevertheless, during the course of November and December, the Germans shot down more than 200 Soviet aircraft against 17 of their own losses.[67]

German General Allmendinger now likely realized the landing at Eltigen was

---

[63] Forcyzk, *Where the Iron Crosses Grow*, 254.
[64] Ivanov gives the numbers as 420 guns and two regiments of rocket mortars. See Ivanov, "Crimean Landing," 59.
[65] Forcyzk, *Where the Iron Crosses Grow*, 255.
[66] Ivanov, "Crimean Landing," 59.
[67] Forcyzk, *Where the Iron Crosses Grow*, 257.

not a mere diversion but a significant, if contained, threat. He ordered a reinforced *Grenadier Regiment 282* to eliminate the Eltigen beachhead to free up troops to block the larger and more successful landing near Kerch.[68]

Entitled Operation Komet, *Grenadier Regiment 282* went into action on 7 November. It was a complete failure. *Luftwaffe* support failed to materialize and, under Gladkov, the Soviet troops were able to hold against the numerically inferior German and Romanian forces. Perhaps the saving grace from the German perspective was that the *Kriegsmarine*'s S- (fast boat) and R-boats (minesweepers) were finally brought into action in the narrow straits with decisive results. Vladimirsky later remarked his wooden ships were akin to "fight with carts against tanks" when compared to the swift German craft.[69]

The Germans, despite taking losses, had enough small boats for blockade operations and started to choke off logistics to the forces trapped in the south.[70] Their success, coupled with a Soviet inability to successfully counterattack, marked the decline of the Soviet lodgment. Though Komet failed, the *Kriegsmarine* ramped up operations and "began to erect a fairly impenetrable blockade . . . using light warships, armed MFPs [naval ferry barges], and mines, which gradually starved the Soviet forces in the beachhead."[71] On 8 November, "five S-boats found Russian Task Force F sinking patrol boat *0122*, with the Russian flotilla command aboard," further hampering Soviet naval efforts.[72] Withdrawing *Grenadier Regiment 282*, Allmendinger held the cordon around Eltigen with the Romanian *14th Machinegun Battalion* and *6th Cavalry Division*. The slow death of the 318th Rifle Division began.

On 10 November, Petrov's 56th Army conducted a major attack west of Baksy, driving the Germans back three kilometers. By 12 November, 56th Army was on the outskirts of Kerch as Gareis's *98th Infantry Division* kept giving up ground. The resultant losses prompted the German High Command to dispatch Major Erich Bärenfänger's *Grenadier Regiment 123* to help shore up the *98th Infantry Division*. Bärenfänger was an impressive battlefield commander, having already received the Honor Roll Clasp, the Bulgarian Order for Bravery, and swords to his Knight's Cross of the Iron Cross.[73] He would add diamonds to that award in January 1944, in part as a result of his success near Kerch.[74]

On 10 November, the first Soviet T-34 medium tanks arrived in Crimea, with more arriving the next day, giving Soviet forces much needed firepower and mobili-

---

[68] Forcyzk, *Where the Iron Crosses Grow*, 257.
[69] Zonin, *Loyalty to the Ocean*, 91.
[70] Zonin, *Loyalty to the Ocean*, 91.
[71] Forczyk, *Where the Iron Crosses Grow*, 256.
[72] Paterson, *Schnellboote*, 258.
[73] Franz Thomas and Günter Wegmann, *Die Ritterkreuzträger der Deutschen Wehrmacht 1939-1945 Teil III: Infanterie Band 1: A-Be* (Osnabrück, Germany: Biblio-Verlag, 1987), 177.
[74] Veit Scherzer, *Die Ritterkreuzträger 1939-1945 Die Inhaber des Ritterkreuzes des Eisernen Kreuzes 1939 von Heer, Luftwaffe, Kriegsmarine, Waffen-SS, Volkssturm sowie mit Deutschland verbündeter Streitkräfte nach den Unterlagen des Bundesarchives* (Jena, Germany: Scherzers Militaer-Verlag, 2007), 199.

ty. Ten were landed in the first two days. Petrov deployed them on 13–14 November. *Grenadier Regiment 123* stopped the tanks cold, destroying nine, with Bärenfänger destroying one personally. On 19 November, the presence of Soviet armor, presumed to be a tank brigade, was discussed by Hitler as part of his midday situation update.[75] By 20 November, *Grenadier Regiment 123* had destroyed 24 Soviet tanks and the possibility of a Soviet armored offensive ended.[76]

November dragged on, with neither side able to gain the upperhand, but the Soviets almost inexorably grinding away at the Axis positions. On 15 November, the Soviets renamed 56th Army as the Separate (or Independent) Coastal Army, reviving the name after the previous Coastal Army had been disbanded in July 1942. The name change, however, did little to improve the Soviet tactical position.

In central and western Crimea, partisan operations also took their toll on German and Romanian troops.[77] Partisan operations in Crimea were, as they were across the rest of the Soviet Union, increasingly effective by late 1943. By that December, for example, every known partisan detachment had contact with the Soviet central staff via radio, as opposed to only 10–15 percent nationwide in August 1942. Furthermore, partisans had increasing access to Soviet aviation, which transported radio equipment, sabotage devices, and almost 75 percent of explosives nationwide. Logistics extended beyond arms and ammunition and included vital food. Attempts to requisition food in Crimea appears to have been difficult for pro-Soviet partisans due to pro-German populations and successful German military operations.[78] Soviet aviation, in spite of being unable to establish superiority over the landing beaches, played a crucial role, as "in some cases, [aircraft] brought the only source of food, as occurred in Crimea when starvation had reduced some partisans to cannibalism" previously.[79] Through the remaining months of occupation, partisans remained "largely dependent upon these [airlifted] supplies."[80] After the peninsula's isolation in late 1943, partisan attacks in Crimea surged. Attacks around Simferopol spiked, for example, with more than 100 attacks occurring.[81] The Axis response was swift, brutal, and largely effective but continued to tie down needed troops.

Allmendinger started withdrawing rear echelon troops and supplies, largely un-

---

[75] Heiber and Glantz, *Hitler and His Generals*, 285.
[76] Forczyk, *Where the Iron Crosses Grow*, 257.
[77] Of note, local antipathy against the Romanians may have been higher than against the Germans due "not necessarily because they were given to wholesale extermination actions but, rather, because they aroused widespread hostility by their wanton looting, theft, and abuse." Quote from Theo J. Schulte, *The German Army and Nazi Policies in Occupied Russia* (Oxford, UK: Berg Publishers, 1989), 72. Schulte does not provide specific examples from Crimea but references others from mainland Ukraine.
[78] Earl Ziemke, "Composition and Morale of the Partisan Movement," in John A. Armstrong, ed., *Soviet Partisans in World War II* (Madison: University of Wisconsin Press, 1964), 369.
[79] Kenneth Slepyan, *Stalin's Guerrillas: Soviet Partisans in World War II* (Lawrence: University of Kansas Press, 2006), 123; and Ziemke, "Composition and Morale of the Partisan Movement," 369.
[80] Ziemke, "Composition and Morale of the Partisan Movement," 370.
[81] Harward, *Romania's Holy War*, 190.

hampered by the Soviet Navy or Air Force that deep in the rear. Nevertheless, Soviet reinforcements kept pouring into the beaches near Kerch, running the gauntlet of the *Kriegsmarine's 1st Schnellbootsflottille*, which inflicted heavy losses on Gorshkov's Azov Flotilla. The Germans suffered heavy losses, too, losing more than one-third of their attack boats.

By December, Allmendinger was again ready to eliminate the beachhead at Eltigen. Here, the Romanian *6th Cavalry Division* under General Corneliu Teodorini, reinforced with two full battalions from the *3d Mountain Division*, the *52d Tank Company*, and two batteries of German *Sturmgeschütz* assault guns, plus 12 artillery batteries, received the assignment to wipe the beachhead.[82] The beachhead was viewed as a "plum ripe for the picking."[83]

On 4 December, Teodorini launched his assault. The cavalry squadrons conducted holding or fixing attacks in the south while the mountain troops and assault guns attacked from the west. The trapped Soviets fought hard, inflicting significant casualties on the Romanians and their German allies. On 6 December, the North Caucasus Front issued the order to evacuate the Eltigen bridgehead.[84] By 7 December, the bridgehead was in Romanian hands. The Soviets "left stranded three large harbor boats (with deck), one large launch, three gunboats, and 24 landing boats, seven large and seventeen small fishing cutters, all armed, but without engines."[85] These were heavy losses for an already weak Soviet naval force in the Black and Azov seas. During their assault, the Romanians suffered some 886 casualties but killed more than 1,200 Soviets and captured another 1,570, of whom one-half were wounded. Disconcertingly, approximately 800 Soviet troops succeeded in breaking out to the north in an attempt to reach the Kerch beachhead and another 1,000 were evacuated by the Azov Flotilla.[86] In the process, they overran German artillery positions atop Mount Mithridat and had to be driven out by Romanian troops of the *3d Mountain Division*. Only by 11 December was the mountain retaken, with another 1,500 Soviets captured by the Romanians.[87]

From here, a stalemate settled over the peninsula as both sides figured out how to resume the fight in the spring. Almost 190,000 German and Romanian troops dug in across Crimea. In eastern Crimea, one German infantry division, along with the Romanian *3d* and *6th Cavalry Divisions* held the line while *1st* and *2d Mountain Divisions* watched the southern mountains and coast line.[88]

---

[82] Mark Axworthy, Cornel Scafeș, and Cristian Craciunoiu, *Third Axis Fourth Ally: Romanian Armed Forces in the European War, 1941-1945* (London: Arms and Armour, 1995), 131.
[83] Quoted in Axworthy, Scafeș, and Craciunoiu, *Third Axis Fourth Ally*, 131. Original source unknown.
[84] Ivanov, "Crimean Landing," 59.
[85] Ruge, *The Soviets as Naval Opponents*, 117-18.
[86] Zonin, *Loyalty to the Ocean*, 92.
[87] Axworthy, Scafeș, and Craciunoiu, *Third Axis Fourth Ally*, 131; and Harward, *Romania's Holy War*, 189.
[88] Harward, *Romania's Holy War*, 189.

By the end of December, some 29,000 Soviet troops had become casualties. Numerous others had received awards, including four Crimean Tatars whose fate would be less than glorious several months later at the hands of Levrenty Beria's People's Commissariat for Internal Affairs (NKVD). In contrast, the Germans and Romanians suffered approximately one-half that number of casualties. In the words of Alexander Hill, "at least the gap between enemy and Soviet casualties was far smaller than was typically the case earlier in the war, where total German and Rumanian losses were perhaps in the region of 14,000 for a similar period and where both figures cover the period of the most intensive fighting."[89]

As fighting and the stalemate dragged on, German and Romanian troops increasingly found themselves trapped. As early as mid-October, Romanian desertions reached a point where a regiment was established to " 'reeducate' first-time deserters for two months before sending them back to their units."[90] Soviet propaganda targeted the Romanians in particular, telling them "Your fate in Crimea is sealed, do not believe the Germans."[91] After the Soviet landings, some Romanians around Simferopol deserted and became bandits.[92]

SOVIET SHORTCOMINGS

Doctrinally, the Soviets suffered from a lack of dedicated and coordinated planning model for amphibious operations. Prewar Soviet doctrine, specifically the *Instructions on Marine Operations (1940)*, was still in place in 1943 and did not divide an amphibious operation into stages.[93] While landings like Kerch-Eltigen were by nature phased operations, the planning documents may not have reflected that, which would have had an adverse impact on the operation.[94] New doctrine was forthcoming but it is unclear if Soviet planners had access to it in October 1943.

Naval doctrine, in the form of the *1943 Instructions for Joint Action by the Ground Forces, the Navy, and Military River Flotillas*, divided landing operations into six stages:
- Preparation for the operation,
- Embarkation of the landing force,
- Transit at sea,
- Fighting for landing and landing,

---

[89] Alexander Hill, *The Red Army and the Second World War* (Cambridge, UK: Cambridge University Press, 2019), 471, https://doi.org/10.1017/9781139107785.
[90] Harward, *Romania's Holy War*, 187.
[91] Harward, *Romania's Holy War*, 187.
[92] Harward, *Romania's Holy War*, 188.
[93] Yeshchenko, "Analyzing the Practice of Conducting Amphibious Assault Operations during the Great Patriotic War (1941–1945)," 56.
[94] Yeshchenko, "Analyzing the Practice of Conducting Amphibious Assault Operations during the Great Patriotic War (1941–1945)," 56.

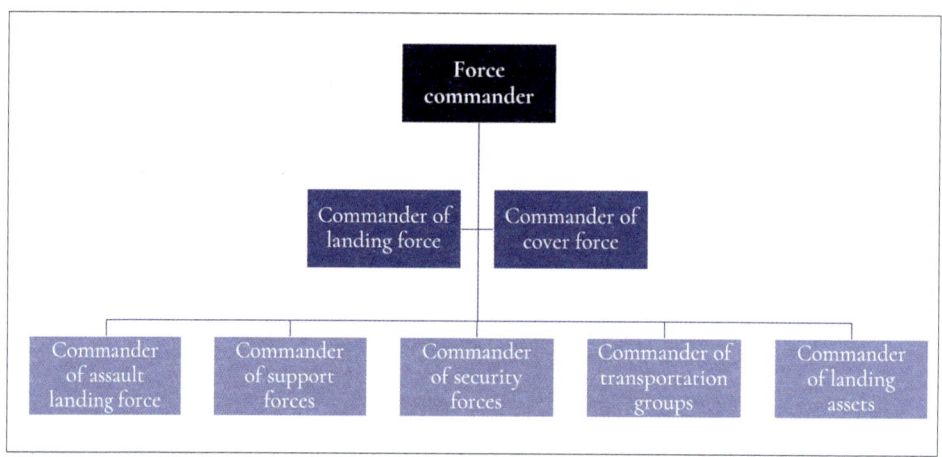

FIGURE 1
Comand and control relationships.
Source: courtesy of the author, adapted by MCUP

- Fulfillment of the mission of the landing ashore, and
- Curtailment of the operation or regrouping for the next operation.[95]

The assault was identified as the most important stage and was further broken down into a series of tactical stages:
- Deploying the landing forces,
- Landing the lead echelon,
- Landing subsequent assault elements comprising the main landing forces dropped in the specified area during the battle,
- Waging the battle ashore to seize the beachhead, and
- Landing logistical support units.[96]

In *Instructions on Marine Operations (1940)*, command and control relationships were identified roughly (figure 1).[97]

For Black Sea operations, likely including Kerch-Eltigen, the inclusion of air force units complicated the chain of command and made achieving unity of command more difficult. The lack of unity "caused inadequate interaction organization, made

---

[95] Yeshchenko, "Analyzing the Practice of Conducting Amphibious Assault Operations during the Great Patriotic War (1941-1945)," 56. These are similar to prewar doctrine as described by Achkasov and Pavlovich, *Soviet Naval Operations in the Great Patriotic War, 1941-1945*, 97.
[96] Achkasov and Pavlovich, *Soviet Naval Operations in the Great Patriotic War, 1941-1945*, 98.
[97] Based on Yeshchenko, "Analyzing the Practice of Conducting Amphibious Assault Operations during the Great Patriotic War (1941-1945)," 57.

impossible a rapid response to situation changes, and reduced the degree of responsibility in the operation."[98]

Additional tasks, including hydrology and meteorology, antitorpedo defenses, and operational deception and security operations, were also required to support landing operations. During Kerch-Eltigen, they were accomplished to varying levels of success.

POSTWAR REFLECTIONS

As German and Romanian soldiers held the line in eastern Crimea, trapping the Soviet amphibious force, Axis leaders identified their overall position as dire. Hitler called Crimea a "second Stalingrad" and believed holding it was a strategic necessity.[99] This belief, however, was likely of little consolation to the troops isolated on the peninsula.

On 8 April 1944, the Stavka launched the third of 10 blows in 1944 when Soviet troops attacked the German divisions at the Perekop Isthmus.[100] The landing at Kerch-Eltigen was a lengthy and costly stalemate that was ultimately broken when the 4th Ukrainian Front pushed into Crimea across the Perekop Isthmus and through the Syvash Sea from the north. The massive and unexpected assault threw the Axis off balance and allowed Petrov's Independent Coastal Army to drive out of their beachheads. In the preceding months, Gorshkov's Azov Flotilla landed approximately 240,000 troops with artillery, armored vehicles, and equipment.[101] With the launch of the third blow, Axis troops across Crimea "began a pell-mell, every-man-for himself race to Sevastopol."[102] On 11 April, Kerch was liberated. By 12 May, all of Crimea was in Soviet hands.

Performance by the Black Sea Fleet was lackluster at best and prewar doctrine, a lack of equipment, and limited specialized troops all hampered the ability of the Soviets to exploit their strength in eastern Crimea. In *The Seapower of the State*, Gorshkov wrote:

> *In the course of the defensive and offensive operations in maritime areas the Soviet fleet, using battleships and ships ill-suited for the landing of troops, put ashore in sea landings over 250,000 men with technical supplies and arms, or some 30 troop divisions. On average the fleet every fortnight of the war disembarked one landing force. At the same time active operations did not allow the Germans to stage a single landing*

---

[98] Yeshchenko, "Analyzing the Practice of Conducting Amphibious Assault Operations during the Great Patriotic War (1941–1945)," 58.
[99] Warlimont, *Inside Hitler's Headquarters, 1939–1945*, 394.
[100] The term was first used in Stalin's speech on the "27th Anniversary of the Great October Socialist Revolution." For more, see Hill, *The Red Army and the Second World War*, chap. 21.
[101] Norman Polmar, Thomas A. Brooks, and George Federoff, *Admiral Gorshkov: The Man Who Challenged the U.S. Navy* (Annapolis, MD: Naval Institute Press, 2019), 71.
[102] Harward, *Romania's Holy War*, 195.

on our coast although they possessed specially constructed landing craft and had the experience of the successful conduct of such operations in the Western European theatre.[103]

Gorshkov's experience was unparalleled within the Soviet Navy, with approximately one-third of all Soviets troops landed during the war doing so in landings he commanded.[104]

Soviet naval experience in the Great Patriotic War was "almost exclusively" limited as the Soviet Navy was " 'the faithful handmaiden' of the Army ground forces."[105] Perhaps nowhere was this more evident than around Kerch-Eltigen in 1943-44. In a postwar analysis, Admiral of the Fleet Ivan Isakov remarked:

Throughout the war, the enemy was constantly menaced by our landing forces at various sectors of the Black Sea coast . . . the Soviet Black Sea Fleet carried out extensive amphibious operations, and frequently and successfully landed diversionist, tactical and operational forces from the sea. This presented a constant threat to the enemy's flanks and rear. . . . Thus operations by our fleet riveted the enemy to the coast and paralyzed large bodies of his men which otherwise could have been hurled into action at the front.[106]

The landings at Kerch-Eltigen in late 1943 certainly pinned down large bodies of Axis troops but also a sizeable portion of the Soviet Black Sea Fleet and the Independent Coastal Army.

---

[103] S. G. Gorshkov, *The Sea Power of the State* (Oxford, UK: Pergamon Press, 1979), 147.
[104] Polmar, Brooks, and Federoff, *Admiral Gorshkov*, 73.
[105] Robert Waring Herrick, *Soviet Naval Theory and Policy: Gorshkov's Inheritance* (Newport, RI: Naval War College Press, 1988), 156.
[106] Quoted in Garthoff, *Soviet Military Doctrine*, 372.

CHAPTER SIXTEEN

Not a Carbon Copy of the U.S. Marine Corps

The Development of the People's Liberation Army Navy Marine Corps since 1979 and What that Means for the Chinese Power Project in the Pacific and Beyond

Edward Salo, PhD

In 2018, the movie *Operation Red Sea* arrived at the theaters in China. Coproduced by the People's Liberation Army Navy (PLAN), the film offered a fictionalized account of the 2015 Chinese military forces' rescue of Chinese and foreign nationals from pirates in Yemen. The director of the film, Dante Lam, commented that the PLAN happily collaborated with the movie's filming and provided all the assistance he required. The Chinese government-run *Global Times* called *Operation Red Sea*, "A patriotic movie about Chinese marines carrying out a daring rescue mission."[1] *Global Times* quoted a "military insider" who commented that the movie was a success because it "showcased the Chinese Navy's capabilities."[2] Li Jie, a Chinese naval expert, suggested that the movie offered an opportunity for the Chinese Navy to show off its overall maritime force that "is even better than the U.S. in some areas."[3] Much like *Top Gun* in the 1980s, *Operation Red Sea* was a commercial film with a strong pro-military propagandistic slant. The movie was designed to highlight the skills and abilities of the different areas of the PLAN. One of those areas were the Chinese Navy's Marines, who were the central heroes of the movie.[4]

---

[1] "China's Military Finances Blockbuster Movie about Heroic Chinese Marines in Yemen," Chinamil.com, 17 February 2018.
[2] "China's Military Finances Blockbuster Movie about Heroic Chinese Marines in Yemen."
[3] "China's Military Finances Blockbuster Movie about Heroic Chinese Marines in Yemen."
[4] *Operation Red Sea*, directed by Dante Lam, starring Zhang Yi, Huang Jingyu, Hai Qing, Du Jiang and Jiang Luxia (Beijing: Bona Film Group, 2018).

The People's Liberation Army Navy Marine Corps (PLANMC) serves as the marine corps of the PLA, and one of the sub-branches of the PLAN.[5] Much like U.S. Marine Corps across the globe, the PLANMC is China's primary force to conduct amphibious warfare and expeditionary operations, but that was not always the military mission of the force. The original PLAN Marine Corps was a single regiment established in 1953 for the specific mission of providing a force to support Communist amphibious operations against islands controlled by the Chinese Nationalist forces. It quickly grew to a force of eight divisions with more than 110,000 members, but the PLAN disbanded the units in 1957 when the Communist leadership walked back their plans to invade Taiwan. For most of the 1960s and the 1970s, the PLA maintained several army units trained in amphibious warfare and small naval infantry units to fulfill those missions for the PLAN.[6]

However, after the Chinese military experienced less than stellar success in the Paracel Islands conflict with South Vietnam, the Chinese military began to reexamine its organizational structure, as well as doctrine.[7] In 1979, the Central Military Commission of China reestablished the PLANMC, placing the 1st Marine Brigade in Hainan, to ensure the new force's ability to bolster the PLA's force projection in the South China Sea. Since its reconstitution in the 1980s, the force has grown as China focused on naval power projection in the Pacific and other areas.[8]

The major question is how does the PLA plan to use their marine corps in future military operations? Is the PLANMC going to be a rapid expeditionary force like the U.S. Marine Corps? Will the PLANMC be responsible for an amphibious invasion of islands as part of a larger military operation? Or do they have another use for the force?

This chapter examines the history of the PLANMC and places its growth and combat role into the model of the Chinese Navy established by Dr. Toshi Yoshihara and James R. Holmes in *Red Star Over the Pacific: China's Rise and the Challenge to U.S. Maritime Strategy*.[9] This chapter explores the development of the Chinese Marine Corps (i.e., equipment, size of the force, and doctrine) and what capabilities the PLANMC will provide the PLA in the future. The author focuses on the development of the force since 1979 as a way to forecast how the PLA will use them in future operations in both war and peace.

---

[5] Within the Chinese military organization, all branches of the military align under the PLA. The PLA consists of five service branches: the ground force, navy, air force, rocket force, and strategic support force.
[6] Dennis J. Blasko, "China's Marines: Less Is More," *China Brief* 10, no. 24 (December 2010).
[7] Toshi Yoshihara, "The 1974 Paracels Sea Battle: A Campaign Appraisal," *Naval War College Review* 69, no. 2 (2016): 41–65.
[8] Blasko, "China's Marines."
[9] Toshi Yoshihara and James R. Holmes, *Red Star Over the Pacific: China's Rise and the Challenge to U.S. Maritime Strategy* (Annapolis, MD: Naval Institute Press, 2018).

THE ORIGINAL PLAN MARINE CORPS, 1950–2016

In 1949, the People's Liberation Army defeated the Nationalist forces and forced them off the mainland, but the Nationalists maintained control of several large islands off the coast, including Formosa (Taiwan) and Hainan, as well as smaller islands. Realizing that the islands would be targets of attack, the Nationalist forces fortified the areas and prepared for invasion from the sea. In 1950, the PLA began amphibious operations against the Nationalist-controlled islands, and the PLA did capture Hainan island in the South China Sea. However, Nationalist forces repulsed the PLA landing in Kinmen and stopped the initial attacks on Taiwan.[10]

Recognizing the errors in their earlier amphibious operations, the People's Liberation Army created its first PLANMC regiment as part of plans to invade Taiwan in April 1953. The initial unit expanded with the addition of other army units, and on 9 December 1954, formed the 1st Marine Division. Soon, the new PLANMC division was deployed successfully during the battle for Yijiangshan Islands during the First Taiwan Strait Crisis (1954–55).[11] After the Korean conflict ended, the PLA expanded the marine corps into eight divisions of approximately 110,000 troops and opened an amphibious warfare school in Fujian in 1955 where the PLANMC and other army units trained for the future amphibious landings and operations.[12]

In October 1957, the Chinese Central Military Commission deactivated the PLANMC after the government decided that invasion of Taiwan was not an immediate goal of the Communist nation. However, the PLAN did maintain several infantry and amphibious tank regiments necessary for basic naval operations. Furthermore, in the 1960s and 1970s, the PLA earmarked an army division for each of the PLAN's fleets to be trained and equipped to conduct amphibious operations if necessary. This strategy soon proved to be inadequate for their needs.[13]

In 1974, a poor performance by the PLA Army against a numerically inferior South Vietnamese force during the Paracel Islands campaign caused the Central Military Commission to reassess the need for a marine corps force in the PLA Navy. On 5 May 1980, the PLAN reconstituted the 1st Marine Brigade, subordinate to the South Sea Fleet (SSF) of the PLA Navy, on Hainan. The force later relocated to the Zhanjiang area in Guangdong Province on the mainland and would serve as the only marine corps unit in the PLAN for almost two decades.[14]

---

[10] Sebastien Roblin, "When America Threatened to Nuke China: The Battle of Yijiangshan Island," *Buzz* (blog), *National Interest*, 19 February 2017.
[11] Christopher P. Isajiw, "China's PLA Marines: An Emerging Force," *Diplomat*, 17 October 2013; and Blasko, "China's Marines."
[12] John Pike, "People's Liberation Navy-Marine Corps-Organizational Structure," Global Security, 1 July 2022.
[13] Pike, "People's Liberation Navy-Marine Corps-Organizational Structure."
[14] Blasko, "China's Marines."

In 1997, as part of a larger reduction in forces and reorganization of the military, the PLA Army's 164th Division, which was also stationed in the vicinity of Zhanjiang, was converted into the 164th Marine Brigade and also placed under the control of the SSF for a force of 10,000-12,000 troops. Retired Army lieutenant colonel Dennis J. Blasko, an expert on the Chinese military and former defense department attaché to Beijing, commented that the location of the two brigades and their training illustrated the PLANMC's primary area of responsibility was the South China Sea.[15]

The primary mission of the PLANMC during this period was to defend the PLAN mainland bases as well as Chinese bases in the Paracel and Spratly island chains. While the Chinese media referred to the PLANMC as the national rapid-reaction force, it appears that this new mission meant the support of natural disaster response, probably because of the small size and ease of deployment of the PLANMC.[16] Steve Ostrosky argued that the PLANMC's focus on natural disaster response was good for the image of the PLANMC, but did little to prepare them for combat. Additionally, the PLANMC remained about the same size from 1980 until 2016.[17]

THE EXPANSION AND REORGANIZATION OF PLANMC, 2017–PRESENT

In 2016, as part of a larger reorganization of the PLA, Chinese leaders announced the PLANMC was to undergo a significant expansion, growing from two to eight brigades. The U.S. Department of Defense's *Annual Report to Congress: Military and Security Developments Involving the People's Republic of China 2018* described the importance of the expansion of the PLANMC as "one of the most significant PLAN structural changes in 2017 was the expansion of the PLAN Marine Corps (PLANMC). . . . By 2020, the PLANMC will consist of 7 brigades, may have more than 30,000 personnel, and will expand its mission to include expeditionary operations on foreign soil."[18] It appears that the newly expanded PLANMC could serve as "the core of the PLA's future expeditionary force"; however, it will take many years before the PLANMC is capable of large-scale amphibious operations.[19]

For the expansion, the PLANMC grew from the two existing brigades to eight brigades by creating four new maneuver brigades (two each in the Northern Theater Command and Eastern Theater Command); expanding the former PLAN "Jiaolong" commando regiment to a brigade and moving it from the PLAN to the PLANMC;

---

[15] Blasko, "China's Marines."
[16] Steve Ostrosky, "The PLANMC: Will the PLA Marine Corps Become Its Own Service," *Marine Corps Gazette* (September 2019): 56.
[17] Ostrosky, "The PLANMC," 56.
[18] *Annual Report to Congress: Military and Security Developments Involving the People's Republic of China 2018* (Washington, DC: Office of the Secretary of Defense, 2018), 28.
[19] Cristina L. Garafola, *The PLA Airborne Corps in a Joint Island Landing Campaign*, China Maritime Report No. 19 (Newport, RI: China Maritime Studies Institute, U.S Naval War College, 2022), 1.

Table 1. The PLANMC brigades

Theater command	Name	Location
Southern	1st Marine Brigade	Zhanjiang, Guangdong
Southern	2d Marine Brigade	Zhanjiang, Guangdong
Southern	Special Operations Brigade	Sanya, Hainan
Eastern	3d Marine Brigade	Jinjiang, Fujian
Eastern	4th Marine Brigade	Jieyang, Guangdong
Northern	5th Marine Brigade	Qingdao, Shandong
Northern	6th Marine Brigade	Qingdao, Yantai, Shandong
Northern	Naval Shipborne Aviation Brigade	Zhucheng, Shandong

Source: Conor Kennedy, *The New Chinese Marine Corps: A "Strategic Dagger" in a Cross-Strait Invasion*, China Maritime Report no. 15 (Newport, RI: China Maritime Studies Institute, U.S Naval War College, 2021), 4.

and finally, creating a PLANMC aviation brigade in Shandong that can support all of the PLANMC brigades (table 1).[20]

To further suggest that the PLANMC is not a current threat to Taiwan, the PLAN based the majority of the expanded PLANMC in the Norther and Southern Theater Commands, and not with the Eastern Theater Command, which is directly across from Taiwan.[21] Alan Burns, a China expert with CNA believes the expansion of the PLANMC also includes changes to the PLANMC's command structure, including the establishment of a single headquarters with a single commander:

> *Previously, the two Chinese marine brigades were subordinate to the PLA Navy South Sea Fleet. Now, the PLANMC commander will likely be subordinate to the PLA Navy headquarters directly, which indicates a significant increase in status and the evolution of the PLANMC into something greater than just one of five equal branches of the PLA Navy. This could indicate that the types of missions that Chinese marines are suited to perform are becoming a higher priority for Beijing.*[22]

In addition to the PLANMC being expanded, the Chinese also reorganized the PLANMC brigades to better suit them for expeditionary operations. Previously, the brigades each contained several infantry battalions and an amphibious armor regiment as the primary assault forces. After 2017, the PLANMC used the combined arms

---

[20] Kennedy, *The New Chinese Marine Corps*, 1.
[21] Kris Osborn, "DIA Report on China Threat: Stealth Fighters, Carrier-Killer Missiles and ICBMs," Warrior Maven: Center for Military Modernization, 24 January 2019.
[22] Kris Osborn, "China Is Tripling the Size of Its Marine Corps," Warrior Maven: Center for Military Modernization, 29 August 2018.

battalion organizational chart, similar to the PLA's table of command, for the new battalions. The typical PLANMC brigade contains nine battalions:[23]

- Amphibious mechanized infantry, 1st Battalion (两栖机械化步兵一营)
- Amphibious mechanized infantry, 2d Battalion (两栖机械化步兵二营)
- Light mechanized infantry, 3d Battalion (轻型机械化步兵营)
- Air assault infantry battalion (空中突击步兵营)
- Reconnaissance battalion (侦察营)
- Artillery battalion (炮兵营)
- Air defense battalion (防空营)
- Operational support battalion (作战支援营)
- Service support battalion (勤务保障营)

Each of the amphibious mechanized infantry battalions contained:
- Four mechanized infantry companies (机步连)
- Firepower company (火力连)
- Reconnaissance platoon (侦察排)
- Air defense element (防空分队)
- Artillery element (炮兵分队)
- Engineer element (工兵分队)
- Repair team (修理小队)

SPECIAL OPERATIONS FORCES IN THE PLANMC

As part of the expansion, the PLANMC also gained a special operations force, the "Jiaolong" Commando Unit (a.k.a. Water Dragons), which is located in Hainan. Additionally, the individual PLANMC brigades have reconnaissance battalions that can carry out special operations like missions but are not considered a special operations force.[24] Founded in 2002, the PLAN created the Jiaolong Commando Unit to serve as a counter to the U.S. Navy's SEALs or the British Royal Navy's Special Boat Squadron. The PLANMC describes the Jiaolong Commando Unit as an "elite special operations force of the People's Liberation Army Navy," and train in "backwater infiltration, jungle search, and urban counter-terrorism among other things." The Jiaolong Commandos might be the best-known Chinese special operation forces because of their success in recapturing a ship from pirates in the Gulf of Aden and the evacuation of civilians from war-torn Yemen in 2018.[25] Gong Kaifeng, a squadron officer of the Jiaolong Commandos commented on the importance of the Water Dragons: "Our

---

[23] Blasko, "China's Marines."

[24] John Chen and Joel Wuthnow, *Chinese Special Operations in a Large-Scale Island Landing*, China Maritime Report no. 18 (Newport, RI: China Maritime Studies Institute, U.S Naval War College, 2022), 6–7.

[25] Stavros Atlamazoglou, "How China's Special Forces Stack Up against the US's Special Operators," *Business Insider*, 1 December 2020.

special operations force is the vanguard in joint operations. We should be the point of the sword in joint operations, to strike terror into the enemy."[26]

Like many maritime special operations forces, the PLANMC Jiaolong Commandos train for more than a year in such skills as parachuting, rappelling, land and sea navigation, special vehicle driving, search and seizure, demolition, and hand-to-hand combat skills, as well as reconnaissance skills like map identification, photography and video recording, and encryption protocols for transmitting intelligence.[27] The addition of the Jiaolong Commandos changes the way the PLANMC can operate on the battlefield.

In December 2020, PLANMC forces conducted a combined arms island landing and seizure exercise involving squad-size mechanized infantry units. Jiaolong Commandos used mine-clearing line charges to destroy landing obstacles, while Jiaolong sniper teams wreaked havoc on the enemy. Also, Jiaolong commandos worked with the conventional forces to destroy vital enemy targets and control the battlefield.[28] This exercise showed how the Jiaolong would support conventional operations, as well as their unconventional warfare mission.

THE MISSION OF THE PLANMC

With the expansion and reorganization of the PLANMC, it is necessary to examine their new mission. However, the PLA's amphibious forces are split between the amphibious combined arms brigades in the army and the PLANMC.[29] Therefore, any amphibious operations would likely employ forces from both services. The U.S. Department of Defense's assessment of the Chinese military in 2020 and 2021 states that

> both PLAA and PLANMC units equipped for amphibious operations conduct regular company- to battalion-level amphibious training exercises, and the PLA continues to integrate aerial insertion training into larger exercises. . . . The PLA rarely conducts amphibious exercises involving echelons above a battalion, although both PLAA and PLANMC units have emphasized the development of combined-arms battalion formations since 2012.[30]

In fact, during a discussion of Chinese military tactics, the U.S. Army describes the PLANMC in terms of its similarity to the U.S. Marine Corps:

> *The People's Liberation Army Navy Marine Corps (PLANMC) is the PLA's expeditionary amphibious warfare capability. Like the U.S. Marine Corps, it falls under administrative control of the navy, but it is equipped and organized in a manner*

---

[26] Han Bin and Huang Xiaodong, "The Jiaolong Commandos," CGTN News, 15 April 2019.
[27] Chen and Wuthnow, *Chinese Special Operations in a Large-Scale Island Landing*, 11.
[28] Chen and Wuthnow, *Chinese Special Operations in a Large-Scale Island Landing*, 13.
[29] Kennedy, *The New Chinese Marine Corps*, 1.
[30] Dennis J. Blasko, *The PLA Army Amphibious Force*, China Maritime Report no. 20 (Newport, RI: China Maritime Studies Institute, U.S Naval War College, 2022), 2.

similar to that of the army. Unlike the U.S. Marine Corps, however, the PLANMC does not have the PLA's heavy amphibious warfare mission—this belongs to the People's Liberation Army Army (PLAA). Instead, the PLANMC should be viewed as a light and strategically mobile force built to conduct expeditionary warfare missions away from Chinese shores.[31]

Alan Burns agrees with this argument that "the PLANMC has been developing into a rapid response force that could be tasked with conducting a variety of expeditionary missions to defend China's overseas interests. . . . While it looks like the PLANMC has increased in status . . . it is still not likely to reach quite the position that the USMC has in the U.S. military."[32] However, it appears that the PLA Army would conduct major amphibious operations such as assaulting Taiwan, while the PLANMC would be tasked with smaller operations such as seizing small islands.[33]

TRAINING THE PLANMC

In addition to expanding the force, PLANMC has also expanded its training regime to become more of an expeditionary force. Chinese media have highlighted that the PLANMC has been active in becoming "an all-weather, multirole special amphibious fighting force able to fight in highlands, jungles, water, and other extreme environments."[34]

In addition to the training for amphibious landings, the PLANMC has been active in developing its vertical envelopment capabilities as well as training in winter, desert, mountain, and jungle environments. For example, PLANMC Jiaolong Commandos have been experimenting with long-range parachuting infiltration methods, very similar to U.S. Navy SEALs. The PLANMC is also developing its own air assault units.[35] Likewise, PLANMC is also working on developing better doctrine for air assault operations including "overcoming difficulties such as obstacles in low-altitude flight and few reference objects in night flight . . . [avoiding] enemy radar reconnaissance and . . . anti-aircraft fire power with their all-weather combat capability."[36] In addition to training its force, since 2005, PLANMC has been active in overseas military exercises with Russia and other nations. In addition, PLANMC has been an active part of China's antipiracy patrols in the Gulf of Aden.[37]

---

[31] *Chinese Tactics*, ATP 7-100.3 (Washington, DC: Department of the Army, 2021), 3-4.
[32] Osborn, "China Is Tripling the Size of Its Marine Corps."
[33] Osborn, "China Is Tripling the Size of Its Marine Corps."
[34] Minnie Chan, "Beijing Marks 73rd Navy Anniversary with Video of Island-Control Drills," *South China Morning Post*, 21 April 2022.
[35] Garafola, *The PLA Airborne Corps in a Joint Island Landing Campaign*, 14.
[36] "Marine Commandos Conduct Armed Parachuting Training," China Military, 5 March 2022; and "PLA Navy Marine Corps Organizes Air Assault Training," China Military, 27 May 2022.
[37] Osborn, "China Is Tripling the Size of Its Marine Corps."

EQUIPMENT AND VEHICLES OF THE PLANMC

Since the PLANMC has shifted to a combined arms battalion, they also require vehicles and equipment to support those forces.[38] The PLANMC uses standard Chinese military equipment (i.e., small arms, field gear, etc.) and weapons that are similar to the PLA Army fields. The marines have had their own ocean pattern camouflage uniform since 2013.[39] Because PLANMC now has combined arms, it also has been improving its armored vehicle components with tanks and infantry fighting vehicles.[40] The PLANMC fields the Norinco Type 63 amphibious light tank, which was an army design but was capable of some small-scale amphibious operations. The PLANMC later fielded an upgraded Type 63A, which could operate in the open ocean. Both designs lacked heavy armor and were vulnerable to antitank guided missiles and heavy armor. The Type 63 is being replaced by the ZLT-05 amphibious tank, a combination of infantry fighting vehicle (IFV), tank, and assault gun. Still lacking in armor, it has superior maneuverability and weaponry. Alongside, the ZLT-05, is the ZBD-05 IFV, PLANMC's amphibious fighting vehicle that carries eight passengers and employs a 30-mm autocannon, rather than the 105-mm main gun. The PLANMC also currently fields standard PLA transport and attack helicopters. Most experts assume that as the PLANMC continues to expand, it will begin to get specialized helicopters and other vehicles that are designed for its specific missions.[41]

AMPHIBIOUS TRANSPORT FOR THE PLANMC

Of course, any marine corps requires amphibious transport ships to transport the force to the beaches. In 2000, the PLAN only had approximately 20 small landing craft (LST type), which indicated it had no real way of conducting large-scale amphibious operations. To support the expansion of the PLANMC, the PLAN has focused an active ship-building effort on the Type 071 Landing Platform, Dock (LPD) and the Type 075 Landing Helicopter, Dock (LHD).

First built in 2002, the PLAN constructed eight of the Type 071 LPDs between 2002 and 2019. The ships have a length of 210 meters, a beam of 28 meters, and a draft of 7 meters. The amphibious warfare ships support the necessary command and control facilities to direct amphibious operations and can field a force that includes:

- amphibious assault vehicles, including the ZBD-05 amphibious IFV and the ZTD-05 amphibious assault tracked armored vehicle

---

[38] Gabriel Dominguez, Samuel Cranny-Evans, and J. Michael Cole, "PLANMC May Be Re-Equipping for Combined Arms, Multidomain Operations," Janes, 3 June 2021.
[39] "People's Liberation Army Equipment and Gear," Far East Tactical, accessed 6 October 2023.
[40] Gordon Arthur, "PLA Marines Bulk Up with Tanks," Shephard Media, 9 June 2021.
[41] Arthur, "PLA Marines Bulk Up with Tanks."

- four Landing Craft, Air Cushion (LCAC);
- two Changhe Z-8 (SA 321 Super Frelon) transport helicopters; and
- a marine battalion of up to 800 personnel and their associated equipment and supplies.[42]

While the Type 071 LPD provided the PLANMC the ability to conduct small-scale expeditionary missions, it did not offer the necessary capabilities for a large-scale beach assault.

In 2011, the Chinese began development on the Type 075 LHD similar to the U.S. Navy's *Wasp*-class landing helicopter dock, displacing more than 30,000 tons. Analysts suggested that the Type 075 was "likely to increase the 'vertical' amphibious assault capability with the very mountainous East Coast of Taiwan in mind."[43] The new ship would be able to handle up to 30 helicopters, along with ship-to-shore amphibious craft and a complement of marines.[44] By 2021, the PLAN had launched three Type 075 ships, due to the fact that the first Type 075 was constructed in record time and the PLAN averaged one ship every six months. It appears that the PLAN intends to construct eight Type 075 LHDs, as well as introduce the new Type 076 helicopter carriers that would have a full flight deck and operate UAVs.[45]

In addition to the large amphibious ships, PLANMC fields two types of LCAC: the Type 726 (*Yuyi*-class) and the *Zubr*-class.
- A *Zubr*-class LCAC can carry up to 3 main battle tanks, 10 armored vehicles, or 500 marines.
- A Type 726 LCAC can carry one main battle tank or 80 Marines.

Coupled with the large amphibious warfare ships, the PLANMC should be able to move troops and armored vehicles to shore with ease.[46]

The Chinese military contends that the "new generation of large amphibious assault vessels . . . will strengthen the navy as it plays a more dominant role in projecting the nation's power overseas."[47] The Department of Defense also suggests that "the

---

[42] "Type 071 Landing Platform Dock (LPD)," *Naval Technology*, 9 February 2016.
[43] Xavier Vavasseur, "China: End of the Type 071 LPD Program, Start of the Type 075 LHD One?" *Naval News*, 5 August 2019.
[44] "Photos: China's First Type 075 Amphib Heads out on Sea Trials," *Maritime Executive*, 5 August 2020.
[45] Andrew Tate, "China Launches Third Type 075 LHD for PLAN," Janes.com, 29 January 2021; Xavier Vavasseur, "China's 2nd Type 075 LHD Guangxi 广西 Commissioned with PLAN," *Naval News*, 30 December 2021; and H. I. Sutton, "Stealth UAVs Could Give China's Type-076 Assault Carrier More Firepower," *Forbes*, 23 July 2020.
[46] Capt Michael A. Hanson, USMC, "China's Marine Corps Is on the Rise," U.S. Naval Institute *Proceedings* 146, no. 4 (April 2020).
[47] Minnie Chen, "China Building Navy's Biggest Amphibious Assault Vessel," *South China Morning Post*, 29 March 2017.

PLAN's investment in LHAs signals its intent to continue to develop its expeditionary warfare capabilities."[48]

A PLACE IN THE WORLD FOR THE PLANMC

So, what is the role of the new PLANMC? Are they a copy of the U.S. Marine Corps or something different? As China expands its role in the world with bases in the Middle East and Africa, we find PLANMC forces at those bases. The PLANMC is conducting training in different climates and environments than the South China Sea, including desert, jungle, and Arctic environments. In terms of partners, the PLANMC is working with the armed forces of other nations from peer-to-peer operations with Russia to smaller nations. The U.S. Department of Defense states that

> *the PLANMC's presence in Djibouti provides the PRC with the ability to support a military response to contingencies affecting the PRC's investments and infrastructure in the region and the approximately 1 million PRC citizens in Africa and 500,000 in the Middle East. The PLANMC also embarks a contingent of marines with the PLAN's Gulf of Aden counterpiracy-focused naval escort task force that supports the PRC's trade interests. Additionally, the PLANMC supports the PRC's military diplomacy. For example, it has trained with Russian and Thai forces and participated in exchanges with the United States and Australia.*[49]

Maybe the PLANMC is becoming like the U.S. Marine Corps. It is not going to be the Marine Corps of World War II, capable of large-scale amphibious operations, but it can be used for various missions in lands far from China to serve as the "tip of the spear." Alan Burns contends that "having a marine force that can conduct expeditionary operations is one part of Beijing's efforts to build a strong military appropriate for what Chinese leaders see as China's ambitions to be a maritime great power."[50]

---

[48] Xavier Vavasseur, "US DOD's 2021 China Military Power Report: PLAN Is the Largest Navy in the World," *Naval News*, 5 November 2021.
[49] *Annual Report to Congress: Military and Security Developments Involving the People's Republic of China, 2021* (Washington, DC: Office of the Secretary of Defense, 2021), 85.
[50] Osborn, "China Is Tripling the Size of Its Marine Corps."

CONCLUSION

Timothy Heck, B. A. Friedman, and Walker D. Mills

The first volume of *On Contested Shores* was inspired by a combination of related developments. A changing global security environment required a hard look at capabilities across the U.S. military and a reevaluation of operational concepts. In part due to this reevaluation, former Commandant of the Marine Corps general David H. Berger issued his *Commandant's Planning Guidance* (2019), which kicked off a major transformation of the Marine Corps.[1] These editors felt strongly then, as we still do, that rigorous scholarship on amphibious operations, which had been dormant or secondary within the Marine Corps due to ongoing commitments to the Global War on Terrorism, is important for informing current and future developments.

At the time *On Contested Shores* released in 2020, it had been nearly 30 years since the publication of Lieutenant Colonel Merrill L. Bartlett's edited volume *Assault from the Sea: Essays on the History of Amphibious Operations*. As such, it was past time for a similar, follow-on work of scholarship with modern relevancy.[2] We took an approach similar to Bartlett, wanting to pull in a wide range of amphibious case studies and authors who would cover relevant topics beyond the classic Second World War amphib-

---

[1] Gen David H. Berger, *Commandant's Planning Guidance: 38th Commandant of the Marine Corps* (Washington, DC: Headquarters Marine Corps, 2019).
[2] LtCol Merrill L. Bartlett, USMC (Ret), *Assault from the Sea: Essays on the History of Amphibious Warfare* (Annapolis, MD: Naval Institute Press, 1993).

ious assaults in Europe and the Pacific that most readers are already familiar with, while simultaneously giving those examples their due recognition as the epitome of one type of amphibious warfare. The 23 chapters in the first volume of *On Contested Shores* covered topics that were both thematic and operational. We believe it achieved our intent of broadening the scope of English-language scholarship on amphibious operations and providing an academically rigorous foundation to inform decisions about the future of the Marine Corps, while still being accessible enough to attract a general audience and educate any interested reader. But as any author or editor realizes when undertaking a project of this scope and breadth, there was so much that we could not include or had to leave out due to space constraints and the limitation of time. Almost immediately, we became interested in publishing a second volume that continued the original intent of broadening the scholarship and filling what we still believe is a gap of non-Western examples in the English-language literature on amphibious operations.

Now, it seems somehow fitting, that four years after the release of General Berger's *Commandant's Planning Guidance* with its subsequent sea-change in how the Marine Corps views its future role, we are finishing the second volume of *On Contested Shores*. The Marine Corps is still in the thrall of the transformation kicked off in 2019 and then a year later with the publication of *Force Design 2030*. *Force Design 2030* has been the subject of major debate but the impacts and changes are immediately apparent force-wide.[3]

Outside of the Marine Corps, the global security outlook is even darker than it was in 2019. We have witnessed war in Nagorno-Karabakh, war in Ukraine, war in Ethiopia, war in Sudan, and war in Israel and Gaza, all while anxiously watching tensions ratchet up in the Strait of Taiwan and the South China Sea. Since 2019, the possibility of an invasion of Taiwan by the People's Republic of China has increasingly preoccupied much of the defense establishment for the United States and its allies. Such an invasion would be, at its core, a massive amphibious assault. Indeed, should it occur, it will likely be the largest amphibious invasion in recorded history, dwarfing even the landings in Normandy or on Iwo Jima and Okinawa in scale and overall operating area. Various staffs and headquarters should be mining the lessons of history for insight into how the People's Liberation Army (PLA) will approach executing any potential amphibious operation, and how they may be defeated.

If the United States became directly involved in any of these conflicts, even far inland, it is relatively certain that at least some of the forces deployed would be amphibious, that is deployed from naval vessels at sea. But it is also relatively certain that the Marine Corps forces within those deployments will not look like lines of amphibious assault vehicles storming a beach. Those deployments will require new and innovative approaches to amphibious warfare developed by a new generation

---

[3] Gen David H. Berger, *Force Design 2030* (Washington, DC: Headquarters Marine Corps, 2020); and Tim Barrick, "Future Wars and the Marine Corps: Asking the Right Questions," *War on the Rocks*, 1 April 2022.

of amphibious warfare practitioners. Like all forms of warfare, amphibious warfare has been forced to change over time. Little more than a decade after the famous amphibious assaults during the Second World War and only a few years after the assault at Inchon, British and French forces using similar tactics and doctrine faced major difficulties in their combined amphibious assault on Port Said, Egypt, in 1956 during the Suez Crisis. According to historian Ian Speller, "the amphibious landing at Port Said was more reminiscent of the slow, methodical approach required during World War II than the type of rapid and flexible operation that might have brought success within an acceptable timescale."[4] One participant later wrote that it was "a lash-up of half-forgotten ideas of the Second World War, more apt to an old comrades parade than to modern war."[5] While the chapters in this book might be analyses of old operations, the idea remains to prevent the next amphibious operations from looking like an "old comrades parade."

Just as the Marine Corps transforms today, practitioners of amphibious warfare around the world need to develop new approaches and operational concepts if they want to execute effective amphibious operations, which are as important as ever, and the intent of this volume is to help inform that development by providing examples and case studies from which to draw lessons and principles. We believe that both of these volumes contribute to the efforts of tacticians, planners, strategists, and policymakers alike, and also serve as a bridge to literature on amphibious operations if readers want to dig deeper or explore the literature further.

We wanted the second volume to offer the continuity of the first but also take a slightly new approach. So, we decided to organize the second volume thematically, as opposed to the largely chronological organization of the first volume. We made this change so that the reader could more easily draw conclusions from and parallels between the by-design diversity within the chapters. Readers will notice that this organization is similar to the commonly used military acronym DOTMLPF-P, which stands for doctrine, organization, training, materiel, leadership, personnel, facilities, and policy; and it is an analytical tool to examine military capability from a wholistic perspective. Here, it helped the editors organize a volume that looked at amphibious operations from a range of different angles and perspectives and is also presented in a sequence that will feel familiar to practitioners.

This organization is particularly useful for informing the ongoing transformation of the Marine Corps, which is making changes across the spectrum of DOTMLPF-P, and also for other amphibious forces around the world that are grappling with the same challenges and dilemmas surrounding amphibious warfare in the twenty-first century and doing the same thing.

---

[4] Ian Speller, "The Seaborne/Airborne Concept: Littoral Manoeuvre in the 1960s," *Journal of Strategic Studies* 29, no. 1 (February 2006): 59, https://doi.org/10.1080/01402390600566357.
[5] MajGen J. L. Moulton, "Bases or Fighting Forces?," in *Brassey's Annual: The Armed Forces Year-Book, 1964* (London: William Clowes, 1964), 149.

While the main purpose of this volume and its predecessor is to provide a more diverse resource for practitioners of amphibious warfare to inform force development, it also provides an accessible resource for armies, navies, and air forces. Armies like the U.S. Army, the Australian Army, and the Japanese Ground Self-Defense Force (JGSDF) are appropriately seeking to enhance their amphibious or maritime capabilities. Airborne forces have been used in concert with amphibious forces in a variety of conflicts since the advent of aviation, and air forces have likewise been used in support of amphibious warfare. Amphibious warfare is the oldest mission for navies, transporting land forces since at least the mysterious Sea People and the Trojan War of the Bronze Age, before ships were able to directly fight each other. Amphibious operations have always been inherently joint and multidomain operations and are often combined operations between militaries from multiple countries. The 2019 edition of *Amphibious Operations*, Joint Publication 3-02, reminds us that "amphibious operations, no matter their makeup or application, are complex and inherently joint or multi-Service."[6] The study of amphibious warfare should not be limited to just those who practice amphibious warfare, but should be undertaken by any military practitioner who might be involved in one, civilian leadership who might be involved in planning or ordering an operation, or even a general audience and interested public.

The first section, Doctrine and Logistics, has two chapters: one that discusses the intersection of geography, strategy, and logistics at Veracruz in the Mexican American War by Christopher Menking; and another that covers the development of landing craft and the resultant doctrinal evolution by Stephen Strahan. Both of these topics have continued relevance today. The second section, Technology and Innovation, is perhaps the most directly relevant to ongoing efforts. It features chapters by Douglas E. Nash Sr. and Walker D. Mills that discuss the relationship between technology and amphibious warfare in the past and future, respectively. The third section, Organization and Training, is both the largest and most diverse.

Not one of the chapters are focused on American amphibious history. Xiaobing Li, Benjamin Claremont, Isabella Ginor, and Gideon Remez focus their chapters on either the People's Liberation Army or the Soviet Red Army. Eric Sibul and David Katz focus theirs on German amphibious operations from the First World War, and Lance R. Blyth focuses his geographically, on amphibious warfare in Arctic regions.

The fourth section, Policy and Interoperability, is more focused on challenges. Darren Johnson and Shaun Mawdsley both cover historical amphibious operations from the Second World War, albeit on very different scales and in different theaters. But they help demonstrate the challenges in conducting amphibious operations between the Services or even different militaries; challenges to which there is no easy fix and that still exist today. The fifth section, Military Materiel and Personnel, offers another batch of diverse chapters by Howard Fuller, Zachary Ota, Timothy G. Heck, and Edward Salo. The chapters cover more than 100 years of amphibious history and

---

[6] *Amphibious Operations*, Joint Publication 3-02 (Washington, DC: Joint Chiefs of Staff, 2019), I-1.

present American examples along with Soviet and Chinese ones. Ota's chapter on indigenous contributions to amphibious warfare in the Pacific and Salo's chapter on the development of the PLA Navy Marine Corps are particularly relevant for military practitioners today, as the United States and its allies focus on Pacific security, but they also cover topics that are often overlooked in the literature on amphibious operations if they are covered at all.

As we wrote in the introduction, this volume is a testament to our belief that amphibious warfare in all of its forms is as relevant today as it was when it the Allies landed in Fortress Europe and campaigned across the Pacific during the Second World War. We do not need to recount all the times that amphibious warfare was declared irrelevant, outdated, or impossible only for it to return to the forefront of a conflict. However, we believe that in a world becoming increasingly dangerous and uncertain, careful study of amphibious warfare by military practitioners, academics, and informed citizens is as important as ever. Ongoing conflicts in Ukraine and Israel have maritime and amphibious dimensions, and potential conflicts in the Pacific would be inherently amphibious.[7] Further, the ongoing force design and transformation—seen most dramatically in the U.S. Marine Corps but also in the British Royal Marines, the PLA Navy Marine Corps, and other marine corps and amphibious forces around the world—gives further reason why new and updated study and analysis of amphibious operations is important, and we hope to have contributed in a small way toward that end.

---

[7] Walker D. Mills and Timothy Heck, "What Can We Learn about Amphibious Operations from a Conflict that Has Had Very Little of It? A Lot," Modern War Institute, 22 April 2022; and Walker D. Mills, "The Maritime Dimension to the Conflict in Israel," Irregular Warfare Initiative, 30 November 2023.

SELECT BIBLIOGRAPHY AND SUGGESTED FURTHER READING

OFFICIAL HISTORIES AND GOVERNMENT DOCUMENTS

2022 Report to Congress of the U.S.-China Economic and Security Review Commission, 117th Cong., 2d Sess. (November 2022).

Achkasov, V. I., and N. B. Pavlovich. *Sovetskoe voenno-morskoe iskusstvo v Velikoĭ Otechestvennoĭ voĭne* [Soviet Naval Operations in the Great Patriotic War, 1941–1945]. Translated by U.S. Naval Intelligence Command Translation Project. Annapolis, MD: Naval Institute Press, 1981.

Aerology and Amphibious Warfare: The Invasion of Sicily. NAVAER 50-30T-1. Washington, DC: Aerology Section, Chief of Naval Operations, 1944.

Alexander, Col Joseph H., USMC (Ret). *Across the Reef: The Marine Assault of Tarawa*. Marines in World War II Commemorative Series. Washington, DC: Historical Center, Headquarters Marine Corps, 1993.

The Aleutians Campaign, June 1942–August 1943: Combat Narratives. Washington, DC: Office of Naval Intelligence, U.S. Navy, 1945.

Amphibious Operations. Joint Publication 3-02. Washington, DC: Joint Chiefs of Staffs, 2019.

Anderson, Charles A. *Algeria–French Morocco: The U.S. Army Campaigns of World War II*. Washington, DC: U.S. Army Center of Military History, 2003.

Annual Report to Congress: Military and Security Developments Involving the People's Republic of China 2018. Washington, DC: Office of the Secretary of Defense, 2018.

Annual Report to Congress: Military and Security Developments Involving the People's Republic of China, 2021. Washington, DC: Office of the Secretary of Defense, 2021.

Arctic Manual. 2 vols. Washington, DC: Army Air Corps, U.S. Army, 1940.

Army Climate Strategy: Implementation Plan, Fiscal Years 2023–2027. Washington, DC: Department of the Army, 2022.

Bailey, Maj Alfred D., USMC (Ret). *Alligators, Buffaloes and Bushmasters: The History of the Development of the LVT Through World War II.* Washington, DC: History and Museums Division, Headquarters Marine Corps, 1986.

Barbey, VAdm Daniel E., USN (Ret). *MacArthur's Amphibious Navy: Seventh Amphibious Force Operations 1943–1945.* Annapolis, MD: Naval Institute Press, 1969.

Becker, Capt Marshall O. *The Army Ground Forces.* Amphibious Training Center Study no. 22. Washington, DC: Historical Section, Army Ground Forces, 1946.

Berger, Gen David H. *Commandant's Planning Guidance: 38th Commandant of the Marine Corps.* Washington, DC: Headquarters Marine Corps, 2019.

———. *Force Design 2030: Annual Update, June 2023.* Washington, DC: Headquarters Marine Corps, 2023.

———. *Sustaining the Force in the 21st Century: A Functional Concept for Future Installations and Logistics Development.* Washington, DC: Headquarters Marine Corps, 2022.

Birtle, Andrew J. *Sicily, 9 July–17 August 1943.* Washington, DC: U.S. Army Center of Military History, 2021.

Blasko, Dennis J. *The PLA Army Amphibious Force.* China Maritime Report no. 20. Newport, RI: China Maritime Studies Institute, U.S Naval War College, 2022.

Bulkley, Capt Robert J., Jr. *At Close Quarters: PT Boats in the United States Navy.* Washington, DC: Naval History Division, 1962.

Carter, Adams, trans. *Manual for Service in the Mountains.* Vienna: War Ministry, 1917 and 1918.

Chen, John, and Joel Wuthnow. *Chinese Special Operations in a Large-Scale Island Landing.* China Maritime Report no. 18. Newport, RI: China Maritime Studies Institute, U.S Naval War College, 2022.

Chinese Tactics. ATP 7-100.3. Washington, DC: Department of the Army, 2021.

Clifford, LtCol Kenneth J. USMCR. *Progress and Purpose: A Developmental History of the United States Marine Corps, 1900–1970.* Washington, DC: History and Museums Division, Headquarters Marine Corps, 1973.

Climate Action 2030. Washington, DC: Department of the Navy, 2022.

Cold Injury, Ground Type. Washington, DC: Medical Department, Office of the Surgeon General, Department of the Army, 1958.

Collyer, BGen J. J. *The Campaign in German South West Africa, 1914–1915.* London: Government Printing and Stationery Office, 1937; Nashville, TN: Battery Press, 1997 reprint.

A Concept for Stand-in Forces. Marine Corps Doctrinal Paper. Washington, DC: Headquarters Marine Corps, 2021.

Conn, Stetson, Rose C. Engelman, and Byron Fairchild. *The Western Hemisphere: Guarding the United States and Its Outposts.* United States Army in World War II, CMH Pub 4-2. Washington, DC: Center of Military History, 2000.

Craven, Wesley Frank, and James Lea Cate, eds. *The Army Air Forces in World War II,* vol. 4, *The Pacific: Guadalcanal to Saipan, August 1942 to July 1944.* Washington, DC: Government Printing Office, 1950.

Cureton, LtCol Charles H., USMCR. *U.S. Marines in the Persian Gulf, 1900–1991: With the First Marine Division in Desert Shield and Desert Storm.* Washington, DC: History and Museums Division, Headquarters Marine Corps, 1993.

Darden, Capt T. F., USN (Ret). *Historical Sketch of the Naval Administration of the Government of American Samoa, April 17, 1900–July 1, 1951.* Washington, DC: Department of the Navy, 1952.

Demma, Vincent H. *Department of the Army Historical Summary, Fiscal Year 1989.* Edited by Susan Carroll. Washington, DC: U.S. Army Center of Military History, 1998.

Department of the Air Force Climate Action Plan. Washington, DC: Department of the Air Force, 2022.

Department of the Navy. *Building the Navy's Bases in World War II: History of the Bureau of Yards and Docks and the Civil Engineer Corps, 1940–1946*, vol. 2. Washington, DC: Government Printing Office, 1947.

Documents Relating to New Zealand's Participation in the Second World War, 1939–45, vol. 3. Wellington: Department of Internal Affairs, War History Branch, 1963.

Dyer, VAdm George C. *The Amphibians Came to Conquer: The Story Admiral Richmond K. Turner*. FMFRP 12-109-I. Washington, DC: Headquarters Marine Corps, 1991.

Expeditionary Operations. Marine Corps Doctrinal Publication 3, with change 1. Washington, DC: Headquarters Marine Corps, 2018.

Fighting on Guadalcanal. FMFRP 12-110. Washington, DC: Department of the Navy, 1991.

Fuller, Capt Stephen M., USMCR, and Graham A. Cosmas. *Marines in the Dominican Republic, 1916–1924*. Washington, DC: History and Museums Division, Headquarters Marine Corps, 1974.

Garafola, Cristina L. *The PLA Airborne Corps in a Joint Island Landing Campaign*. China Maritime Report no. 19. Newport, RI: China Maritime Studies Institute, U.S Naval War College, 2022.

Garland, LtCol Albert N., and Howard McGaw Smyth. *Sicily and the Surrender of Italy: The Mediterranean Theater of Operations*. U.S. Army in World War II. Washington, DC: U.S. Army Center of Military History, 1993.

Gebhardt, Maj James F. *The Petsamo-Kirkenes Operation: Soviet Breakthrough and Pursuit in the Arctic, October 1944*. Leavenworth Paper no. 17. Leavenworth, KS: Combat Studies Institute, U.S. Army Command and General Staff College, 1989.

Gillespie, Oliver A. *The Pacific*. Official History of New Zealand in the Second World War, 1939–45. Wellington: War History Branch, Department of Internal Affairs, 1952.

Greenfield, Kent R. *Army Ground Forces and the Air-Ground Battle Team Including Organic Light Aviation*. Forces Study no. 35. Fort Monroe, VA: Historical Section, Army Ground Forces, 1948.

Guide to United States Naval Administrative Histories of World War II. Compiled by William C. Heimdahl and Edward J. Marolda. Washington, DC: Naval History Division, Department of the Navy, 1976.

Harrison, Gordon A. *Cross-Channel Attack: The European Theater of Operations*. U.S. Army in World War II. Washington DC: U.S. Army Center of Military History, 1993.

Hines, John G., and Ellis Mishulivich. *Soviet Intentions, 1965–85*, vol. 2, *Soviet Post-Cold War Testimonial Evidence*. Washington, DC: Office of Net Assessment, Department of Defense, 1993.

Hough, LtCol Frank O., Maj Verle E. Ludwig, and Henry I. Shaw Jr. *Pearl Harbor to Guadalcanal*, vol. 1, *History of U.S. Marine Corps Operations in World War II*. Washington, DC: Historical Branch, G-3 Division, Headquarters Marine Corps, 1970.

Installations and Logistics 2030. Washington, DC: Headquarters Marine Corps, 2023.

Jay, John C. *History of the Mountain Training Center*. Study no. 24. Fort Monroe, VA: Historical Section, Army Ground Forces, 1948.

Jianguo yilai Mao Zedong wengao, 1949–1976 [Mao Zedong's Manuscripts since the Founding of the State, 1949–1976], vol. 1. Beijing: CCP Central Archival and Manuscript Press, 1993.

Joint Warfighting. Joint Publication 1. Washington, DC: Joint Chiefs of Staff, 2023.

Kennedy, Conor. *The New Chinese Marine Corps: A "Strategic Dagger" in a Cross-Strait Invasion*. China Maritime Report no. 15. Newport, RI: China Maritime Studies Institute, U.S Naval War College, 2021.

Land Operations, vol. 5, *Operational Techniques under Special Conditions*, pt. 1, *Mountainous Country*. London: Ministry of Defence, 1972.

Landing Operations Doctrine. FTP 167. Washington, DC: Office of Naval Operations, Division of Fleet Training, 1938.

Landing Operations Doctrine. FTP-167, change 1. Washington, DC: Office of Naval Operations, Division of Fleet Training, U.S. Navy, 1942.

Logistics. Marine Corps Doctrinal Publication 4. Washington, DC: Headquarters Marine Corps, 2023.

Lowrey, Col Nathan S., USMCR. *U.S. Marines in Afghanistan, 2001–2002: From the Sea*. U.S. Marines in the Global War on Terrorism. Washington, DC: History Division, Headquarters Marine Corps, 2011.

MajGen David R. Nimmer Oral History, vol. 3. Washington, DC: History and Museums Division, Headquarters Marine Corps, 1970.

Messages of the Presidents of the United States, with the Correspondence, therewith Communicated, between the Secretary of War and Other Officers of the Government, on the Subject of the Mexican War. House Executive Documents no. 60, 30th Cong., 1st Sess., Serial Set 520. Washington, DC: Wendell and Van Benthuysen, 1848.

Miller, John, Jr. *Cartwheel: The Reduction of Rabaul*. U.S. Army in World War II: The War in the Pacific. Washington, DC: Office of the Chief of Military History, Department of the Army, 1959.

———. *Guadalcanal: The First Offensive*. U.S. Army in World War II: The War in the Pacific. Washington, DC: U.S. Army Center of Military History, 1995.

Ministry of Defense Climate Change and Sustainability Strategic Approach. London: UK Ministry of Defense, 2021.

Murphy, Paul J., ed. *Naval Power in Soviet Policy*. Studies in Communist Affairs vol. 2. Washington, DC: Government Printing Office, 1978.

Nalty, Bernard C. *The United States Marines in Nicaragua*. Washington, DC: Historical Branch, G-3 Division, Headquarters Marine Corps, 1958.

Naval Warfare. Naval Doctrine Publication 1. Washington, DC: U.S. Navy, Marine Corps, and Coast Guard, 2020.

Navy Readiness: Actions Needed to Maintain Viable Surge Sealift and Combat Logistics Fleets. Washington, DC: Government Accountability Office, 2017.

Operational Maneuver from the Sea. Marine Corps Concept Paper 1. Washington, DC: Headquarters Marine Corps, 1996.

Operations. Field Manual 100-5. Washington, DC: War Department, 1941.

Operations in Snow and Extreme Cold. FM 31-15. Washington, DC: Government Printing Office, 1941.

O'Rourke, Ronald. *Navy Light Amphibious Warship (LAW) Program: Background and Issues for Congress*. Washington, DC: Congressional Research Service, 2022.

Pathfinder: Recollections of Those Who Served, 1942–1971. Silver Spring, MD: Office of National Oceanic and Atmospheric Administration Corps Operations, 1994.

Prepositioning Programs Handbook: Appendix F to Marine Corps Installations & Logistics Roadmap (MCILR). Washington, DC: Headquarters Marine Corps, 2015.

Proceedings of the American-British Joint Chiefs of Staff Conferences, 2 pts. Washington, DC: Joints Chiefs of Staff, 1941.

Ramirez, Byron, and Robert J. Bunker. *Narco-Submarines: Specially Fabricated Vessels Used for Drug Smuggling Purposes*. Fort Leavenworth, KS: U.S. Army Foreign Military Studies Office, 2015.

Reddel, Col Carl W. USAF, ed. *Transformation in Soviet and Russian Military History: Proceed-*

ings of the Twelfth Military History Symposium. AFD-101028-004. Washington, DC: U.S. Air Force Academy, Office of Air Force History, U.S. Air Force, 1986.

Rentz, Maj John N. *Bougainville and the Northern Solomons*. Washington, DC: Historical Section, Division of Public Information, Headquarters Marine Corps, 1948.

Risch, Erna. *Quartermaster Support of the Army: A History of the Corps, 1775–1939*. Washington, DC: Quartermaster Historian's Office, Office of the Quartermaster General, 1962.

Santelli, Gabrille M. Neufeld. *Marines in the Mexican War*. Edited by Charles R. Smith. Washington, DC: History and Museums Division, Headquarters Marine Corps, 1991.

Second Marine Division Report on Gilbert Islands Tarawa Operation. FMFRP 12-90. Washington, DC: Headquarters Marine Corps, 1991.

Securing Defense-Critical Supply Chains: An Action Plan Developed in Response to President Biden's Executive Order 14017. Washington, DC: Department of Defense, 2022.

Shaw, Henry I. Jr., and Maj Douglas T. Kane. *History of the U.S. Marine Corps in World War II*, vol. 2, *Isolation of Rabaul*. Washington, DC: Historical Branch, G-3 Division, Headquarters Marine Corps, 1963.

The Soviet Army: Operations and Tactics. FM 100-2-1. Washington, DC: Department of the Army, 1990.

Stockman, Capt James R., USMC. *The Battle for Tarawa*. Washington, DC: Historical Section, Division of Public Information, Headquarters Marine Corps, 1947.

Strategy. Joint Doctrine Note 2-19. Washington, DC: Joint Chiefs of Staff, 2019.

Tentative Manual for Expeditionary Advanced Base Operations, 2d ed. Washington, DC: Headquarters Marine Corps, 2023.

U.S. Army Climate Strategy. Washington, DC: Department of the Army, 2022.

The U.S. Army in Multi-Domain Operations 2028. TRADOC Pamphlet 525-3-1. Washington, DC: U.S. Army, 2018.

U.S. Civilian Production Administration. *Official Munitions Production of the United States by Months, July 1, 1940–August 31, 1945*. Washington, DC: War Department Production Board, 1947.

von Tschischwitz, Erich. *The Army and Navy in the Conquest of the Baltic Islands in October 1917*. Translated by Henry Hossfield. Fort Leavenworth, KS: Command and General Staff School Press, 1933.

War Department. *German Mountain Troops*. Washington, DC: Military Intelligence Division, 1944.

Warfighting. Marine Corps Doctrinal Publication 1. Washington, DC: Department of the Navy, 1997.

Ziemke, Earl F. *The German Northern Theater of Operations, 1940–1945*. Army Pamphlet 20-271. Washington, DC: Department of the Army, 1959.

Zimmerman, Maj John L., USMCR. *The Guadalcanal Campaign*. Washington, DC: Historical Division, Headquarters Marine Corps, 1949.

SECONDARY SOURCES

Achkasov, V. I., and N. B. Pavlovich. *Soviet Naval Operations in the Great Patriotic War, 1941–1945*. Translated by U.S. Naval Intelligence Command Translation Project. Annapolis, MD: Naval Institute Press, 1981.

Across the Taiwan Strait: Mainland China, Taiwan, and the 1995–1996 Crisis. Edited by Suisheng Zhao. London: Routledge, 1999.

Adams, Charles Francis, ed. *Memoirs of John Quincy Adams*, vol. 6, *Comprising Portions of His Diary from 1795 to 1848*. Philadelphia, PA: J. B. Lippincott, 1875.

———. *The Monroe Doctrine and Mommsen's Law*. Boston, MA: Houghton Mifflin, 1914.

Adams, Ephraim Douglass, ed. *British Diplomatic Correspondence Concerning the Republic of Texas, 1838–1846*. Austin: Texas State Historical Association, 1918.

Alexander, Col Joseph H. *Utmost Savagery: The Three Days of Tarawa*. Annapolis, MD: Naval Institute Press, 1995.

Alman, David. "Extend Air Wing Range with Seaplane Tankers." U.S. Naval Institute *Proceedings* 147, no. 5 (May 2021).

———. "Seaplanes Go to War." *Naval History Magazine* 35, no. 4 (August 2021).

Anderson, Edgar. "The Military Situation in the Baltic States." *Baltic Defence Review* 6, no. 2 (2001).

Andronic, Benone. "Warfare Actions of the Large Romanian Military Units for Defense and Evacuation of Crimea in World War II." *Annals-Series on Military Sciences* 13, no. 1 (2021).

Armstrong, John A., ed., *Soviet Partisans in World War II*. Madison: University of Wisconsin Press, 1964.

Ashton, Nigel J., ed. *The Cold War in the Middle East: Regional Conflict and the Superpowers 1967-73*. London: Routledge, 2007. https://doi.org/10.4324/9780203945803.

Asprey, Robert B. *Once a Marine: The Memoirs of General A. A. Vandegrift, U.S.M.C.* New York: W. W. Norton, 1964.

Atkinson, Rick. *An Army at Dawn: The War in North Africa, 1942–1943*, vol. 1. New York: Owl Book, an imprint of Henry Holt, 2003.

———. *The Day of Battle: The War in Sicily and Italy, 1943–1944*, vol. 2. New York: Henry Holt, 2007.

Axworthy, Mark, Cornel Scafeș, and Cristian Craciunoiu. *Third Axis Fourth Ally: Romanian Armed Forces in the European War, 1941–1945*. London: Arms and Armour, 1995.

Badsey, Stephen. "An Overview of the Falklands War: Politics, Strategy and Operations." *NIDS Military History Annual* (2013).

Ball, Rhys, and Shaun Mawdsley. "Australasian Special Operations in the Second World War." In *The Routledge History of the Second World War*. Edited by Paul Bartrop, 608–22. Oxon, UK: Routledge, 2022.

Bar-Siman-Tov, Yaacov. *The Israeli-Egyptian War of Attrition, 1969–70: A Case Study of Limited Local War*. New York: Columbia University Press, 1980.

Bates, 1stSgt Cecil R. "The Fita-Fita Guard." *Leatherneck*, October 1940.

Bauer, K. Jack. *The Mexican War, 1846–1848*. New York: Macmillan, 1974.

———. *Surfboats and Horse Marines: U.S. Naval Operations in the Mexican War, 1846–48*. Annapolis, MD: U.S. Naval Institute, 1969.

Bentinck, Mark. *Vertical Assault: The Story of the Royal Marines Mountain Leaders' Branch*. Hants, UK: Royal Marines Historical Society, 2008.

Bioletti, H. L. *Pacific Kiwis: Being the Story of the Service in the Pacific of the 30th Battalion, Third Division, Second New Zealand Expeditionary Force*. Wellington: A. H. & A. W. Reed, 1947.

Brezhnev, Leonid Il'ich. *How It Was: The War and Post-war Reconstruction in the Soviet Union*. Oxford, UK: Pergamon Press, 1979.

Bolia, Robert S. "The Bluff Cove Disaster." *Military Review* (November–December 2004).

Booth, Christopher D. "Overcome the Tyranny of Distance." U.S. Naval Institute *Proceedings* 146, no. 12 (December 2020).

Boswell, Rod. *Mountain Commandos at War in the Falklands: The Royal Marines Mountain and Arctic Warfare Cadre in Action during the 1982 Conflict*. Philadelphia, PA: Pen & Sword Military, 2021.

Browning, Robert S., III. *Two If by Sea: The Development of American Costal Defense Policy*. Westport, CT: Greenwood Press, 1983.

Buchner, Alex. *Narvik: The Struggle of Battle Group Dietl in the Spring of 1940*. Translated by Janice W. Ancker. Philadelphia, PA: Casemate, 2020.

Burke, Ryan Patrick. *The Polar Pivot: Great Power Competition in the Arctic and Antarctic*. Boulder, CO: Lynne Rienner, 2022.

Burke, Ryan, and LtCol Jahara Matisek. "The Polar Trap: China, Russia, and American Power in the Arctic and Antarctica." *Journal of Indo-Pacific Affairs* (October 2021).

Cagle, Malcolm W. "The Strategic Danish Straits." U.S. Naval Institute *Proceedings* 86, no. 10 (October 1960).

Carter, Ashton B., and William J. Perry. *Preventive Defense: A New Security for America*. Washington, DC: Brookings Institution Press, 2000.

Chokshi, Niraj. "Air Force Receives Its First Electric Air Taxi." *New York Times*, 25 September 2023.

Churchill, Randolph S., and Winston S. Churchill. *The Six-Day War*. London: Heinemann/Penguin, 1967.

Churchill, Winston S. *A History of English-Speaking Peoples*, vol. 4. New York: Dodd, Mead, 1958.

Clapp, Michael, and Ewen Southby-Tailyour. *Amphibious Assault Falklands: The Battle of San Carlos Water*. Barnsley, UK: Pen & Sword Military, 1996.

Clemens, Martin. *Alone on Guadalcanal: A Coastwatcher's Story*. Annapolis, MD: Naval Institute Press, 1998.

Collingham, Lizzie. *The Taste of War: World War Two and the Battle for Food*. London: Penguin Books, 2011.

Corbett, Julian. *The Spectre of Navalism*. London: Darling & Son, 1915.

Crawford, Neta C. *Pentagon Fuel Use, Climate Change and the Costs of War*. Providence, RI: Watson Institute, Brown University, 2019.

Croizat, Victor J. *Across the Reef: The Amphibious Tracked Vehicle at War*. Quantico, VA: Marine Corps Association, 1989.

Dallin, Alexander. *German Rule in Russia, 1941–1945: A Study in Occupation Policies*. London, UK: MacMillan, 1957.

David, Saul. *The Force: The Legendary Special Ops Unit and WWII's Mission Impossible*. New York: Hachette Books, 2019.

Deletant, Dennis. *Hitler's Forgotten Ally: Ion Antonescu and His Regime, Romania, 1940–1944*. New York: Palgrave Macmillan, 2006.

D'Este, Carlo. *Bitter Victory: The Battle for Sicily, July–August 1943*. New York: Harper Collins, 1988.

DiNardo, R. L. *Mechanized Juggernaut or Military Anachronism?: Horses and the German Army of WWII*. Mechanicsburg, PA: Stackpole Books, 1991.

Donnelly, Christopher. *Red Banner: The Soviet Military System in Peace and War*. London: Jane's Information Group, 1988.

Dougherty, Chris. *Buying Time: Logistics for A New American Way of War*. Washington, DC: Center for a New American Security, 2023.

Dunning, William Archibald. *The British Empire and the United States: A Review of Their Relations during the Century of Peace Following the Treaty of Ghent*. New York: Charles Scriber's Sons, 1914.

Einarsson, Niels, Joan Nymand Larsen, Annika Nilsson, and Oran R. Young. *Arctic Human Development Report*. Akureyri, Iceland: Arctic Council, 2004.

Elemia, Camille. "How a Decaying Warship Beached on a Tiny Shoal Provoked China's Ire." *New York Times*, 11 November 2023.

Ellis, E. H. "Bush Brigade." *Marine Corps Gazette* 6, no. 1 (March 1921).

Ennes, James M., Jr. *Assault on the Liberty: The True Story of the Israeli Attack on an American Intelligence Ship*. New York: Random House, 1979.

Erickson, John. *The Soviet High-Command: A Military-Political History, 1918–1941*, 3d ed. Oxford, UK: Frank Cass, 2001.

Evans, Maj Peter, RM. "The Value of Amphibious Raiding in the Twentieth Century: A Historical Perspective." *Defence Studies* 1, no. 3 (Autumn 2001). https://doi.org/10.1080/714000047.

Ferris, Norman B. *The Trent Affair: A Diplomatic Crisis*. Knoxville: University of Tennessee Press, 1977.

Filonik, A. O., ed. *Blizhniy Vostok: Komandirovka na voyn: Sovetskie voennye v Egipte* [Middle East: Mission to War: Soviet Military in Egypt]. Moscow: Academy of Sciences and Moscow State University, 2009.

Fish, Carl Russell. *American Diplomacy*. New York: Henry Holt, 1916.

Fitzgerald-Black, Alexander. *Eagles over Husky: The Allied Air Forces in the Sicilian Campaign, 14 May to 17 August 1943*. Solihull, UK: Helion, 2018.

Forczyk, Robert. *Where the Iron Crosses Grow: The Crimea, 1941–44*. Oxford, UK: Osprey Publishing, 2014.

Fowler, William, and Michael Chappell. *Battle for the Falklands (1): Land Forces*. London: Osprey, 1982.

Freedman, Lawrence. *The Official History of the Falklands Campaign*, vol. 2, *War and Diplomacy*. New York: Routledge, 2005.

Fremont-Barnes, Gregory. *The Falklands 1982: Ground Operations in the South Atlantic*. New York: Osprey Publishing, 2012.

Friedman, Norman. *U.S. Amphibious Ships and Craft: An Illustrated Design History*. Annapolis, MD: Naval Institute Press, 2002.

Foreign Relations of the United States, 1964–1968, vol. 19, Arab-Israeli Crisis and War, 1967, doc. 253. "Telegram from the Joint Chiefs of Staff to the Commander-in-Chief European Command (Lemnitzer)." Recorded Date 10 June 1967, 1522Z. National Security File, Country File, Middle East Crisis, vol. 9. Lyndon B. Johnson Library, Austin, TX.

Fry, Capt Nathan. "Survivability, Sustainability, and Maneuverability: The Need for Joint Unity of Effort in Implementing the DOD Arctic Strategy at the Tactical and Operational Levels." *Military Review* 94, no. 6 (November–December 2014).

Fuller, Howard J. *Clad in Iron: The American Civil War and the Challenge of British Naval Power*. Westport, CT: Praeger/Greenwood Press, 2007.

———. "'The Whole Character of Maritime Life': British Reactions to the U.S.S. *Monitor* and the American Ironclad Experience." *Mariner's Mirror* 88, no. 3 (August 2002).

Garcia, Antonio. *The First Campaign Victory of the Great War: South Africa, Manoeuvre Warfare, the Afrikaner Rebellion and the German South West African Campaign, 1914–1915*. Warwick: Helion & Company Limited, 2019.

Gardiner, Ian. *The Yompers: With 45 Commando in the Falklands War*. Havertown, PA: Pen & Sword, 2012.

Gardner, Bruce, and Barbara Stahura. *Seventh Infantry Division, 1917–1992: World War I, World War II, Korean and Panamanian Invasion—Serving America for 75 Years*, rev. ed. Nashville, TN: Turner Publishing, 1997.

Garfield, Brian. *The Thousand-Mile War: World War II in Alaska and the Aleutians*. Fairbanks: University of Alaska Press, 1995.

Garthoff, Raymond L. *Soviet Military Doctrine*. Santa Monica, CA: Rand, 1953.

Giangreco, D. M. *Hell to Pay: Operation Downfall and the Invasion of Japan, 1945–47*. Annapolis, MD: Naval Institute Press, 2009.

Ginor, Isabella, and Gideon Remez. *Foxbats over Dimona: The Soviets' Nuclear Gamble in the Six-Day War*. New Haven, CT: Yale University Press, 2007.

———. " 'Shestidnevnaya voyna' 1967 g. i pozitsiya SSSR [The 'Six-Day War' and the position of the USSR]." *USA and Canada*. Moscow: Russian Academy of Sciences, USA and Canada Institute, 2002.

———. "The Six-Day War as a Soviet Initiative: New Evidence and Methodological Issues." *Middle East Review of International Affairs* 12, no. 3 (September 2008).

———. *The Soviet-Israeli War, 1967-1973: The USSR's Military Intervention in the Egyptian-Israeli Conflict*. London: Oxford University Press, 2017. https://doi.org/10.1093/oso/9780190693480.001.0001.

———. "Veterans' Memoirs as a Source for the USSR's Intervention in the Arab-Israeli Conflict: The Fluctuations in Their Appearance and Character with Political Change in Post-Soviet Russia." *Slavic Military Studies* 29, no. 2 (2016).

Glantz, David M. "Forgotten Battles of the German-Soviet War (1941-45), Part 6: The Winter Campaign (5 December 1941-April 1942): The Crimean Counteroffensive and Reflections." *Journal of Slavic Military Studies* 14, no. 1 (2001). https://doi.org/10.1080/13518040108430472.

———. *The Soviet Conduct of Tactical Maneuver: Spearhead of the Offensive*. London: Frank Cass, 1991.

Goldstein, Lyle J., and Yury M. Zhukov. "A Tale of Two Fleets: A Russian Perspective on the 1973 Naval Standoff in the Mediterranean." *Naval War College Review* 57, no. 2 (Spring 2004).

Gorlaski, Robert, and Russel W. Freeburg. *Oil & War: How the Deadly Struggle for Fuel in WWII Meant Victory or Defeat*. New York: William Morrow and Company, 1987; Quantico, VA: Marine Corps University Press, 2022 reprint. https://doi.org/10.56686/9780160953613.

Gorshkov, S. G. *The Sea Power of the State*. Oxford, UK: Pergamon Press, 1979.

Granger, Rob. "British Army Cold Weather and Mountain Warfare Training in the Second World War." *British Journal for Military History* 8, no. 1 (2022). https://doi.org/10.25602/GOLD.bjmh.v8i1.1606.

Greenwood, John T. "The U.S. Army and Amphibious Warfare during World War II." *Army History*, no. 27 (Summer 1993).

Gudmundsson, Bruce I. *On Armor*. Westport, CT: Greenwood Publishing, 2004.

Halsey, FlAdm William F., and LtCdr J. Bryan III. *Admiral Halsey's Story*. New York: McGraw-Hill, 1947.

Hameiri, Yehezkel *Mishnei evrei harama* [On both sides of the heights]. Tel Aviv: Lewin-Epstein, 1970

Hamilton, Stanislaus Murray, ed. *The Writings of James Monroe*, vol. 7, *1824-1831*. New York: G. P. Putnam's Sons, 1903.

Harward, Grant T. *Romania's Holy War: Soldiers, Motivation, and the Holocaust*. Ithaca, NY: Cornell University Press, 2021.

Hastings, Max, and Simon Jenkins. *The Battle for the Falklands*. New York: W. W. Norton, 1983.

Heiber, Helmut, and David M. Glantz, eds. *Hitler and His Generals: Military Conferences, 1942-1945*. London: Greenhill Books, 2002.

Herrick, Robert Waring. *Soviet Naval Strategy: Fifty Years of Theory and Practice*. Annapolis, MD: U.S. Naval Institute, 1968.

———. *Soviet Naval Theory and Policy: Gorshkov's Inheritance*. Newport, RI: Naval War College Press, 1988.

Herring, George C. *Aid to Russia, 1941-1946: Strategy, Diplomacy, and the Origins of the Cold War*. New York: Columbia University, 1973.

Hess, Wilhelm. *Arctic Front: The Advance of Mountain Corps Norway on Murmansk, 1941.* Translated by Linden Lyons. Havertown, PA: Casemate Publishers, 2021.

Hill, Alexander. *The Red Army and the Second World War.* Cambridge, UK: Cambridge University Press, 2019. https://doi.org/10.1017/9781139107785.

Hines, John. *Soviet Intentions, 1965–1985,* vol. 1, *An Analytical Comparison of U.S.-Soviet Assessments during the Cold War.* McLean, VA: BDM Federal, 1995.

Hoffman, Jon T. *Once a Legend: "Red Mike" Edson of the Marine Raiders.* Novato, CA: Presidio Press, 1994.

Hoppe, Jon. "The Measure of the Sierra Madre: The Extensive History of the Sierra Madre, Originally the USS LST-821." *Naval History Magazine,* vol. 36, no. 1, February 2022.

Isakov, Adm I. S. *Red Fleet in the Second World War.* London: Hutchinson, 1947.

Isely, Jeter A., and Philip A. Crowl. *The U.S. Marines and Amphibious War: It's Theory, and Its Practices in the Pacific.* Princeton, NJ: Princeton University Press, 1951.

Jenkins, McKay. *The Last Ridge: The Epic Story of America's First Mountain Soldiers and the Assault on Hitler's Europe.* New York: Random House, 2003.

Johanson, Art. *General Nikolai Reek Writings Including Operation Albion and Battle of Cēsis.* Tartu: Baltic Defence College, 2021.

Johnson, Darren, and Claudio Innocenti. *The West Point Guide to the Campaigns of World War II: Sicily.* New York: Rowan Technology Solutions, 2022.

Jones, Michael. *Total War: From Stalingrad to Berlin.* London: John Murray, 2011.

Kapitanets, Adm Ivan M. (Ret). *Na sluzhbe okeanskomu flotu, 1946–1992: zapiski komandujuschego dvumja flotami* [In the service of the oceanic fleet, 1946–1992: Notes of the commander of two fleets]. Moscow: Andreyevsky Flag, 2000.

Kennedy, Greg, ed. *Imperial Defence: The Old World Order, 1856–1956.* London: Routledge, 2008.

Kiaupa, Zigmantas. *The History of the Baltic Countries.* Tallinn, Estonia: Avita, 1999.

Kiefer, Chester L. *Maligned General: The Biography of Thomas Sidney Jesup.* San Rafael, CA: Presidio Press, 1979.

Kipp, Jacob W. *Naval Art and the Prism of Contemporaneity: Soviet Naval Officers and the Lessons of the Falklands Conflict.* Stratech Studies Series. College Station: Center for Strategic Technology, Texas A&M University, 1983.

Kravchenko, Adm V. A., ed. *Podvodnye sily Chernomorskogo flota* [Submarine forces of the Black Sea Fleet]. Simferopol, Crimea: Tavrida, 2004.

Kress, Moshe. *Operational Logistics: The Art and Science of Sustaining Military Operations.* Boston, MA: Kluwer Academic Publishers, 2002. https://doi.org/10.1007/978-3-319-22674-3.

Krulak, LtGen Victor H., USMC (Ret). *First to Fight: An Inside View of the U.S. Marine Corps.* Annapolis, MD: Naval Institute Press, 1984.

Kwai, Anna Annie. *Solomon Islanders in World War II: An Indigenous Perspective.* Acton: Australian National University Press, 2017. http://doi.org/10.22459/SIWWII.12.2017.

Lewis, Emanuel Raymond. *Seacoast Fortifications of the United States: An Introductory History.* Annapolis, MD: Naval Institute Press, 1979.

Lewis, Jack A. "The Forgotten Battalion." *Leatherneck,* March 2005.

Li, Xiaobing. *The Cold War in East Asia.* New York: Routledge, 2018.

Li, Xiaobing, and Hongshan Li, eds. *China and the United States: A New Cold War History.* Lanham, MD: University Press of America, 1998.

Lunde, Henrik O. *Hitler's Pre-Emptive War: The Battle for Norway, 1940.* Philadelphia, PA: Casemate, 2009.

Macintyre, Capt Donald. "A Forgotten Campaign—IV: Forlorn Hope." *RUSI Journal* 106, no. 624 (1961). https://doi.org/10.1080/03071846109420730.

Macksey, Kenneth. *Commando Strike: The Story of Amphibious Raiding in World War II*. London: Guild Publishing, 1985.
Malkasian, Carter A. *Charting the Pathway to OMFTS: A Historical Assessment of Amphibious Operations from 1941 to the Present*. Alexandria, VA: CNA, 2002.
"The Marines Amphibian." *Marine Corps Gazette* 37, no. 6 (June 1953).
Marsh, A. R. "A Short but Distant War–The Falklands Campaign." *Journal of the Royal Society of Medicine* 76 (November 1983). https://doi.org/10.1177/014107688307601119.
Mawdsley, Evan. *The War for the Seas: A Maritime History of World War II*. New Haven, CT: Yale University Press, 2020.
May, Robert E. *Slavery, Race and Conquest in the Tropics: Lincoln, Douglas, and the Future of Latin America*. Cambridge, UK: Cambridge University Press, 2013.
McCormick, Gordon H. *The Soviet Presence in the Mediterranean*. Santa Monica, CA: Rand, 1987.
McGee, William L. *The Amphibians Are Coming!: Emergence of the 'Gator Navy and Its Revolutionary Landing Craft*. Napa, CA: BMC Publications, 2000.
McManners, Hugh. *Falkland Commando*. London: William Kimber, 1984.
Mills, Capt Walker D. "Contested Logistics: Look to the Drug Trade." U.S. Naval Institute *Proceedings* (August 2021).
Mills, Capt Walker D., and Christopher D. Booth. "Marines Need a Few Good Mules." U.S. Naval Institute *Proceedings* 148, no. 4 (April 2022).
Mills, Capt Walker D., USMC, and Erik Limpaecher. "Sustainment Will Be Contested." U.S. Naval Institute *Proceedings* 146, no. 11 (November 2021).
Mills, Walker D., Joshua Taylor, and Dylan Phillips-Levine. "Modern Sea Monsters: Revisiting Wing-in-Ground Effect Aircraft for the Next Fight." U.S. Naval Institute *Proceedings* (September 2020).
Mills, Capt Walker D., USMC, and LCdr Dylan Phillips-Levine, USN. "Give Amphibians a Second Look." U.S. Naval Institute *Proceedings* 146, no. 12 (December 2020).
Mills, Capt Walker D., Maj Jacob Clayton, and Erik R. Limaecher. "Powering EABO: Aluminum Fuel for the Future Fight." *Marine Corps Gazette* (August 2022).
Mills, Walker, and Timothy Heck. "What Can We Learn about Amphibious Operations from a Conflict that Has Had Very Little of It? A Lot." Modern War Institute, 22 April 2022.
Mitcham, Samuel W., Jr., and Gene Mueller. *Hitler's Commanders*. Lanham, MD: Scarborough House, 1992.
Monick, S. *A Bugle Calls: The Story of the Witwatersrand Rifles and Its Predecessors, 1899–1987*. Johannesburg: Witwatersrand Rifles Regimental Council, 1989.
Morison, Samuel Eliot. *History of United States Naval Operations in World War II*, vol. 6, *Breaking the Bismarcks Barrier: 22 July 1942–1 May 1944*. Boston, MA: Little, Brown, 1989.
Newell, Reg. *Operation Squarepeg: The Allied Invasion of the Green Islands, February 1944*. Jefferson, NC: McFarland, 2017.
Nichols, Edward J. *Zach Taylor's Little Army*. Garden City, NY: Doubleday, 1963.
Nicholson, Maj Dustin, USMC. "Marines Need Regenerative Logistics." U.S. Naval Institute *Proceedings* 148, no. 11 (November 2022).
O'Brien, Phillips Payson. *How the War Was Won: Air-Sea Power and Allied Victory in World War II*. Cambridge, UK: Cambridge University Press, 2015.
O'Connell, Aaron B. *Underdogs: The Making of the Modern Marine Corps*. Cambridge, MA: Harvard University Press, 2012.
Ostrosky, Steve. "The PLANMC: Will the PLA Marine Corps Become Its Own Service." *Marine Corps Gazette* (September 2019).

Oswandel, J. Jacob. *Notes of the Mexican War, 1846-1848.* Knoxville: University of Tennessee Press, 2010.

Panicacci, Paul S. "How to Do Logistics in EABO: It's a MAGTF, Not a MAGLTF." *Marine Corps Gazette* 104, no. 12 (December 2020).

Parker, Capt William Harwar. *Recollections of a Naval Officer, 1841-1865.* New York: Charles Scribner's Sons, 1883.

"Peace with Russia May Be German Goal: Operations in Baltic Possibly Have This End in View as Well as the Influencing of Sweden by Seizing Aland Islands." *New York Times*, 21 October 1917.

Polmar, Norman. *Guide to the Soviet Navy*, 3d ed. Annapolis, MD: Naval Institute Press, 1983.

Polmar, Norman, Thomas A. Brooks, and George Federoff. *Admiral Gorshkov: The Man Who Challenged the U.S. Navy.* Annapolis, MD: Naval Institute Press, 2019.

Privratsky, Kenneth L. *Logistics in the Falklands War: A Case Study in Expeditionary Warfare.* Philadelphia, PA: Pen & Sword, 2014.

Reddaway, W. F. *The Monroe Doctrine.* New York: G. E. Stechert, 1924.

Reek, Nikolai. *Saaremaa Kaitsmine Ja Vallutamine A. 1917* [The Defense and Conquest of Saaremaa in 1917]. Tallinn, Estonia: Kindralstaab IV Osakond, 1937.

Rennie, Frank. *Regular Soldier: A Life in the New Zealand Army.* Auckland, NZ: Endeavour Press, 1986.

Reynolds, David, and Vladimir Pechatnov, eds. *The Kremlin Letters: Stalin's Wartime Correspondence with Churchill and Roosevelt.* London: Yale University Press, 2018. https://doi.org/10.2307/j.ctv7cjvz5.14.

Richter, Donald. *Where the Sun Stood Still: The Untold Story of Sir Jacob Vouza and the Guadalcanal Campaign.* Calabasas, CA: Toucan Publishing, 1992.

Rigby, David. *Allied Master Strategists: The Combined Chiefs of Staff in World War II.* Annapolis, MD: Naval Institute Press, 2012.

"Roebling's 'Alligator' for Florida Rescues." *Life*, 4 October 1937.

Ross, Robert, Anne Kelk Mager, and Bill Nasson, eds. *The Cambridge History of South Africa.* Cambridge, UK: Cambridge University Press, 2011. https://doi.org/10.1017/CHOL9780521869836.007.

Rottman, Gordon L. *U.S. Marine Corps World War II Order of Battle: Ground and Air Units in the Pacific War, 1939-1945.* Westport, CT: Greenwood Press, 2002.

———. *U.S. World War II Amphibious Tactics: Army and Marine Corps, Pacific Theater.* New York: Osprey, 2004.

The Routledge History of the Second World War. Edited by Paul Bartrop. Oxon, UK: Routledge, 2022.

Ruge, Fredrich. *The Soviets as Naval Opponents, 1941-1945.* Annapolis, MD: Naval Institute Press, 1979.

Ryan, Mark A., David M. Finkelstein, and Michael A. McDevitt, eds. *Chinese Warfighting: The PLA Experience since 1949.* Armonk, NY: M. E. Sharpe, 2003.

Sage, Clive B. *Pacific Pioneers: The Story of the Engineers of the New Zealand Expeditionary Force in the Pacific.* Wellington: A. H. & A. W. Reed, 1947.

Samson, William H., ed. *Letters of Zachary Taylor from the Battle-Fields of the Mexican War.* Rochester, NY: Genesee Press, 1908.

Schlossmann, Karl. *Estonian Curative Sea-Muds and Seaside Health Resorts.* London: Boreas, 1939.

Schroden, Jonathan. "Lessons from the Collapse of Afghanistan's Security Forces." *CTC Sentinel* 14, no. 8 (October 2021).

Scott, Winfield. *Memoirs of Lieut.-General Winfield Scott.* Edited by Michael Gray and Timothy D. Johnson. Knoxville: University of Tennessee Press, 2015.

Seaton, Albert. *The Russo-German War, 1941–45*. New York: Praeger Publishers, 1971.

Semmes, Adm Raphael. *Memoirs of Service Afloat during the War between the States*. Baltimore, MD: Kelly, Piet, 1869.

Simmons, Edwin H., and J. Robert Moskin, eds. *The Marines*. Quantico, VA: Marine Corps Heritage Foundation, 1998.

Schwirtz, Michael, Anton Troianovski, Yousur Al-Hlou, Masha Froliak, Adam Entous, and Thomas Gibbons-Neff. "Putin's War." *New York Times*, 16 December 2022.

Shirokorad, A. B. *Flot, kotory unichtozhil Khrushchev* [The fleet that Khrushchev destroyed]. Moscow: Vzoi-AST, 2004.

Shugart, Cdr Thomas, USN. *First Strike: China's Missile Threat to U.S. Bases in Asia*. Washington, DC: Center for New American Security, 2017.

Slepyan, Kenneth. *Stalin's Guerrillas: Soviet Partisans in World War II*. Lawrence: University of Kansas Press, 2006.

Smith, Gen Holland M. "The Development of Amphibious Tactics in the U.S. Navy, Part IV." *Marine Corps Gazette* 30, no. 9 (November 1946).

Smith, Holland M., and Percy Finch. *Coral and Brass* New York: Charles Scribner's Sons, 1949.

Smith, Thomas T. *The U.S. Army and the Texas Frontier Economy, 1845–1900*. College Station: Texas A&M University Press, 1999.

Spencer, Ivor D. *The Victor and the Spoils: A Life of William L. Marcy*. Providence, RI: Brown University Press, 1959.

St. Clair Golden, Francis, Thomas James Roose Francis, Deborah Gallimore, and Roger Pethybridge. "Lessons from History: Morbidity of Cold Injury in the Royal Marines during the Falklands Conflict of 1982." *Extreme Physiology & Medicine* 2 (2013). https://doi.org/10.1186/2046-7648-2-23.

SSRC Soviet Amphibious Warfare. The Hague: Soviet Studies Research Center, 1985.

Stewart, Nora Kinzer. *Mates & Muchachos: Unit Cohesion in the Falklands/Malvinas War*. Washington, DC: Brassey's, 1991.

Strekhnin, Yuriy Fedorovich. *Commandos from the Sea: Soviet Naval Spetsnaz in World War II*. Translated by James F. Gebhardt. Annapolis, MD: Naval Institute Press, 1996.

Sweeney, Capt Michael. "Sleeper Cell Logistics: Sustaining New Warfighting Concepts." *Marine Corps Gazette* 105, no. 1 (January 2021).

Temperley, Harold. *The Foreign Policy of Canning, 1822–1827: England, the Neo-Holy Alliance, and the New World*. London: G. Bell and Sons, 1925; Routledge, 2006 reprint.

Thompson, Julian. *No Picnic: 3 Commando Brigade in the South Atlantic, 1982*. London: L. Cooper with Secker & Wargurg, 1985.

Thompson, Mark. *The White War: Life and Death on the Italian Front, 1915–1919*. New York: Basic Books, 2010.

Till, Geoffery. *Seapower: A Guide for the Twenty-First Century*. New York: Routledge, 2013.

Tomblin, Barbara Brooks. *With Utmost Spirit: Allied Naval Operations in the Mediterranean, 1942–1945*. Lexington: University Press of Kentucky, 2004.

Tooze, Adam. *The Wages of Destruction: The Making and Breaking of the Nazi Economy*. New York: Penguin, 2006.

Turner, Jobie. *Feeding Victory: Innovative Military Logistics from Lake George to Khe Sanh*. Lawrence: University Press of Kansas, 2020.

Twining, Merrill B. *No Bended Knee: The Battle for Guadalcanal*. Edited by Neil Carey. New York: Presidio Press, 1996.

Tyler, Patrick. *A Great Wall: Six Presidents and China—An Investigative History*. New York: Public Affairs, 1999.

van Creveld, Martin. *Supplying War: Logistics from Wallenstein to Patton*, 2d ed. Cambridge, UK: Cambridge University Press, 2004.

van der Bijl, Nick, and David Aldea. *5th Infantry Brigade in the Falklands*. Barnsley, UK: Pen & Sword, 2014.

Vaux, Nick. *Take that Hill!: Royal Marines in the Falklands War*. Washington, DC: Pergamon-Brassey's International Defense Publishers, 1986.

Vego, Milan N. *Naval Strategy and Operations in Narrow Seas*. New York: Frank Cass, 1999.

———. "On Major Naval Operations." *Naval War College Review* 60, no. 2 (2007).

von Kobinski, Capt G., German Navy (Ret). "The Conquest of the Baltic Islands." U.S. Naval Institute *Proceedings* 58, no. 7 (July 1932).

von Manstein, Erich. *Lost Victories: The War Memoirs of Hitler's Most Brilliant General*. Minneapolis, MN: Zenith, 2004.

Walton, Timothy A., Harrison Schramm, and Ryan Boone. *Sustaining the Fight: Resilient Maritime Logistics for a New Era*. Washington, DC: Center for Strategic and Budgetary Assessments, 2019.

War by Land, Sea, and Air: Dwight Eisenhower and the Concept of Unified Command. New Haven, CT: Yale University Press, 2010.

Waring Herrick, Cdr Robert. *Soviet Naval Strategy: Fifty Years of Theory and Practice*. Annapolis, MD: U.S. Naval Institute, 1968.

Warlimont, Gen Walter. *Inside Hitler's Headquarters, 1939-1945*. Novato, CA: Presidio Press, 1991.

Warren, Gordon H. *Fountain of Discontent: The Trent Affair and Freedom of the Seas*. Boston, MA: Northeastern University Press, 1981.

Watling, Jack, and Nick Reynolds. *Operation Z: The Death Throes of an Imperial Delusion*. London: Royal United Services Institute, 2022.

White, Geoffrey M., ed. *Remembering the Pacific War*. Occasional Paper no. 36. Honolulu: University of Hawai'i at Manoa, 1991.

Wilson, Michael. *Baltic Assignment: British Submariners in Russia, 1914-1919*. London: Leo Cooper, 1985.

Wright, Matthew. *Pacific War: New Zealand and Japan, 1941-45*. Auckland, NZ: Reed Publishing, 2003.

Yoshihara, Toshi. "The 1974 Paracels Sea Battle: A Campaign Appraisal." *Naval War College Review* 69, no. 2 (2016).

———. *Chinese Lessons from the Pacific War: Implications for PLA Warfighting*. Washington, DC: Center for Strategic and Budgetary Assessments, 2023.

Yoshihara, Toshi, and James R. Holmes. *Red Star Over the Pacific: China's Rise and the Challenge to U.S. Maritime Strategy*. Annapolis, MD: Naval Institute Press, 2018.

Zhang, Xiaoming. *Red Wings over the Yalu: China, the Soviet Union, and the Air War in Korea*. College Station: Texas A&M University Press, 2002.

INDEX

1st Estonian Infantry Regiment, 148
15-inch Rodman Gun, 271

Aberdeen, George Hamilton-Gordon (4th Earl), 260, 263–66
Adams, Charles Francis, Jr., 257–59
Adams, John Quincy, 251–55, 257–62, 267, 274
additive manufacturing, 77
A History of the English-Speaking Peoples, 248. *See also* Churchill, Winston S.
Aiping, Zhang, 238, 240–41
aircraft,
    bombers, 190, 241, 243, 245, 306, 312
    F-84 Thunder, 239
    fighters, 60, 241, 243, 245
    MiG-15, 239
    Sukhoi Su-17, 60
    Sukhoi Su-27, 60
    Tupolev Tu-22M Backfire, 60
air operations, 142–62
airships, 154–55
Alexander I, 252
Aleutians, 163, 197–75
Alexander, Gen Harold, 200
Alexandria, Egypt, 188, 192, 195
Algiers, 198–99

Allied/Allies, 41, 48 51, 89, 165–66, 198–99, 202–7, 211–12, 215, 276–77, 292, 337
Alligator, 90–92, 184. *See also* Landing Vehicle, Tracked 1; and Roebling, Donald
American Civil War, 20, 247–75, 235
amphibious operations,
    command and control, 58, 312, 318, 329
    deception, 154, 173, 319
    doctrine, 7–88, 94–95, 101, 115, 118–19, 143, 182, 202, 206, 217, 244, 276–300, 310, 317, 319, 322
    intelligence, 12, 42, 129, 137, 141, 155, 191, 194, 199–200, 207, 213, 217–19, 230, 234, 277, 282, 292–95, 305, 311
    planning, 8, 19, 46, 58, 66, 68, 70, 98, 100–1, 107–9, 112, 127, 133–34, 139–40, 152–54, 162, 174, 199–200, 203–6, 209–16, 220, 226, 306–7, 317
    reconnaissance, 4, 56, 58–59, 73, 86, 89, 138, 144, 154, 157, 167, 180, 204, 208, 213–14, 217–20, 222, 225, 294, 309, 311–12, 326–28
Amphibious Training Center (ATC), 202, 204
Amtank (LVT[A]-1), 107
Amtrack, 91, 110
Antarctic, 163, 176–77, 181
antiaccess/area-denial (A2/AD), 162, 238, 247–75

Antikythera anchorage, Greece, 184, 187
Arcadia Conference, 198
Arctic, 55n81, 163-64, 167, 169, 171-72, 175, 179-81, 331, 336
Attu, 169-75, 181
Australian units
    Army, 296, 336
    Royal Air Force, 296
    Royal Navy, 277
Axis, 197-200, 203, 205-10, 303-9, 312-15, 319-20
amphibious raid, 58, 187-88, 198n7, 213, 218-27, 295-98
austere environment, operating in, 81
auxiliary personnel destroyer (APD), 223, 225-26

Bagot, Charles, 258
Baring, Alexander (1st Baron Ashburton), 260
Barrowclough, Harold E., 215-220, 226
bases, military
    Camp Lejeune, NC, 102
    Camp Pendleton, CA, 102, 113
    Norfolk, VA, 98, 103
    Pearl Harbor, HI, 65-66
    Quantico, VA, 102
bataille conduite (methodical battle), 161-62
battleships (German)
    SMS *Bayern* (1915), 156
    SMS *König* (1913), 154, 158
    SMS *Kronprinz* (1914), 158
battleships (Russian)
    SMS *Grazhdanin* (1903), 151, 159
    SMS *Grom* (1916), 159
    SMS *Slava* (1905), 151, 159
Beijing, 230, 236-38, 243-44, 324-25
Berger, Gen David H., 4, 65, 70-72, 74, 300, 333-34
Berkeley, Maurice Frederick (1st Baron Fitz Hardinge), 262-63
Biao, Cdr Lin, 231-35
Black Sea, 62n129, 183-84, 187-90, 194, 303, 305-13, 318-20
Bolívar, Simon, 254
Bonaparte, Napoléon, 270
Bonin Islands, 101
boots, cold weather, mountain, 167, 172, 174, 179, 181
Bougainville Island, 107, 213-16, 226
Bourne, Kenneth, 250n3, 263, 265
Bradley, LtGen Omar N., 206
Brazos Santiago, TX, 11-14
Brezhnev, Leonid, 183, 192, 309
Buchanan, James, 257n28, 263, 266-67
Buffalo, 119. *See also* Landing Vehicle, Tracked 4

Bureau of Construction and Repair (BuC&R), 23, 27-32, 35-41

campaigns,
    Baltic Islands, 142-62
    Civil War (U.S.), 247-75
    Kerch-Eltiger, 301-20
    Mediterranean, 197-212
    Mexico, 7-20
    Pacific, 64-120, 213-27, 276-300, 321-31
    People's Liberation Army (PLA), 228-48
    Polar, 163-81
    South West Africa, 121-41
    Soviet, 42-63
    Taiwan, 228-46
Canning, George, 248, 250-59, 267, 274
Canning, Stratford (1st Viscount Stratford de Redcliffe), 258
Caroline affair, 265
Cass, Lewis, 256-60
Central Pacific, 100-1, 107, 116, 119, 287
Camp Edwards, MA, 202
Casablanca, Morocco, 197, 199
Casablanca Conference, 197, 199
Charleston, SC, 247, 261, 271-72
China, People's Republic of (PRC), 48, 228, 234, 334, 331
    nuclear doctrine of, 246
China, Republic of (ROC), 228, 233, 236n33, 237-45
    Army, 230n8-9
Navy, 240
Chinese Civil War, 230, 235-36
Chinese Communist Party (CCP), 228, 230n10, 235-37, 246
    Central Military Commission (CMC), 229n5, 230n10, 232n11, 235-36, 238n40, 240, 322-23
    National Congress, 228, 236, 246
Chris Craft Corporation, 28, 35-36
Churchill, Winston S., 107, 197-98, 204n37, 248, 275
City Park Plant, 30-33
Clarendon, George William Frederick Villiers (4th Earl), 262-65
clothing, cold weather, mountain, 167, 171-72, 175, 179, 181
coalition warfare, 44, 206, 264
cold injuries, 169n37, 180
Columbia River, 258-59
Combined Chiefs of Staff (CCS), 199-200, 204
commando raid, 213-27
Corbett, Julian S., 255
Corn Laws, 261, 266-67

Cornwall, LtCol Frederick C., 219-25
Crampton affair, 262-63
Crampton, John (2d Baronet), 262, 263n45
Crimean War, 263n45, 264, 270-71, 307
Cunningham, Andrew B., 200, 208

Daggett, Cdr Ross B., 31
Dahlgren, Lt John A., 262n42
Dalrymple-Hay, John (3d Baronet), 272
Dayan, Moshe, 195
Disney, Paul A., 211
Disraeli, Benjamin (1st Earl of Beaconsfield), 274-75
distributed maritime operations (DMO), 65
Dmitriev, V. I., 192-93
DUKW (6-wheeled duplex drive amphibious truck), 99, 103-5, 106n41, 112-13, 208n59, 308
Dunning, William, 258
Du Pont, Samuel Francis, 268n64

Earle, Capt Ralph, 219-20, 223
Eisenhower, Dwight D., 198, 200-1, 204
Egypt, 126, 182-94, 335
electric vehicles, 78-82
Emancipation Proclamation, 269
Erie Canal, 262n44, 268
Eureka boat, 22-33
expeditionary advanced base operations (EABO), 65, 73-78, 81, 83, 300

Falkland Islands, 69, 141, 163, 176-81, 255
Fei, Gen Yue, 230
Fengzhi, Nie, 241, 243
Ferdinand VII, 250
First World War (Word War I), 21, 53, 121, 142, 161-62, 166-67, 202, 255, 336
Fitch, VAdm Aubrey W., 215
Fleet Development Board (FDB), 25-29
Fleet Landing Exercise (FLEX)
    FLEX 4, 25
    FLEX 5, 27
    FLEX 6, 27
Forsyth, John, 267
Fort, RAdm George H., 216-17
Fort Schuyler, 269n66
Fort Sumter, 247, 261
French invasion of Mexico (1862), 255, 269-70
Friedrichshafen FF41A (seaplane), 154, 159

Geiger, MajGen Roy S., 215
Gela, Sicily, 200n18, 211
George IV (George Augustus Frederick), 250

German military units
    6th Army, 304
    7th Army, 303-4
    8th Army, 146
    11th Army, 302
    17th Army, 304-7
    9th Flak Division, 305
    42d Infantry Division, 154
    98th Infantry Division, 314
    3d Mountain Division, 163-64, 168
    139th Mountain Infantry Regiment, 164, 166
    18th Shock Company, 157
    Army Group A, 303-4
    Battle Group Dietl, 164, 166
    Division Group van der Hoop, 50, 52
    Fliegerkorps, 305
    Grenadier Regiment, 123, 282, 314-15
    Kriegsmarine, 301, 305, 314, 316
    Luftwaffe, 169, 313-14
    Panzer Division, 211
    Schutztruppe, 125, 127-28, 132, 136
    V Armeekorps, 306
Ghormley, RAdm Robert, 198
Gladstone, William Ewart, 274
Gloire (1860), 270
Glorious Revolution, 250
Golan Heights, Syria, 188, 191
Golovko, Adm Arseni, 50
Gorokhov, Col V. I., 194
Gorshkov, Adm Sergey G., 52-53, 182-83, 195, 302, 310-11, 319
Grechko, Andrei, 183
Green Islands, 213-27
Guadalcanal, 66-67, 76, 97-101, 116-17, 215-17, 277, 284, 288, 290-300
Guam, 104, 118, 277, 299
Guantánamo Bay, Cuba, 93, 96, 100
Guizot, Francois Pierre Guillaume, 260
Guzzoni, Cdr Alfredo, 206-7

Haddock, Graham, 38
Haifa, Israel, 187-91
Hainan Island, 229, 234-36, 244, 322-25
Halsey, William F., 215-16, 226-27
Harmon, LtGen Millard F., 216-17
Helms, Richard, 191
Higgins, Andrew J., 21-41, 51, 95, 110
Holy Alliance, 250, 253
Hope, Adm James, 272
horse boats, 150, 152-53
Hua, Deng, 234-35
Huet, George, 38

hydrogen, 79–81, 248n1

Inchon, South Korea, 120, 335
Industrial Canal, 36–37
Israel, 182–96, 334, 337
Ivliev, RAdm Nikolay, 194
Iwo Jima, 112, 118–19, 334

Jackson, Andrew, 13, 255–56, 259, 262n43
James II, 250
Jefferson, Thomas, 253, 256, 259
Jenkinson, Robert (2d Earl of Liverpool), 254
Jesup, Gen Thomas S., 8, 11–13, 19
Jianying, Ye, 244
Jinguang, Adm Xiao, 233, 236
Jinping, Xi, 228–29, 246
Johnson, Lyndon B., 191
Joint Chiefs of Staff, Joint Planning Staff, 100, 108, 111
Joint War Planning Committee, 107–8
Jones, RAdm Claude A., 30

Kai-shek, Chiang, 229, 233, 236n33, 239, 240n43
Kaliningrad, USSR, 186–87
Kaohsiung, Taiwan, 245–46
Kapitanets, Ivan, 187
Keelung, Taiwan, 245–46
Kerensky, Aleksandr, 145
Kesselring, Albert, 206–7
Khripunkov, Capt Yuri, 185, 187, 189
Khrushchev, Nikita, 182
Kiska, Aleutian Islands, 174–75
Kislov, Aleksandr, 185, 187
Knox, Frank, 28, 40–41
Knüpfer, Capt M. G., 150
Korea, North (DPRK), 237
Korea, South (ROK), 237
Korean War, 237–39, 240n43
Krulak, LtGen Victor H., 21, 31, 102, 108, 111–12
Kuomintang (KMT, Nationalist Party), 229–30, 233, 235–36
Kwajalein Atoll, 116, 118

Laaman, Eduard, 159–60
Landing Boat Board (LBB), 25–29, 32
Landing Craft, Infantry (LCI), 206, 223
Landing Craft, Medium (LCM), 33–41, 83, 118, 120
Landing Craft Personnel, Large (LCPL), 23, 28–29, 41
Landing Craft, Personnel (Ramp) (LCP[R]), 23, 36, 41
Landing Craft, Tank (LCT), 51, 104, 118, 120, 206

Landing Craft, Vehicle and Personnel (LCVP, or Higgins Boat), 23, 30, 32–33, 35–36, 95–96, 101, 109, 111, 114–17, 206
Landing Ship, Dock (LSD), 107, 110, 120
Landing Ship, Medium (LSM), 51, 83, 118n78
Landing Ship, Tank (LST), 4, 51, 89, 97–99, 102–20, 184, 329
Landing Vehicle, Tracked (LVT-1, -2, -3 and -4), 51, 89–120
Latakia, Syria, 195–96
Latvian Rifle Regiments, 160, 311
Lawson, Leonard Axel, 259
Lend-Lease Act, 98, 197n1, 248, 308
Lenin, Vladimir I., 44, 160
Leopold II, 269
Licata, Sicily, 201, 205, 210
Lincoln, Abraham, 250n3, 271
Lind, William S., 161–62
Linsert, Maj Ernest E., 31, 284
Little Creek Naval Base, VA, 103–4
Louis-Philippe I, 259
Lucas, MajGen John P., 204
Lyons, Richard B. (1st Viscount), 274

Madison, James, 251, 253, 256
Makin Island, 110, 115–6, 218
Mallin, Valery, 190–92, 195
Manifest Destiny, 275
Marcy, William L., 8–12, 263
Mariana Islands
Marine Corps Amphibious Warfare School (later Expeditionary Warfare School), 162
Marshall, Gen George C., 64, 174, 199, 204
Marshall Islands, 101, 107–9, 115–18
Martin, VAdm William I., 184
Maximilian, Ferdinand, 269–70
McDowell, LtCdr Ralph S., 25–26
McNair, LtGen Lesley, 93, 204
McNamara, Robert, 191
Mediterranean, 107, 184–87, 191–99, 202, 204, 208–9
Mediterranean Air Command, 202–3
Meretskov, Gen Kirill A., 50
Mexican-American War, 7–20, 257, 267
military mountain guides (*heeresbergfuhrer*) (German), 168
Minto, Gilbert Elliot-Murray-Kynynmound (2d Earl), 265
Mommsen, Theodor, 257
Monroe Doctrine, 248, 250–59, 263, 265, 269–70, 274–75
Monroe, James, 248, 251, 253–58, 261–62
Montgomery, Gen Bernard, 200–1, 205

Mosquito Coast, 263-64
Moulton, Cdr H. Douglas, 215
Mountain Training Center, 171
mountain warfare, 163-81
mountaineering, 168, 180
Muhu (Moon) Causeway, 142n1, 146, 148, 150-53, 157-59
Muhu Sound Fortified Position (*Moonzundskaya ukreplennaya positziya*), 146, 148, 151, 159
multidomain operations (MDO), 65, 87, 336

Narvik, Norway, 163-69, 177, 181
Nesselrode, Karl Vasilyevich (Count), 257
New Caledonia, 102, 108, 216, 288
New Georgia Island, 98, 107, 109
New Guinea, 67, 101, 215, 290
New Zealand, 111, 213-27, 300
New Zealand military units
    14th New Zealand Brigade, 218, 220, 222
    3d New Zealand Division, 215, 223, 226
Nissan Island, 218, 221
Norfolk Naval Base, VA, 26, 28-29, 33, 35, 39, 90, 98-99, 103
North Africa, 37n71, 197-200, 204, 206, 212
North Atlantic Treaty Organization (NATO), 48, 178, 184
North Cape, Norwegian theater, 63, 271n75

offshore islands, 229, 230n8, 238, 240n43, 244
Okinawa, 66, 117-19, 334
operational maneuver from the sea (OMFTS), 5, 66
operations
    Albion, 142-62
    Barclay, 207-8
    Coronet, 68
    Desert Storm, 56
    Ferdinand, 289
    Galvanic, 97, 105, 109-11, 114
    Goldrush, 102, 105
    Husky, 103, 197-212
    Mincemeat, 207-8
    Olympic, 68
    Squarepeg, 213
    Torch, 197-99
    Uranus, 303-4
Ostend Manifesto, 257

Pago Pago, American Samoa, 113, 277, 286
Pakenham, Richard, 266
Palmerston, Henry John Temple (3d Viscount), 248-50, 260, 262-67, 270-72

Panama Canal, 258
Paracel Islands, 229, 244, 322-24
patrol-torpedo boat (PT), 186, 191, 220, 223-26, 314
Patton, Gen George S., 205, 210
Peel, Robert (2d Baronet), 260, 265-67
People's Liberation Army (PLA), 321-31
    amphibious campaigns, 228-46, 334
    East China Military Region (ECMR), 238, 240
Petsamo-Kirkennes Offensive, 49-0
Pitt, William (the Younger), 248
PLA Air Force (PLAAF), 233-34, 239-45
    2d Division, 239
PLA Army (PLAA), 327-28
    7th Army Group, 232, 236
    8th Army Group, 236
    9th Army Group, 232, 236
    10th Army Group, 230, 232, 233n14, 237n35
    13th Army Group, 236
    15th Army Group, 234-35
    19th Army Group, 236
    3d Field Army, 230-31, 233, 236
    4th Field Army, 231, 233-36
PLAN Marine Corps (PLANMC), 322-31, 337
PLA Navy (PLAN), 4, 122, 125, 128, 233-36, 240, 243, 245, 321-46,
    East Sea Fleet (ESF), 240, 245
Polignac, Jules de, 253
Polk, James K., 8, 265-66
Port Fuad, Egypt, 193, 195
Port Said, Egypt, 183, 186, 190, 192-95, 335
Potter, Leslie, 218, 225
Provisional Training Instruction for Mountain Troops of 1935 (German), 166

Quemoy, Taiwan (Kinmen or Jinmen), 229-34, 237n38

Ramsey, Adm Bertram, 208
Rawlings, Capt Norborne L., 30
Reek, Nikolai, 148, 157, 161
Riggs, Capt W. F. Jr., 215
Riley, Col William E., 215
Riva-Aguero, Jose de la, 254
Rongzhen, Nie, 236
Roosevelt, Franklin D., 107, 197, 204n37, 248
Roosevelt, Theodore, 257
Rush, Richard, 251, 256n25, 259
Russell, John (1st Earl), 265, 269, 271, 274

Saipan, 94, 106, 116, 118, 299

Salisbury, Robert Gascoyne-Cecil (3d Marquess), 255
San Juan (Greytown), Puerto Rico, 267
Scott, Gen Winfield, 8-19
seaplanes, 85-86, 154, 157
Second World War (World War II), 42-43, 48, 51, 54, 56, 62, 65-70, 77-78, 80, 86, 88, 213, 227, 248, 301, 333, 335-37
Semenov, Evgeny, 195
Sevastopol, USSR, 48, 62, 194, 262, 263n48, 264, 271, 306-7, 310, 319
Seward, William H., 40-41, 250n3
Shevchenko, Viktor, 186, 189-90
ships
    British
        C-class, 144, 151
        E-class, 144
        HMHS *Dover Castle*, 133
        HMS *Astrea*, 133-34
        HMS *E8*, 144
        HMS *E19*, 144
        HMS *Black Prince*, 270
        HMS *Warrior*, 270
        RMS *Trent*, 250n3
    German
        SMS *Bayern*, 156
        SMS *Gneisenau*, 141
        SMS *Konig*, 158
        SMS *Kronprinz*, 158
        SMS *Leipzig*, 141
        SMS *Nurnberg*, 141
        SMS *Prinz Adalbert*, 144
        SMS *Scharnhorst*, 141
    monitors, 247, 249, 273
    Russian
        Admiral Makarov, 159
        Bayan, 159
        Grazhdanin, 151, 159
        Grom, 159
        Slava, 151, 159, 187
    United States
        USS *Bonita*, 16
        USS *Harris*, 111
        USS *Falcon*, 16
        USS *Liberty*, 191
        USS *Liscome Bay*, 111
        USS *LST-34*, 114
        USS *LST-242*, 114
        USS *LST-243*, 113-14
        USS *LST-486*, 113
        USS *LST-831*, 112
        USS *Massachusetts*, 14
        USS *Petrel*, 16
        USS *Petrita*, 15
        USS *President Hayes*, 92
        USS *Raritan*, 16
        USS *Reefer*, 16
        USS *San Jacinto*, 250n3
        USS *Spitfire*, 16
        USS *Tampico*, 16
        USS *Vixen*, 16
        USS *Washington*, 280
Sicily, 98, 103, 107, 198-209, 212
siege of Kars, Turkey, 263n48
Six-Day War, 1967, 182, 184, 185n8, 186
ski, training, 167-68, 179-80
Small Boat Desk, 27, 29, 32, 35-36, 41
Smith, Gen Eric M., 70, 76
Smith, Gen Holland M. 27, 29, 31-32, 36, 41, 93-96, 99-101, 106, 108, 110-11, 114, 119-20
Smith, J. McDonald, 219. 225
Solomon Islands, 104, 109, 213-14, 276-77, 285, 290-91, 295-300
Somerset, Edward Adolphus Seymour (12th Duke), 270, 272
South China Sea, 76, 229, 231, 244-45, 322-24, 331, 334
Soviet Army Studies Office, 42
Soviet concepts
    advanced force, 47, 58, 60
    armed conflict, 44, 58, 86n11
    military affairs, 45
    military art, 46
    deep battle/operations, 57
    war, 45-6
Soviet military doctrine, 42-63
Soviet military units
    12th Army, 145-46, 233
    2d Army Corps, 144
    4th Coastal Defense Artillery Battalion, 149, 150n16
    107th Infantry Division, 158
    425th Infantry Regiment, 158
    12th Naval Infantry Brigade, 50
    61st Naval Infantry Brigade, 55-56
    63d Naval Infantry Brigade, 50
    Baltic Fleet, 194
    Fifth *Eskadra* (squadron), Navy, 48, 53, 59, 62, 184-91, 195
    Forward Detachment (*Peredvoy Otriad*), 47
    Marine Guards, 183, 192-193, 148n15, 149-50
    Naval Aviation (*Morskaya Aviatsya*), 42, 60, 151
    Naval Infantry (*Morskaya Pekhota*, Black Death), 5, 42-63, 182-83, 186-95, 302, 308, 313, 322
    Northern Fleet, 49-50, 55

Operational Mobile Group (OMG), 47
special autonomous sector (Muhu Sound Fortified Position), 148-50
Spratly Islands, 4, 229, 245, 324
Suez Canal, 183, 186, 189, 192, 335
stand-in forces, 65, 73-75
St. Charles Avenue plant, 33
St. Lawrence River, 268
surfboat, 10, 13-18
Sveshnikov, RAdm Dimitry Aleksandrovich, 148, 157
Swedish iron ore, 142, 144, 164
synthetic fuel, 80-81
Syracuse, Sicily, 201, 205, 209n66, 210
Sysoev, VAdm Viktor, 187

Tachen (Dachen) Islands, 238-41
Taga Bay (Tagalaht), Tinian, 152-57
Taipei, 238, 245
Taiwan, 228-29, 233, 236-39, 243, 322, 325, 328, 330
Taiwan Strait, crises, 228-29, 231, 234, 237-38, 244-46, 323, 334
Tampico, 11-12, 14
Tarawa Atoll, 97, 105, 109-12, 114-17
Tartus, Syria, 195
Task Force Papa Bear, 55-56
Taylor, Gen Zachary, 8, 11-13
Tedder, Arthur W., 200, 202, 212
Temperley, Harold, 251, 254
Third System (U.S. fortifications), 247
Tinian Atoll, 106, 118, 299
Totten, Lt George M., 10, 262n43
Trent affair (1861), 247, 250, 265
Trident Conference, 99, 107-9
Triple Entente, 142, 144-45, 161, 255
Trotsky, Leon, 160-61
Truman, Harry S., 38-41, 237
Truscott, MajGen Lucian, 206
Tunisia, 199-201
twenty-first century foraging, 76-77, 87

United Kingdom military units
    3 Commando Brigade, 175-80
    42 Commando, 176, 178
    45 Commando, 176, 178-79
    5th Infantry Brigade, 176, 178-79
    2 Parachute Battalion, 176
    3 Parachute Battalion, 176
    Commando Logistics Regiment, 180
    Eighth Army, 200, 205
    Gurkhas, 176, 178
    Mountain and Arctic Warfare Cadre, 180
    Mountain Leaders, 180
    Scots Guards, 176
    Welsh Guards, 176
unmanned aerial vehicle (UAV), 80, 83-85, 87
U.S. Air Force, 68, 71, 79-81, 84, 86, 101, 174
U.S. Army military units
    7th Infantry Division, 169, 171-72, 175
    9th Infantry Division, 93
    27th Infantry Division, 110, 115
    43d Infantry Division, 109
    45th Infantry Division, 103, 206, 208n59, 209, 211
    87th Mountain Infantry Regiment, 174-75
    1st Special Service Force (FSSF), 174
    Seventh Army, 200n18, 205, 210-12
U.S. Marine Corps military units
    1st Amphibious Tractor Battalion, 91
    2d Amphibious Tractor Battalion, 111, 113
    1st Battalion, 7th Marines, 284
    7th Defense Battalion, 286
    9th Defense Battalion, 100
    1st Marine Amphibious Brigade, 93, 99
    2d Marine Brigade, 288
    3d Marine Brigade, 288
    1st Marine Division, 56, 91, 93, 95-96, 100, 102, 120, 240, 288, 290-91, 293, 295-97
    2d Marine Division, 93, 109-12, 114
    4th Marine Division, 102, 106, 119
    5th Marine Division, 292
    1st Marine Regiment, 240
    2d Parachute Battalion, 102
    1st Provisional Marine Brigade, 284
    1st Samoan Battalion, 286-87
    I Marine Amphibious Corps, 101, 102n33, 106, 215
    III Marine Amphibious Corps, 102n33
    V Marine Amphibious Corps, 109
    I Marine Expeditionary Force (I MEF), 70
    Fleet Marine Force, Atlantic, 90
    Fleet Marine Force, Pacific, 106
    Task Force 58, 58, 87
U.S. Navy Bureau of Ships, 21, 29, 37n71, 39-41, 91, 98, 112
U.S. Navy military units
    Atlantic Fleet, Amphibious Command, 35-36, 93, 98-99, 102-3, 106
    Pacific Fleet, Amphibious Training Command, 106, 109, 111, 113, 215, 276, 287
    Seventh Fleet, 237-38
    Sixth Fleet, 184, 189, 191, 194
    Task Force 31, 215, 217, 223

Van Buren, Martin, 262n43, 265
Vella Lavella, Solomon Islands, 215-16, 222, 226

Veracruz, Mexico, 7–20, 336
Vietnam, Republic of, 4, 69, 244–45
von Hutier, LtGen Oskar, 146
von Pohl, Gen Maximillian, 207
von Tschischwitz, Col Erich, 161

War of 1812, 247, 251, 259n35
Webster-Ashburton Treaty (1842), 259, 265
Webster, Daniel, 267
Wei-kuo, Chiang, 233
Wilhelm II, German emperor, 146
Wilkinson, RAdm T. S., 215–16, 220
Welles, Gideon, 271–72
Wellesley, Arthur (1st Duke of Wellington), 250–51
Wood, Charles (1st Viscount Halifax), 262–64

Wyly, Col Michael D., 162

Xiamen (Amoy), China, 230, 233
Xiaoping, Deng, 243–44

Yijiangshan Island, 229, 241–44, 323
Yom Kippur War (1973), 182

Zakharov, Gennady, 188
Zakharov, Matvei, 192
Zedong, Mao, 230, 234, 246
Zemin, Jiang, 229, 245
Zhejiang, China, 238, 240
Zhoushan, China, 229, 236

ABOUT THE AUTHORS

Dr. Lance R. Blyth is the command historian at North American Aerospace Defense Command (NORAD) and U.S. Northern Command and adjunct professor of history at the U.S. Air Force Academy in Colorado. A backcountry skier with a history problem, Blyth is currently completing a manuscript, "10th Mountain and Mountain Warfare," looking at U.S. mountain warfare efforts from World War II (WWII) to the present day. All views are his own and do not necessarily represent the views of the Department of Defense or its components. https://orcid.org/0009-0004-1621-115X

Benjamin Claremont is an American historian and current PhD candidate in international relations at the University of St. Andrews in Scotland. Claremont's passion for military history was sparked by his grandfather's stories of serving aboard USS *Robely D. Evans* (DD 552) during the Pacific campaign and his grandmother's experiences as an Royal Air Force radar operator during WWII. His research interests include twentieth- and twenty-first-century large-scale high-intensity war, the Soviet military in theory and practice (especially post–WWII), Cold War U.S. and UK threat doctrine development, naval combat in the missile age, and the Russo-Ukrainian War. Claremont has previously published articles with the Center for International Maritime Security, and his forthcoming doctoral thesis explores the history of U.S. Army efforts to develop threat doctrine concerning the Soviet Army and what lessons these historical efforts can teach current threat doctrine practitioners. https://orcid.org/0009-0002-7614-1611

B. A. Friedman is a retired U.S. Marine Corps artillery officer currently employed as a strategic assessment analyst and as a PhD student at the University of Reading in the United Kingdom. He is the author of numerous works, including *On Tactics: A Theory of Victory in Battle* (2017) and *On Operations: Operational Art and Military Disciplines* (2021).

A reader in war studies at the University of Wolverhampton, **Dr. Howard J. Fuller** completed his BA in history at the Ohio State University, and his MA and PhD in war studies at King's College London. In 2002–3, he was the Rear Admiral John D. Hayes Fellow in U.S. Naval History through the U.S. Naval Historical Center (now Naval History and Heritage Command) in Washington, DC; he was also a West Point Fellow in Military History, Caird Research Fellow, and is a Fellow of the Royal Historical Society. Dr. Fuller is a frequent contributor to conferences and journals on both sides of the Atlantic, associate editor for the *International Journal of Naval History*, and a managing editor for the *British Journal of Military History*. His first book, *Clad in Iron: The American Civil War and the Challenge of British Naval Power* (2007), was runner-up for the 2008 John Lyman Book Award in U.S. Naval History; followed by the critically acclaimed *Empire, Technology and Seapower: Royal Navy Crisis in the Age of Palmerston* (2013); and most recently, *Turret vs. Broadside: An Anatomy of British Naval Prestige, Revolution and Disaster, 1860–1870* (2020) as part of the Wolverhampton Military Studies Series. https://orcid.org/0000-0001-8993-6531

Isabella Ginor, former Soviet/Russian affairs specialist at Israel's newspaper of record *Haaretz*, and **Gideon Remez**, former head of foreign news at Voice of Israel Radio, are associate fellows of the Harry S. Truman Research Institute, Hebrew University of Jerusalem, specializing in Soviet/Russian involvement in the Middle East. Their coauthored publications include *Foxbats over Dimona: The Soviets' Nuclear Gamble in the Six-Day War* (2007) and *The Soviet-Israeli War, 1967–1973: The USSR's Military Intervention in the Egyptian-Israeli Conflict* (2017). https://orcid.org/0000-0003-2593-9965

Timothy G. Heck is an artillery officer by training. Currently employed as a historian with the U.S. Navy, he previously coedited *On Contested Shores: The Evolving Role of Amphibious Operations in the History of Warfare* (2020) and *Armies in Retreat: Chaos, Cohesion, and Consequences* (2023). https://orcid.org/0000-0002-5270-2341

Darren Johnson is a U.S. Army infantry officer who is currently serving as an instructor of history at the U.S. Military Academy at West Point. He has served in the U.S. Army for 13 years and has two combat deployments to Afghanistan as an infantry platoon leader and company commander. He earned a bachelor's degree in accounting and finance from Corban University and a master's degree in history from Florida State University, where he focused his research on the Holocaust and the Allied liberation experience. He has published numerous book reviews, presented his research at national and international conferences, and is currently researching how the U.S. Army adapted at the tactical and operational levels of war during World War II. https://orcid.org/0009-0000-9912-9274.

David Brock Katz obtained a bachelor's in commerce and accounting at the University of the Witwatersrand, South Africa, in the early 1980s and practiced as a chartered accountant thereafter. After obtaining his master of military science (cum laude) in 2014, he qualified for a doctor of philosophy in military science in 2021 at Stellenbosch University. Katz currently serves as an active member of the South Africa National Defence Force with the Andrew Mlageni Regiment and is a research fellow at Stellenbosch University with the faculty of military science. He is also the editor-in-chief of the *South African Army Journal*. He lectures at the Army, War, and Defence Force Colleges and has published numerous papers in several academic journals. His book publications include *South Africans versus Rommel: The Untold Story of the Desert War in World War II* (2019) and *General Jan Smuts and His First World War in Africa, 1914-1917* (2022). He coauthored *20 Battles: Searching for a South African Way of War, 1913-2013* (2023). https://orcid.org/0000-0003-0689-0668

Dr. Xiaobing Li is a professor at the Department of History and Geography and Don Betz Endowed Chair in International Studies at the University of Central Oklahoma. He is the executive editor of the *Chinese Historical Review* and editorial board member of *The Journal of Military History* and *Journal of Chinese Military History*. He was born in China and served in the People's Liberation Army (PLA). Among his recent books are *China's New Navy: The Evolution of PLAN from the People's Revolution to a 21st Century Cold War* (2023), coedited *Sino-American Relations: A New Cold War* (2022), *The Dragon in the Jungle: The Chinese Army in the Vietnam War* (2020), *Attack at Chosin: The Chinese Second Offensive in Korea* (2020), coauthored *East Asia and the West: An Entangled History* (2020), coedited *A Century of Student Movements: The Mountain Movers, 1919-2019* (2020), *China's War in Korea: Strategic Culture and Geopolitics* (2019), *Building Ho's Army: Chinese Military Assistance to North Vietnam* (2019), *The History of Taiwan* (2019), coedited *Corruption and Anticorruption in Modern China* (2019), *The Cold War in East Asia* (2018), coedited *Urbanization and Party Survival in China: People vs. Power* (2017), coauthored *Power versus Law in Modern China: Cities, Courts, and the Communist Party* (2017), coedited *Ethnic China: Identity, Assimilation, and Resistance* (2015), *China's Battle for Korea: The 1951 Spring Offensive* (2014), coedited *Evolution of Power: China's Struggle, Survival, and Success* (2014), coedited *Oil: A Cultural and Geographic Encyclopedia* (2014), coedited *Modern Chinese Legal Reforms: New Perspectives* (2013), edited *China at War: An Encyclopedia* (2012), and *Voices from the Vietnam War: Stories from American, Asian, and Russian Veterans* (2011).

Dr. Shaun Mawdsley is a military historian and defence and security researcher currently employed as an analyst with the New Zealand Police. From 2016 to 2022, Dr. Mawdsley taught at the Centre for Defence and Security Studies and the College of Humanities and Social Sciences at Massey University, New Zealand. In 2023, he received a PhD in history from Massey University, where his dissertation focused on Australian and New Zealand divisional combat effectiveness in the Western Desert campaign of 1941–42. He has previously published on matters of intelligence oversight, defense transparency, and Australasian special operations in WWII. His other research areas include transnational organized crime and New Zealand's adaptation to amphibious operations and jungle warfare in WWII. He is currently adapting his dissertation for publication. https://orcid.org/0009-0004-3060-5454

Christopher Menking is a professor of history at Tarrant County College in Fort Worth, Texas. His research focuses on the U.S. military experience during the Mexican-American War and in the South Texas borderlands. He has published chapters and articles related to food during the war with Mexico, South Texas, and the Mexican-American War. He received his undergraduate degree at Texas A&M University in College Station and completed his graduate work at the University of North Texas in Denton. His current research looks at music during the Mexican-American War and the experience of average soldiers during the war. https://orcid.org/0000-0003-3961-2445

Walker D. Mills is a U.S. Marine Corps infantry officer training to be a General Atomics MQ-9A unmanned aerial vehicle pilot. He previously coedited *Armies in Retreat: Chaos, Cohesion, and Consequences* (2023) and contributed a chapter to the first volume of *On Contested Shores: The Evolving Role of Amphibious Operations in the History of Warfare* (2020). He has written extensively on the Marine Corps and future warfare. https://orcid.org/0000-0002-7047-555X

Douglas E. Nash Sr. is a West Point graduate and a retired U.S. Army colonel with 32 years of service, including assignments in Kosovo, Iraq, Afghanistan, Germany, Cuba, and Uzbekistan. He served in a variety of armored cavalry, armor, and special operations units, including civil affairs and psychological operations. He recently retired after serving 10 years as the senior supervisory historian at History Division, Marine Corps University in Quantico, Virginia. His works include *Hell's Gate: The Battle of the Cherkassy Pocket* (2002), *Victory Was Beyond Their Grasp: With the 272nd Volks-Grenadier Division from the Huertgen Forest to the Heart of the Reich* (2008), and the *From the Realm of a Dying Sun* trilogy, as well as numerous articles in *Marine Corps History*, *Army History*, *Armchair General*, *Iron Cross*, *Special Warfare*, and *World War II* magazines. https://orcid.org/0009-0001-0142-437X

Evan Zachary Ota is from Kealakekua, Hawai'i, and earned bachelor's degrees in history and political science from the University of Hawai'i at Manoa and a master's degree in regional security studies from the Naval Postgraduate School in Monterey, California. He is an infantry officer and an international affairs officer in the U.S. Marine Corps and served in Iraq, Afghanistan, Japan, South Korea, the Philippines, Indonesia, Thailand, Singapore, Australia, Solomon Islands, and at sea. He is a recipient of the Lieutenant Colonel Earl "Pete" Ellis award for innovative operational approaches and is a nonresident fellow at the Marine Corps University's Krulak Center for Innovation and Future Warfare. https://orcid.org/0009-0003-9538-2354

Dr. Edward Salo, FRHistS, is an associate professor of history and the associate director of the Heritage Studies PhD Program at Arkansas State University. He was in the initial Normandy Scholars class at the University of Tennessee. His work has been published in the *Journal of America's Military Past*, *1945*, *Inkstick*, *National Interest*, *Eunomia Journal: The Journal of the U.S. Army Civil Affairs Association*, Modern War Institute, and the U.S. Army Corps of Engineers, and he has presented at numerous conferences including the USMA Social Studies Department Security Seminar (2023). He is currently studying the weaponization of heritage as part of irregular warfare. https://orcid.org/0000-0003-2365-793X

Dr. Eric A. Sibul has a BA in international relations from Penn State University, an MA in history from San Jose State University, and a PhD in history from the University of York in the United Kingdom. He served as an assistant professor of military theory and history at the Baltic Defence College in Tartu, Estonia, from 2006 to 2015. He also served as principal lecturer on seapower theory, naval history, and strategy at the Baltic Naval Command and Staff

Course in Riga, Latvia. Prior to his academic appointment in Estonia, Sibul served as a military English instructor at the Army Intelligence School in Songnam, Republic of Korea. As a member of the U.S. Navy Reserve, he served in an Office of Naval Intelligence unit and Amphibious Construction Battalion Two. Currently, Dr. Sibul is an instructor of history, business, and economics at the City Colleges of Chicago. https://orcid.org/0009-0003-3223-2332

Jerry E. Strahan holds a BA in education and an MA in history from the University of New Orleans and is the author of *Andrew Jackson Higgins and the Boats that Won World War II* (1994). He has spent 50 years researching Andrew Higgins and more than 17 years helping restore Higgins boats for the National World War II Museum.